STATISTICS COMPANION

SUPPORT FOR INTRODUCTORY STATISTICS

STATISTICS COMPANION

SUPPORT FOR INTRODUCTORY STATISTICS

Roxy Peck
California Polytechnic State University, San Luis Obispo

Tom Short
West Chester University of Pennsylvania

Contributors:
Paul D. Nolting
Hillsborough Community College

Kimberly Nolting
Academic Success Press, Inc.

Sue Ann Jones Dobbyn
Pellissippi State Community College

Australia • Brazil • Mexico • Singapore • United Kingdom • United States

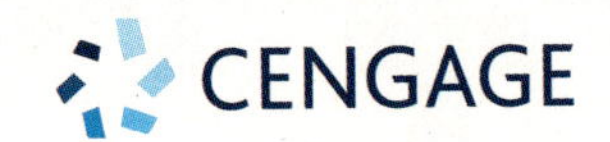

Statistics Companion: Support for Introductory Statistics
Roxy Peck, Tom Short

Product Director: Mark Santee

Product Manager: Catherine Van Der Laan

Senior Learning Designer: Elinor Gregory

Product Assistant: Amanda Rose

Marketing Manager: Mike Saver

Senior Content Manager: Michael Lepera

IP Analyst: Reba Frederics

IP Project Manager: Carly Belcher

Production Service: MPS Limited

Compositor: MPS Limited

Art Director: Vernon Boes

Text Designer: Diane Beasley

Cover Designer: Cheryl Carrington

Cover Image: simon2579/DigitalVision Vectors/Getty Images

Library of Congress Control Number: 2018950693

ISBN: 978-1-337-70559-2

Cengage
20 Channel Center Street
Boston, MA 02210
USA

Cengage is a leading provider of customized learning solutions with employees residing in nearly 40 different countries and sales in more than 125 countries around the world. Find your local representative at **www.cengage.com.**

Cengage products are represented in Canada by Nelson Education, Ltd.

To learn more about Cengage platforms and services, register or access your online learning solution, or purchase materials for your course, visit **www.cengage.com.**

Printed in the United States of America
Print Number: 01 Print Year: 2018

To Uri Treisman, for introducing
me to the joyful conspiracy.
Roxy Peck

To Allan Rossman and Roxy Peck, for their
friendship and mentorship.
Tom Short

Author Bios

ROXY PECK is a professor emerita of statistics at California Polytechnic State University, San Luis Obispo. She was a faculty member in the Statistics Department for thirty years, serving for six years as Chair of the Statistics Department and thirteen years as Associate Dean of the College of Science and Mathematics. Nationally known in the area of statistics education, Roxy was made a Fellow of the American Statistical Association in 1998, and in 2003 she received the American Statistical Association's Founders Award in recognition of her contributions to K-12 and undergraduate statistics education. In 2009, she received the USCOTS Lifetime Achievement Award in Statistics Education. In addition to coauthoring the textbooks *Statistics: Learning from Data, Introduction to Statistics and Data Analysis*, and *Statistics: The Exploration and Analysis of Data*, she is also editor of *Statistics: A Guide to the Unknown*, a collection of expository papers that showcases applications of statistical methods. Roxy served from 1999 to 2003 as the Chief Faculty Consultant for the Advanced Placement Statistics exam and she is a past chair of the joint ASA/NCTM Committee on Curriculum in Statistics and Probability for Grades K-12 and of the ASA Section on Statistics Education. Outside the classroom, Roxy enjoys travel and has visited all seven continents. She collects Navajo rugs and heads to Arizona and New Mexico whenever she can find the time.

TOM SHORT is an Associate Professor in the Statistics Program within the Department of Mathematics at West Chester University of Pennsylvania. He previously held faculty positions at Villanova University, Indiana University of Pennsylvania, and John Carroll University. He is a Fellow of the American Statistical Association and received the 2005 Mu Sigma Rho Statistics Education Award. Tom is part of the leadership team for readings of the Advanced Placement (AP) Statistics Exam, and was a member of the AP Statistics Development Committee. He has also served on the Board of Directors of the American Statistical Association. Tom treasures the time he shares with his four children and the many adventures experienced with his wife, Darlene.

Contributor Bios

Dr. Paul Nolting for over 30 years worked at State College of Florida in Bradenton, FL, as an undergraduate instructor, a learning specialist, an institutional test administrator, Title III director, and math lab coordinator. He also has been a graduate instructor for assessment and measurement courses. He is an expert on assessing math learning problems, developing effective student learning strategies, math study skills, assessing institutional variables that affect math success, math redesigns, and tutor training. He has consulted nationally and internationally with over a hundred college/university campuses on their QEPs and to improve math success and retention. He is now an adjunct professor at Hillsborough Community College in Tampa, FL.

Ms. Kim Nolting currently works as Research and Development Vice President for Academic Success Press, Inc. In addition, she consults with higher education institutions, focusing on faculty development and intervention workshops for improving classroom teaching and learning, as well as enriching academic support services. Her MAT focused on teaching English and communications, while her doctoral coursework specialized her in the area of measurement and research in higher education with a special focus on teaching and learning. Her multifaceted career includes college instruction, learning assistance supervision, and manager of grant initiatives. In addition, she worked in mid-level administration, which allowed her to work closely with faculty in the academic disciplines of general education to improve student academic performance.

Sue Ann Jones Dobbyn graduated from the University of Tennessee in 1974 with a BS in Chemical Engineering. After working for Proctor and Gamble and Ford Motor Company, Sue Ann took a Masters' degree in Curriculum and Instruction with the primary goal of having a working schedule which better matched that of her three young daughters. Falling in love with the role of educator was a wonderful surprise. Sue Ann spent the next twenty years as a high school teacher in an alternative high school in Las Cruces, New Mexico. In this position, Sue Ann and her students participated in national and international educational programs, including *Project del Rio*, an ecological partnership with schools in New Mexico, Texas, and Mexico, and *Critical Issues Forum*, a nuclear nonproliferation program which involved students in the United States and the Russian Federation sponsored by the US Department of Energy. In addition, she received grants from GTE and Toyota to create curriculum in science and mathematics. An online version of one of these programs, *The Physics of Sports*, received an Eisenhower Clearinghouse award. Returning to Tennessee in 2007, Sue Ann accepted a position at Pellissippi State Community College in the Mathematics Department. When the state of Tennessee in 2014 decided to mandate that remediation in math be provided in a co-requisite model, Sue Ann volunteered to help construct the program for the entry level statistics course. The resulting curriculum was piloted in Fall 2015. The success of the co-requisite model in Introductory Statistics resulted in the implementing of this model for other entry level math courses.

Brief Contents

Contents

Preface

Introductory statistics is the required college-level mathematics course for a large and growing number of students. As a consequence, two-year and four-year colleges nationwide are exploring ways to broaden access to introductory statistics courses by rethinking placement and prerequisite policies. To accommodate a group of students that is more diverse with respect to mathematics preparation, many have chosen to implement structures that include a pre- or a co-requisite course for students who need additional support to be successful. These courses are focused on the mathematics background and foundation needed for success in introductory statistics.

This companion text is designed to provide introductory statistics students with the support that they need to be successful by addressing mathematical content, study skills, and a productive mindset. And, because statistics is a subject in which context plays a critical role, a chapter on strategies for reading and understanding statistics problems in context is also included in the text.

The *Statistics Companion* is written to align with the text *Statistics: Learning from Data*, 2nd edition, but can be adapted for use with any introductory statistics textbook or resource.

Organization of the *Statistics Companion*

Chapter 0 is a chapter written by nationally recognized math study skills experts Paul Nolting and Kimberly Nolting. It is designed to help students develop skills necessary for success in college and for the introductory statistics course. This chapter covers content that statistics faculty may not have previously included in their courses, and the authors of this chapter have written helpful suggestions for how to incorporate this content into a prerequisite or a co-requisite support course. The suggestions are included in the "Advice for Instructors" section that follows later in this preface. It is recommended that the material in Chapter 0 be covered during the first two weeks of a semester-long course, teaching it alongside the early mathematics content of Chapters 1 and 2.

Chapters 1 through 6 cover the basic mathematics that provides the foundation needed for the introductory college-level statistics course. These chapters are organized around the ordering of topics in a typical introductory statistics course. This enables a review of the necessary mathematics to be addressed in a supporting, parallel course as it is encountered in the statistics course. If descriptive analysis of bivariate data (correlation and linear regression) is covered before inference, most of the mathematics prerequisites are needed in the first half of the introductory statistics course. With this in mind, there is an assessment of the mathematics introduced in Chapter 1 through Chapter 6 available on the instructor resource site for this text. This assessment can be used midsemester to confirm student mastery of the mathematics content and to identify any areas where additional review might be needed.

Chapter 7 introduces a strategy for reading and understanding statistics problems. Midway through an introductory statistics course, the content transitions from descriptive statistics and probability to inferential statistics. This is where the reading load increases, as nearly every problem students encounter involves context, interpretation, and communication of conclusions. Chapter 7 provides students with a systematic strategy for reading and understanding the types of problems that they will encounter throughout the second half of their statistics course. Depending on how much time is devoted to the co-requisite course (which may range from two hours to four hours per week), if time permits, the material in this chapter could also be covered earlier because the reading strategies introduced here also apply to problems that students encounter in the descriptive statistics part of the course.

Chapters 8 through 20 are short chapters that focus on support for the material on confidence intervals and hypothesis testing. Because not all statistics textbooks present this material in the same order, we have chosen to break this material up into short chapters that can be taught in any order. Each of these chapters focuses on a specific inference topic, such as estimating a population proportion or testing a hypothesis about a difference in means using independent samples. These chapters open with a section that walks students through evaluating the mathematical expressions they will encounter in the statistics course. The sections that follow provide students with guided practice in applying the reading strategy introduced in Chapter 7, and then solving the types of problems they will see in their statistics course.

Advice for Instructors

In this section, we are fortunate to be able to include contributions from Sue Ann Jones Dobbyn, Paul Nolting, and Kimberly Nolting.

Sue Ann Jones Dobbyn is a faculty member at Pellissippi State Community College in Tennessee, where she teaches a co-requisite support course for introductory statistics. Tennessee was an early adopter of the co-requisite strategy at both two-year colleges and four-year universities, and has seen a significant increase in the number of students passing the college-level statistics course in their first year. Professor Dobbyn shares what she has learned as the co-requisite strategy was implemented and offers recommendations for instructors teaching the co-requisite support course.

Following Sue Ann Jones Dobbyn's recommendations, Paul and Kimberly Nolting offer insight on how the material in Chapter 0 can be integrated into a co-requisite support course.

On Co-requisite Mathematics: Advice to Instructors

Sue Ann Jones Dobbyn

Nationally, over 50% of entering freshmen at two-year colleges require academic remediation, most frequently in mathematics. At four-year institutions the proportion is smaller, around 20%, but still daunting.[1] The state of Tennessee enacted the Complete College Tennessee Act in 2010, which established a progressive reform agenda of the educational system of the state. One of the initiatives at the college level was to move from prerequisite remediation programs to co-requisite programs.

As an Associate Professor at Pellissippi State Community College, I have been involved in the development of co-requisite math courses for the past four years. The Tennessee Board of Regents gave us the following mandate: The co-requisite experience will serve the dual purpose of supporting and illuminating the skills and concepts of the college-level credit-bearing course while also providing instruction for students to remediate those mathematics developmental competencies in which they have a deficiency. In delivering on this mandate, we have consistently focused on the needs of the remedial students, which has affected many of the pedagogical and structural decisions that we made. The results of our efforts have been gratifying: success rates, as defined by receiving an A–C in the college-level course, have increased from 34% (2013) to 60% (Fall 2017). The following recommendations are based on some of the lessons we have learned from piloting the co-requisite format to full-scale implementation.

Recommendation 1: Choose the best materials

At the time that my institution began co-requisite reform, almost no comprehensive materials were available. The original recommendation was simply to pancake remedial algebra modules on top of the college-level course materials. We worried that students would simply feel like they were being forced to take two math courses at the same time. In addition, we were concerned that the order and timing of the remedial algebra topics would not meld with the needs of the college-level class. As a result, we decided to author

1 Complete College America (2012). *Remediation: Higher Education's Bridge to Nowhere*. p. 2.

our own remedial course materials that kept the skill requirements of the college-level course as the driving force.

Ideally, the students in the co-requisite course should see immediately how the content of the remedial course informs and assists them in the college-level course. Examples and exercises should be aligned with the content and timing of the college course but focused on the remedial topic. For example, to construct and analyze frequency tables in a statistics class, students need to be able to convert between fractions, decimals, and percentages, which is a remedial skill. Therefore, questions in the co-requisite class that target this remedial topic should align with the college-level course's treatment of frequency tables.

Fortunately, over the past few years, more authors have developed co-requisite materials that pace the college-level class and reflect its needs. If it is properly used, students in the co-requisite section should see the benefits of this just-in-time approach as they progress through the curriculum. A two-pronged approach that provides supporting instructional materials paired with examples and exercises written in the context of the college-level course has been very effective at my institution. The students in the co-requisite courses have passing rates (grades of A–C) that are almost the same as the college-ready students.

Recommendation 2: Avoid structuring the co-requisite class as a lecture

Although many of us are most comfortable when we are the "sage on the stage," this is probably not the most effective way to conduct the co-requisite classroom. An outline/schedule of topics for the co-requisite mapped to the college-level course should be provided to instructors. However, this type of document shouldn't be used to create lectures on all those topics. While occasional mini-lectures may be needed, they should be in response to some identified issue that is shared by several students. Strategies that work well include class discussion and selective group work on topics of shared difficulty or concern. These active learning techniques also promote positive social interactions that have long-term benefits. The main objectives of the instructor in a mastery-based curriculum are to encourage students to be persistent in their work, to be diligent in their efforts to understand the content problems of individual students, and to help remediate those issues. This job requires that faculty examine carefully and constantly the work of their students. A feedback loop should also be established with the students. A weekly progress report that can be easily filled out by the instructor can effectively communicate progress to the student. The report should reflect the desired schedule of the course in some way and identify where the student stands in relation to the expected pace. In addition, the report may give the student-specific feedback on assessments such as tests.

I think of the role of co-requisite instructor more in terms of coaching. Certainly, some procedures need to be taught when they reflect commonly shared misconceptions or difficulties. But, at most times, the challenge is to encourage students to keep on working and making progress toward completion and help them to see how the co-requisite can help them be successful in the college-level course.

Recommendation 3: Have college-level sections with mixed populations

In reviewing the anecdotal evidence, it seemed clear that one issue encountered by many underprepared students was a feeling of isolation from the college society in general. Of course, this problem was most exacerbated when all remediation was done as prerequisites. However, even in the co-requisite model this isolation can still occur. At Pellissippi State, we made a deliberate choice to offer our co-requisite course linked with college-level sections that include both college-ready students and co-requisite students. This decision has helped us to keep the college-level course at the same level of rigor for both groups. Maintaining the rigor of the college-level course was part of the state's mandate, but it is also simply the right thing to do.

There have been several benefits derived from mixed-population sections. First, it alleviates the possibility of social isolation that plagued remedial students in the prerequisite model. In fact, the co-requisite students often find themselves in the role of peer tutor when their college-ready colleagues have trouble with some underlying algebra skill. This role reversal is not an unusual occurrence in the college-level class with mixed populations.

Another benefit is that both groups begin to see the inherent connections between the prerequisite mathematics and gateway math courses like Introductory Statistics.

Recommendation 4: Assign the same instructor to both courses

Committing college-accredited faculty to both the college-level and co-requisite classes has been somewhat controversial and problematic. Some faculty may object to, or even have difficulty with teaching the co-requisite course topics. Some institutions may have trouble finding enough fully-accredited faculty members to meet the increased teaching load. However, there are many benefits to making this commitment.

Having the same instructor demonstrates that the institution is equally interested in the success of the underprepared student and the college-ready student. The decision makes a subtle, but important, statement about the value of the co-requisite course, as well. In addition, the same instructor can much more easily follow the progress of all the co-requisite students in both courses. The value of this type of knowledge can't be overstated. For example, if the instructor has noted in the college-level class that co-requisite students don't understand how to change a calculator display in scientific notation to standard notation, they can reinforce the procedure needed in the co-requisite session. This type of synthesis is almost impossible if the courses are taught by two different instructors, even if serious efforts are made to formalize communication between these instructors.

Recommendation 5: Use mastery-based learning in the co-requisite course

Since co-requisite students are working off identified deficits, programs that demand mastery at a specifically high level are encouraged. Mastery-based programs can have the benefit of lowering the anxiety associated with testing for these students, since they know in advance that they can review and retest if their first results fall short of the mastery level chosen. In addition, this system helps document that the student has remediated their math deficits.

Of course, there are problems associated with the choice of a mastery-based program. Since the requirements and pace of the college-level course will dictate to some extent the structure of the co-requisite course, the choice to have a mastery basis can result in some students falling behind due to reviewing and retesting. This problem is in some ways unavoidable and can create management issues for the instructor. However, it is also an opportunity to become more involved with the students who struggle. Struggle can be beneficial when it results in encouraging students to use other resources like tutoring centers. The cycle of working toward mastery can help students become more persistent in their efforts to succeed.

In a mastery-based system, students are also empowered to work ahead if they are able. This acceleration can put those students out of sync with the college-level course. However, for ambitious students the problems created usually are not serious. And, of course, the student always has the option to simply slow down the pace in the co-requisite course. Most students who choose to accelerate understand the choice they are making.

Recommendation 6: Make additional resources available

Co-requisite instructors need to understand the resources available to students that can assist them in their efforts to complete their education. At my institution, we require that underprepared students spend time each week in one of our tutoring centers. This requirement forces them to schedule time dedicated to working on their course. It also makes it easier for them to develop relationships with people whose job is to help students succeed. Properly trained tutors help students become better learners by coaching them in problem-solving techniques and encouraging them to be persistent in their efforts. Developing a relationship with math tutors today may lead to the widespread utilization of tutors in other subject areas as well.

One of the initiatives that we have developed at my institution has been the training and use of supplemental instructors (SI) in the co-requisite classrooms. The SIs are current undergraduate students who have previously made an A in the college-level course and agree to conduct regularly scheduled tutoring sessions for the co-requisite students. This system has the advantage of guaranteeing that the tutoring time is focused on the specific course. Also, the students often develop a strong and productive relationship with the supplemental instructor. Peer tutors frequently teach far more than math by discussing and encouraging good study skills and academic discipline.

Virtual tutoring is another strategy in which we are investing time and effort. The state of Tennessee has recently begun a program called *Reconnect Now* that is targeted specifically for adult learners. One of the issues with adult learners is fitting a college educational schedule into an adult life that includes families and full-time jobs. In response, we have begun offering co-requisite courses in a "weekend college" format. A vital addition to this initiative has been to make tutoring available in a virtual setting since many of these adult learners need tutoring during the evening hours or on weekends when the traditional tutoring centers are closed. So far, we have been able to staff the virtual tutoring hours using faculty who have taught or are teaching the college-level course.

Recommendation 7: Consider incorporating other high-impact practices

There are many other high-impact practices that can be positively utilized in co-requisite classes. One initiative at my institution has been the incorporation of service learning projects as part of the student experience. It is possible to identify projects that are relevant to college math courses in many cases. For example, the TSA at our local airport requested help analyzing survey data. This task had clear relevance to a statistics course, and with guidance the students were capable of performing the work. It may require some out-of-the-box thinking, but service projects allow students to apply what they are learning to real-world settings. In addition, many institutions now acknowledge these service learning experiences on student transcripts.

The incorporation of growth mindset training is another opportunity that can enhance co-requisite courses. Underprepared students often have low opinions of their own ability to succeed academically, especially in mathematics. Countering this nonacademic barrier with good information about how the brain works and how persistence and practice can improve skills is possible. In addition, training for instructors and tutors in giving productive feedback can be valuable as we attempt to maximize the effectiveness of our courses. Resources for conducting this type of intervention with faculty and with students are widely available from sources like mindsetworks.com.

In the end, faculty and students have the same goal: success in both the remedial component of the class and the college-level course. Making the transition to the co-requisite model of remediation does require significant effort from all the stakeholders at an institution, from the faculty to the registrars. However, the benefits at Pellissippi State have included both higher retention rates and higher success rates.

Recommendation 8: Use projects to address critical-thinking competencies in the college-level course

In the state of Tennessee, one of the remedial competency points centers around the development of critical thinking. In order to address this competency, we have developed projects for our entry-level courses. In the case of Introductory Statistics, we have utilized large data sets to design semester-long capstone projects with social justice themes that reinforce the concepts and procedures of the course while encouraging critical thinking through the continued use of interpretive questions. In the liberal arts (general studies) math course, themed projects are used that reflect the varied topics of the course, such as logic, consumer finance, probability/statistics, and modeling.

Projects of this type allow us to incorporate group learning and discussion. It can be challenging to manage a group-learning project. Typically, we suggest that the instructor introduce a set of ground rules for productive discussion and interaction. These rules emphasize respect for other's opinions, including being non-judgmental, giving equal time to all group members' expressions, and keeping verbal exchanges civil. Instructors need to be attentive to these interactions and be prepared to intervene if the discussion becomes too heated.

In addition, many of the projects are enhanced by bringing in and sharing other materials like articles, research papers, and maps. As difficult as group projects sometimes can be, they are worth the effort. These projects are an opportunity to show how very relevant mathematics is to our everyday life. They are also a chance to educate students in current events and show how mathematics interacts with other disciplines.

Advice on Incorporating Chapter 0 Smart Study Strategies

Paul D. Nolting and Kimberly Nolting

After looking at any statistics textbook for the first time, students quickly realize that there is a lot to learn and that their statistics course will not be like other mathematics courses they may have taken. There is an immense amount of information to learn. In addition, students must learn how to apply this information immediately. To accomplish these skills, students need to develop study plans that help the brain learn how to gather and organize information to analyze and synthesize. In addition, it is important to learn the concepts of statistical vocabulary and the relationships between them. As important, students must learn how to study in a way that puts all concepts into long-term memory because as the curriculum progresses, it grows in complexity. The easier it is to pull information from long-term memory, the more available mental energy for the brain to understand and organize more complex information. A study plan helps to use time wisely, allows the brain to work efficiently, and minimizes learning anxiety.

The study strategies suggested in this chapter are also innovative teaching instructions. When instructors use study strategies as teaching strategies, students are more likely to also value them. Here are a few examples. (More study strategies can be found in *Math Study Skills Workbook*, 5th edition, by Dr. Paul Nolting, either online or by contacting Cengage.)

Study Strategy	Teaching Strategy
Skim the assigned reading material. Look at the learning objectives listed to get an overview of what you will learn. As you skim the chapter, circle novel words that you do not understand with a pencil. This step is to simply get an overview of what you will be learning. These two skimming exercises should only take 5–10 minutes.	During the first or second class, work through these two steps in class. While modeling the steps, you are also introducing the statistics content.
After skimming, read again to familiarize yourself with the material enough to prepare for the next class. Remember, this step will help you encode and keep new information presented in class in short-term memory long enough to record it in your notes. Mark the concepts and words that you do not understand. Another option is to write the vocabulary words in your vocabulary list. If you do not clearly understand a word or concept, make sure you ask about them in class.	Include at least a portion of the chapter 0 material, so students can see how the skimming helps with taking notes.
Figure out which learning objective(s) each problem is helping you learn. Write down the objective number. This will help you use your homework when reviewing for a test. Also record the concepts and formulas by the homework problem.	Take 10 minutes to model this strategy in class. Assign one homework problem to each small group in class. The small group takes 5 minutes to decide which learning objective the problem is linked to and explain why. Each group reports back.
Make sure your notes have all the correct information. One of the most common challenges for students is cognitive overload during statistics classes. When on overload, it is easy to miss pertinent information. Find at least two classmates willing to share notes. Take a few minutes to take pictures of each other's notes with your phones right after class. There is only one caution: Make sure their notes are correct.	Allow time at the end of class for students to practice this. It also provides time for them to meet other students and begin building study support groups.

Study Strategy	Teaching Strategy
Form a study group of three or four students within the first two weeks of class. Keep it small.	Encourage study groups. Help students form them. For instance, hand out a weekly schedule grid that is hour by hour. They can highlight the times they are most available for study groups. Collect and see which students share similar available times. Let them have 5 minutes at the end of class to meet in selected time slots. Think of places they can meet. If you are teaching a co-requisite section attached to the academic class, provide times for students to study in their small groups. This is particularly good when reviewing for a test.

It is also wise to attach a type of student incentive for following through with study strategies. For instance, ask them to use a vocabulary list during the weeks before the first test. Tell them to follow the guidelines in the study strategy chapter. If they turn in a complete list of vocabulary words, using the guidelines, they can receive extra points on the test. These points are just enough that it could mean the difference between a B+ and A−. If they receive grades for smaller quizzes, perhaps the incentive is dropping the lowest quiz. Be creative.

Acknowledgments

The authors would like to express our thanks and gratitude to the following people who made this book possible:

Catherine Van Der Laan, our product manager at Cengage, for her support of this project.

Michael Lepera and Spencer Arritt, our content managers at Cengage, for their helpful suggestions and for keeping us on track.

Elinor Gregory, the learning designer.

Lori Hazzard and Ed Dionne, our project managers at MPS Limited.

Stephen Miller, for his careful and complete work on the huge task of creating the student and instructor solutions manuals.

Roger Lipsett, for his attention to detail in checking the accuracy of examples and solutions.

Mike Saver, the Marketing Manager.

Richard Camp, the copy editor and Heather Mann, the proofreader for the book.

MPS Limited, for producing the artwork used in the book.

Sue Ann Jones Dobbyn, for her valuable advice to instructors that appears in the preface.

Paul Nolting and Kimberly Nolting, for sharing their expertise in the area of study skills in authoring Chapter 0 of this book.

Many people provided invaluable comments and suggestion as this text was being developed.

Reviewers

Anna Tivy, Ventura College
Adam Heck, Southern Adventist University
Roberto Colson, University of Texas Rio Grande Valley
Sabrina Ripp, Tulsa Community College
Shannon Solis, San Jacinto College

Francis Nkansah, Bunker Hill Community College
Ryan Kasha, Valencia College
Sam Ofori, Cleveland State Community College
Jordan Bertke, Central Piedmont Community College

And last, but certainly not least, we thank our families, friends, and colleagues for their continued support.

Roxy Peck and Tom Short

0 Smart Study Strategies

This chapter was written by Paul Nolting and Kimberly Nolting

This chapter is the starting line for designing a study plan. First, it focuses on how the brain learns and then suggests study strategies that help the brain process, understand, organize, and store and retrieve information. Often you will see page references to ***Math Study Skills Workbook, 5th edition***, by Dr. Paul Nolting, where you can find additional study suggestions in more detail. You can find this resource online. Second, this chapter focuses on the impact that self-image and emotions have on successful academic study. You will learn strategies to help manage self-talk and thoughts that discourage devoting time to studying when it gets tough. You will also learn strategies that develop a positive self-image as a student. Do not wait until you get behind in classes to try these productive study strategies. Start now.

SECTION 0.1 The Learning and Memory Process

The brain is complex. The more neuroscience experts learn, the more they understand how the brain is immensely intricate. Through all their discoveries, several basic processes remain the same. In order to learn, the brain must complete the following process: encoding, short-term memory, working memory, long-term memory, and long-term retrieval. Even though these are discussed in a sequential manner, the brain processes with immense speed and uses these processes almost simultaneously.

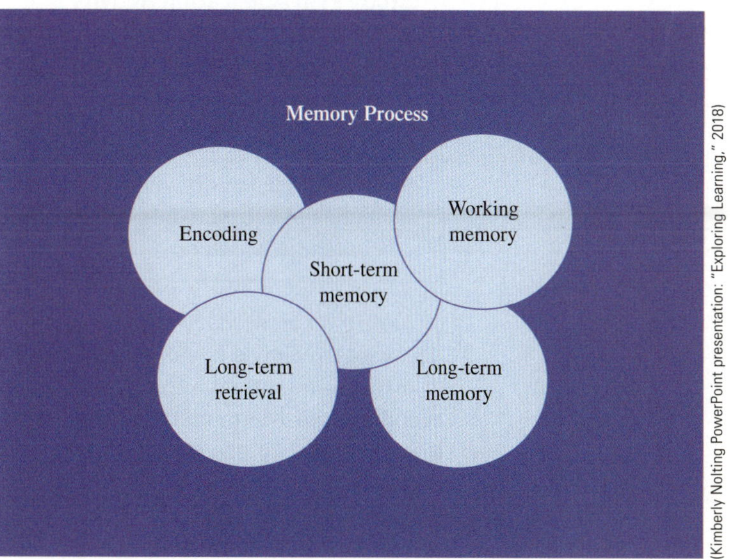

(Kimberly Nolting PowerPoint presentation: "Exploring Learning," 2018)

Encoding is the initial biological step for gathering information through the five senses. For example, the eye sees a new formula on the white board. The retina takes in novel visual information and sends it to the optic nerve, which transfers it to the brain. The brain transforms it into a group of neurons, which are then sent to another area in the brain. Our brain encodes everything but must choose which information to attend to. It lasts for less than a second.

After encoding, the ***short-term memory*** holds information for no more than 30 seconds. It is able to hold small portions of information during these 30 seconds. Another brain function called attention decides whether to discard or send information on to the working memory. The short-term memory also holds on to information that is retrieved from the long-term memory to assist in learning new concepts. The short-term memory depends primarily on acoustic and visual code for storing information.

Study Plan Application Activity

1. Evaluate how well you listen and record information in class. Also, think about how you currently read the textbook and record information during independent study time. Examine how you encode information from online sources and record it. **Write out these descriptions**.
2. What interferes with your concentration during class and study times? **Answer**.
3. What do your class notes look like? **Answer**.

Working memory is the mental storage and work space for the brain to create understanding and to organize the new concepts to send into long-term memory. It is important for reasoning and decision making. New information in working memory is either organized for long-term memory or decays after a short time (10–15 seconds). While in working memory, the new information must be attended to actively or rehearsed.

Working memory requires attention, temporary storage of information, and manipulation of information. The brain must arrange the neurons containing new information into clusters by creating new clusters of information or attaching the neurons of new information with others that are retrieved from the long-term memory.

Working memory is where and when learning takes place. It is important to develop study strategies to ensure that you allow the brain to do what it must in order to create understanding. Most of the working memory activities take place studying outside of class, completing homework, and during class discussions.

Study Plan Application Activity

1. Describe your typical study habits. When? Where? How long are your study sessions? **Answer.**
2. What strategies do you use to organize the information you learn? **Answer.**
3. What strategies do you use to memorize information? **Answer.**
4. What strategies do you use to practice analyzing information? Evaluating information? Understanding concepts of vocabulary and relationships between these concepts? **Answer.**

Long-term memory stores schemas (clusters of neurons that hold information) that were developed in working memory. It has large capacity and time endurance. One of the most important aspects of long-term memory is that it influences the quality of encoding, short-term memory, working memory, and long-term retrieval. This is why it is important to use strategies to organize information while learning it. The more organized a file cabinet is, the easier to find files quickly and when you need them. In the same way, the more organized the information that enters the long-term memory, the easier it is to retrieve.

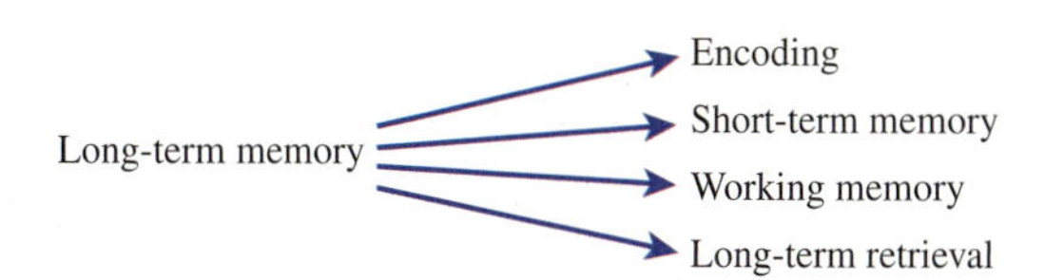

Long-term retrieval involves recalling information from long-term memory into either short-term memory while encoding and/or working memory where the brain needs information from long-term memory to understand new information, answer questions, or apply to new situations. Remember the clusters of neurons formed in working memory and stored in long-term memory? During retrieval, the brain replays these patterns and decides which long-term memories to retrieve given any situation.

Study Plan Application Activity

1. What strategies do you use to organize the information you learn? **Answer.**
2. What strategies do you use to practice retrieving information from long-term memory? **Answer.**

Conclusion

It is now time to see ways each stage of learning affects the other stages. For example, since short-term memory is very limited in time and how much information it can hold, it is important to make sure that the information is recorded when first encoded. Most of your study time should be devoted to working memory, where information is studied long enough and organized to place into long-term memory. Otherwise, you will forget it because it is discarded. Your long-term retrieval is minimal.

The more time you study and ask questions regarding the information you learn, the easier it is to retrieve. It is also important to practice retrieving information for accuracy, effectiveness, and the speed at which you can retrieve it. You retrieve information during all the other stages of memory/learning and during tests. The memory stages, although separate entities when we learn about them, act collaboratively within seconds and minutes.

SECTION 0.2 The Study Plan for Statistics

The following study strategies are based on the way the brain learns. Each study strategy assists in several stages of the learning process. None of the following strategies are busy work, because there is not a single student who has time for busy work! The study strategies are organized based on the primary stage of the learning process we believe it aids. Some strategies will be repeated because they help with several stages of learning.

Strategies to Improve Listening

Active listening is intensive. Your brain uses every stage of learning throughout the entire class. Cognitive overload, also known as "brain fog," may happen in statistics classes. The following strategies will help you encode and hold on to the information long enough to record it in your notes. In addition, these strategies will free up more mental energy for thinking (working memory) while in class.

1. Warm up at least 5–10 minutes before class. Review your notes and homework. This helps with the attention required of you as soon as the instructor begins class. It will also speed up retrieving information from long-term memory that is necessary to learn new statistical concepts in class.

2. If you have questions about the previous class meeting, find time to ask the instructor or colleagues from class during your study times. Otherwise, ask them in class if the instructor provides time for review before moving on with the new material. Prepare them ahead of time, so you won't waste class time with vague questions.

3. The night before class, preview the chapter that contains the material to be covered the next day. Spend some time thinking about the new concepts and how they relate to what you have learned already. You do not have to understand everything. If you note that some concepts appear confusing, plan to ask questions when the instructor presents them in class. Write these questions down, so you won't forget to ask them. Make sure you write the answers down when the instructor answers them.
4. Come to class with completed homework even if you do not turn it in.
5. Make sure you have studied everything up to the next class. At least review it. It may not be easy to retrieve, but at least understand the concepts.
6. Sit where you have the least distractions, can see the board, and can hear the instructor.
7. For more suggestions, read pages 48–49 and 56 in the *Math Study Skills Workbook, 5th edition.*

Strategies to Record Information during Class

Remember that encoding and short-term memory last just moments. Unless you have time to record, rehearse, and discuss it in class, most of the information is out the classroom window.

1. Make sure you complete all the strategies mentioned above.
2. The goal of efficient note-taking is to use the least amount of words to record the greatest amount of information. Never write complete sentences; instead, use short phrases. There are times during statistics classes when mapping is the best technique to use. When the information seems complicated, ask the instructor to design a map or visual to help put the pieces of information together.
3. Always record what is written on the board. And don't forget to write down what the instructor says about the information placed on the board. Not writing down the instructor's comments is a common mistake, but doing so is important because the instructor comments are the explanations of what is on the board.
4. Make sure to capture all the vocabulary words and definitions. If the definitions are long, capture the key words and then fill in after class. If you have time, it is quite helpful to preview the vocabulary before class, even writing the words down along with the textbook definition. Then all you have to do is add what the instructor uses as a definition.
5. Record any questions that the instructor says are important when thinking about different statistical methods. The critical-thinking aspect of statistics is important for solving problems and designing your own research.
6. If you experience cognitive overload, note when it happened in your notes. Then you can make sure you find the information you missed when working with a tutor or on your own.
7. Just because you may understand what is discussed in class does not mean you will remember it, particularly any of the details. Our brain loses information as hours and days pass unless it is rehearsed and studied.

Strategies to Transform Class Notes into Study Tools

In college, students are responsible for the majority of learning. Students are responsible for learning more at a faster pace. Thus, studying involves more than glancing at notes and textbooks. Class notes are important because they contain what the instructor views as significant to learn. They also include the instructor's unique perspective on what you must learn. Organizing class notes into study tools will help in the following ways. First, organizing notes helps to see how the details work with one another. While organizing class notes, you spend time working with the information in the working memory, where the learning takes place. The process of reorganization is itself profitable study time.

Second, the more organized the information, the easier it is for the working memory to organize it and send it into long-term memory. Third, organized study tools help you retrieve information from long-term memory.

1. Make sure your notes have all the correct information. One of the most common challenges for students is cognitive overload during statistics classes. When on overload, it is easy to miss pertinent information. Find at least two classmates willing to share notes. Take a few minutes to take pictures of each other's notes with your phones right after class. There is only one caution: make sure their notes are correct!

2. Use the textbook and other online resources to fill in gaps.

3. Develop a vocabulary resource. The following matrix works well in statistics because it makes you think about how the concepts behind the vocabulary relate to one another and how they are used in statistics. Use a complete 8 × 11 page for it because as you learn more about statistics, you will want to fill in more information. See page 55 in the *Math Study Skills Workbook* for more ideas. When the vocabulary words involve graphs, include a sample graph for each one.

Vocabulary Word	Definition	What Does It Tell Us?*	What Does It Not Tell Us?*	What Situations Is It Best Used In?*
mode				
mean				
median				

*Note that these questions may be different with various vocabulary words.

4. Maintain a journal of statistical formulas. Once again, use an entire 8 × 11 page for each formula because you will learn more information about it throughout the course. You may find that this will take more than one page for each formula. The following is an example of one page from a statistical formula journal. Instructions are included in the example of what to include.

Sample Standard Deviation

Is a measure that is used to quantify the amount of variability in a sample consisting of numerical data.

$$s = \sqrt{\frac{\sum(x - \bar{x})^2}{n - 1}}$$

Leave space for other information instructor presents in class regarding the standard deviation throughout the course.

Answer specific questions that are important for the formula like
What does it tell us? What does it not tell us?

Write out a sample standard deviation word problem.

Complete the standard deviation formula.

Continue to add problems that use the formula but are more involved as you continue to learn more.

5. Use your phone to take pictures of each formula and record information about the formula, so you can review on the go. For more information about reworking your notes, refer to pages 57–59 in the *Math Study Skills Book*.

How to Use the Textbook When Studying

Often students use the textbook as the last resort when they do not understand something. However, it is a useful tool in all the stages of learning. First, previewing the textbook in preparation for the next class assists in encoding, taking notes, and thinking actively during class. Strategies for this have already been discussed. Second, using the textbook to fill in class notes is helpful. Again, this has already been discussed. Third, studying the content of a chapter helps to put the pieces together in the working memory. Auditory learners should read the sections of a chapter out loud. Fourth, reviewing the chapter for 20 minutes each day helps move the information into long-term memory. Memorizing how a concept looks on a page can help in longer-term retrieval, particularly if you are a visual learner. Finally, reviewing the chapter learning objectives, conclusions, and practice tests can make sure you are learning what is important. If you want more suggestions, refer to pages 64–67 in the *Math Study Skills Workbook, 5th edition.*

The following strategies are designed to help you with the statistics textbook you are currently using.

1. Skim the assigned reading material. Look at the learning objectives listed to get an overview of what you will learn. As you skim the chapter, circle novel words that you do not understand with a pencil. This step is to simply get an overview of what you will be learning. These two skimming exercises should only take 5–10 minutes.

2. After skimming, read again to familiarize yourself with the material enough to prepare for the next class. Remember, this step will help you encode and keep new information presented in class in short-term memory long enough to record it in your notes. Mark the concepts and words that you do not understand. Another option is to write the vocabulary words in your vocabulary list. If you do not clearly understand a word or concept, make sure you ask about it in class.

3. After class, study the textbook chapter. Put all your concentration into reading. At this point, you are reading to learn.

   a. During this reading, also rework your notes into some of the learning tools mentioned in this study strategy chapter (such as using the textbook and other online resources to fill in any gaps in your notes, and creating a vocabulary resource and journal of statistical formulas).
   b. Sometimes, it is best to study the textbook chapter in sections because you are using the working memory at this stage of learning. Working memory requires a respectable amount of mental energy.
   c. Pay attention to the examples. Do not skip them. The examples help conceptualize what you need to learn. Copy and work out the examples in the learning tool you have selected.
   d. Erase the circles around the words you now understand.

4. Remember the learning objectives you previewed while skimming the chapter? It is time to reread them. Make sure you understand what the verbs mean in each objective. For example, consider the following learning objective:

   Evaluate *whether conclusions drawn from a study are appropriate, given a description of the statistical study.*

   Achieving this objective requires knowing the different types of statistical studies and thinking about the relationship between how data are collected and the types of conclusions that can be drawn.

   Make sure your learning tools contain all the information you need to understand, memorize, and use for answering specific questions that the instructor might ask on a test to make sure you have attained this learning objective.

5. Finally, skim and review the chapter several times throughout the week to keep the information fresh and to practice retrieving it from long-term memory. Take only 5–10 minutes to do this. Maybe just use this step for the sections that you do not completely understand.

6. If you get completely lost in a chapter, do the following:

   a. Go back to the previous page and reread the information to maintain a train of thought.

   b. Read ahead to the next page to discover if any additional information better explains the misunderstood material.
   c. Locate and review any diagrams, examples, or rules that explain the misunderstood material.
   d. Read the misunderstood paragraph(s) several times aloud to better understand the meaning.
   e. Refer to your class notes for a better explanation.
   f. Refer to another statistics textbook or resource.
   g. Define exactly what you do not understand and call a classmate.
   h. Visit a tutor.

7. Recall important material. If you cannot recall the material, look for the information in the textbook.

8. Reflect on what you have read. Combine what you already know with the latest information that you just read.

9. Write anticipated test questions. Use the learning objectives as a guide to writing appropriate test questions.

The textbook is your friend. The more time you spend studying with the textbook, the more comfortable it will be. Make sure you read the textbook before the next class and identify what to listen for in class. This strategy is mentioned in the previous section on improving listening in class. Also, take time now and then just to read it with no goal in mind except to refresh.

Completing Homework

Homework is not limited to completing assigned exercises. All the suggestions so far are study strategies that are vital in preparation for completing homework problems. Completing assigned exercises is not the time to learn the concepts for the first time. The purpose of homework problems is to increase critical-thinking skills such as application, synthesis, and evaluation. Each problem is meant to help you learn specific concepts and applications.

The following strategies will help you turn homework time into study and test preparation time.

1. Figure out which learning objective(s) each problem is helping you learn. Write down the objective number. This will help you use your homework when reviewing for a test.

2. Also record the concepts and formulas by the homework problem.

3. Write out the questions you must answer to complete a problem. For example, if you are using the textbook *Statistics: Learning from Data, 2nd edition,* there are four questions on page 15 that you should be able to answer to evaluate whether a sample has been selected in a reasonable way. There is also a discussion of types of bias. The descriptions of these types of bias can be turned into questions you can ask as you complete homework problems. Here is an example of how to do this with one of the four questions mentioned on page 15.

Was the sample selected in a reasonable way?	Does the method of selecting a sample exclude important groups in the population, making it so that the sample does not represent the entire population to be studied?
	Are members of the sample volunteers or self-selected?

It always helps to have specific questions to ask when completing homework. This helps you be flexible if a test question is slightly different than what you have completed before. While studying statistics, always ask, "What questions need to be answered to solve this problem?" Check the discussions for information that should be turned into questions to ask while completing homework. Always write these questions down.

This strategy is extremely important because it helps the working memory organize and make sense of the information you must learn.

4. Here are basic steps for completing homework.

   a. Review the textbook before beginning homework.
   b. Review class notes before beginning homework.
   c. Complete homework as neatly as possible.
   d. Write down every step of a problem.
   e. Understand why each step in a problem is taken.
   f. Always finish homework with successful completion of a problem.
   g. Make up note cards containing hard-to-remember problems.
   h. Do not fall behind.

Online Homework

Completing online homework is similar. The steps described above are also important for online homework. However, students tend to ignore writing down notes about important concepts they discover while solving online problems. Take notes!

Also, write down important online problems, so you will have easy access to them when you find a few extra moments to study. If you show your work online, take pictures of the steps with your phone. For difficult problems, record how you solved them in your notes.

Making Time to Study

Finding time to study for your classes is one of the major challenges in college, particularly when a test is looming over you. This is one more reason students studying statistics need to spend more time studying and completing homework as it is due. It is almost impossible to cram for a statistics exam.

1. Compile your vocabulary resource on a regular basis. Use the chart system as illustrated earlier. You can use large note cards if you want.

2. Make time to study and review throughout the weeks, not just before tests. Use a large note card and write one learning objective from the textbook on each card. Write the vocabulary words (without definitions) that relate to the learning objective. Write a sample problem and other pertinent information. These are review cards, so use short phrases that can be used just as reminders. While reviewing, if you forget some of the information, refer to your reworked class notes and problems.

3. Keep up with your formula journal. Then make review cards with just the formula on one side and a brief descriptive reminder on the back. This is just for review to keep information in long-term memory.

Developing Review Strategies for Statistics Exams

Preparing for a test starts after the first class, not a few days before the test. Ideally, the week before a test should involve practicing test questions within the time limit of the exam. Understanding all the information is not enough. You must be able to answer all the test questions within a certain amount of time. The following strategies will help you get to that point a week before the test.

1. As you study between classes, study as if preparing for a test or quiz. Design a study plan using the above strategies, and review the resource tools and homework with notes on a regular basis.

2. It is also necessary to practice unfamiliar problems on a regular basis. Ask your instructor for resources for additional practice problems.

3. When you are confident that you know how to complete a certain type of problem, begin solving unfamiliar problems under a timed setting. Ask your instructor for reasonable time allotments for certain types of problems. Do this throughout the weeks, not a few days before the exam.

4. Ask the instructor what type of questions will be on the test. Many statistics exams include complicated multiple-choice questions. These questions usually require more than recall, and many involve application and critical thinking.

5. The first test is always a challenge because students do not know what it will be like. One of the best strategies is to do the following:

   a. Form a study group of three or four students within the first few weeks of class. Keep it small.
   b. Let the instructor know about the study group.
   c. Ask for some examples of the types of questions that he/she typically asks. Tell the instructor you do not want questions that will appear on the exam. Tell the instructor you want just a few examples, so that the study group can make up its own practice questions.
   d. As a study group, review the material that the exam will cover and design similar test questions. This will help you begin thinking like the instructor thinks, which is a great advantage, particularly on the first exam.
   e. If you are using the textbook *Statistics: Learning from Data, 2nd edition,* review the *Explorations in Statistical Thinking* sections at the end of each chapter.
   f. If you are using the textbook *Statistics: Learning from Data, 2nd edition,* continually review the *Are You Ready to Move On* questions at the end of each chapter. If your exam will include a take-home portion to assess how well you can use the statistical technology, review the *Technology Notes* section. It is important to review how to use the appropriate technology so that you do not waste time looking up how to do something while completing a take-home exam.

6. It is difficult to design a complete practice test because of all the information and level of sophisticated thinking the exams require. There is a solution.

   a. If you concentrate on the *Critical Thinking* portions of each chapter and work with a study group, you will acquire the mindset of a statistician while reviewing material.
   b. Meet with your study group at least once a week to review what you have learned. Design quiz questions every week. Several recent studies in the growing area of neuro-education have discovered that taking quizzes a short time after initial learning significantly improves subsequent retrieval of facts and ideas, as well as overall understanding of topics and the ability to solve related problems.
   c. Save these quizzes and review them the week before the test.
   d. Understand the concept of number sense. This means if the answer does not look right, you may have made a mistake. In other words, test answers should make sense. For example, you were asked to calculate the mean of ten numbers that range from 1 to 15. If your answer is 65, number sense tells you that the answer is outside the range of data values and so it is not a possible value for the mean. This does not make sense. Maybe the answer is 6.5. In any case, you should check your calculations to see where the error occurred. This is number sense.
   e. For more test-taking suggestions, refer to pages 103–114 in the *Math Study Skills Workbook, 5th edition.*

Conclusion

Now that you have read all these suggestions, you may feel like the process of studying statistics is as challenging as the actual statistics course material. To follow through with a study plan, incorporate a few new components each week. For example:

Week One	Strategies to Improve Listening Strategies to Record Important Information in Notes Turning Class Notes into Study Tools
Week Two	How to Use the Textbook Form a Study Group and Incorporate All Strategies Mentioned for Study Groups
Week Three	Completing Homework
Week Four	Develop Review Tools for Statistics Exams

The components for this study plan will increase the efficiency of studying and help process information the way the brain requires for learning. Always take time to evaluate the methods you use to study statistics. Study time is precious, and you must make the most of it. If you are using the textbook *Statistics: Learning from Data, 2nd edition,* there are excellent learning objectives to help you identify what to learn and how to think like a statistician.

SECTION 0.3 Productive Self-concept through Mindfulness

Self-concept

Usually we associate intelligence and "hard work" with academic success. The first section of this chapter indicates that hard work is not enough; rather, smart study that allows the brain to work efficiently is necessary. In recent years, however, experts have learned that affective qualities also influence academic success. One such affective quality is self-concept. **Self-concept** is the set of beliefs people have about themselves. The self-concept is formed by the following:

1. Experiences
2. Your personal perceptions of past experiences
3. Thoughts you have about yourself (self-dialogue)
4. Your perceptions of how others view you
5. Your perceptions of your image and abilities

There are many ways to think about self-concept. For the purposes of this chapter, the discussion is organized according to three components identified by Dr. Carl Rogers, a renowned psychologist: self-image, self-ideal, and self-esteem.

Self-image

Self-image is how you perceive yourself in the present moment. For example, in the academic setting, think about how you would answer the following types of questions, either consciously or subconsciously:

1. How do I label myself in the setting of a specific class? While studying?
2. What does the instructor think of me? How about my family? My friends?
3. What are other students thinking of me in class?

Answers to these questions defines self-image. What is important to realize? Self-perceptions are often incorrect or exaggerated.

Self-ideal

Self-ideal is how you wish you were in the moment and in the future, the future as in the next few hours, year, or even decade. Self-ideal can be positive if the following circumstances exist:

1. The self-ideal is realistic.

2. The person accepts who he/she is currently while striving for higher goals.
3. The self-ideal motivates productive and harmless action.

However, many students often have unrealistic self-ideals. Sometimes, these self-ideals are created by others close to them, like family and friends. Students sometimes begin to own the ideals others have for them. Unrealistic self-ideals often create harmful behaviors such as (a) avoidance, (b) giving up too soon, or (c) sabotage. In the academic setting, these behaviors may resemble the following:

1. Waiting until the last minute to study for a test
2. Studying for other classes first and then math or statistics
3. Dropping the course before the first test
4. Dropping the course after performing poorly on the first test
5. Making excuses like poor instruction, a busy schedule, or a hard class load

What you must always remember is that self-image and self-ideal are perceptions, and perceptions may be incorrect.

Self-esteem

Self-esteem encompasses your emotional status in the present moment. Your self-concept and self-ideal shape your self-esteem. Here is an example of the interaction of the three components in an academic setting.

Since his toddler days of constructing awesome buildings with Legos, his family constantly mentions he should be an architect. Throughout his first 12 years of school, Steven always heard, "If you want to be an architect, you need to get A's in math." Steven earned some A's in those 12 years, but more B's, even a couple of C's. He constantly doubted himself when it came to his math classes. He did not see himself as an exceptional math student like his friends and family. During his senior year, conversations during every family gathering focused on what he wants to do for a career. Everyone agreed he has a talent for architecture, but they also agreed he would have to excel in math courses. The family dreamt big for him. Steven understood that they all meant well. Actually, Steven aspired to be a city planner for a large city one day, close to an architectural career but not quite. During his first semester, he enrolled in Calculus I and Statistics I, both required courses for his degree. At the end of the first week, he reflected on his classes, reread his syllabi, and started reading the textbooks. As he stared at each page, he felt pressure like before, only much worse. He could not shake "what if" thoughts and doubted that he could handle the situation. He panicked and decided to go shoot some hoops.

As a math student, what was his self-image?
What self-ideal does Steven struggle with?
What is Steven's self-esteem after the first week of college regarding math and statistics?

Recent research has demonstrated that self-concept and motivation are key affective traits for successful persistence. After examining how to maintain a positive, realistic self-concept and motivation, you will be able to create a realistic and positive ending for Steven in the above scenario.

Mindfulness

This section explores a couple of components of *mindfulness,* a psychological process that helps people focus on the present moment, not the past nor the future. When practicing mindfulness, you do not judge your emotions as either good or bad. You accept the emotions you are experiencing in the moment. Then you learn how to interact with these emotions in a positive way so that you may focus on what you are doing in the present, like studying statistics.

The Past, Present, and Future Interactions

Why focus in the moment? We cannot change the past and we cannot predict the future. We can influence the present. The following diagram overviews this component of *mindfulness*.

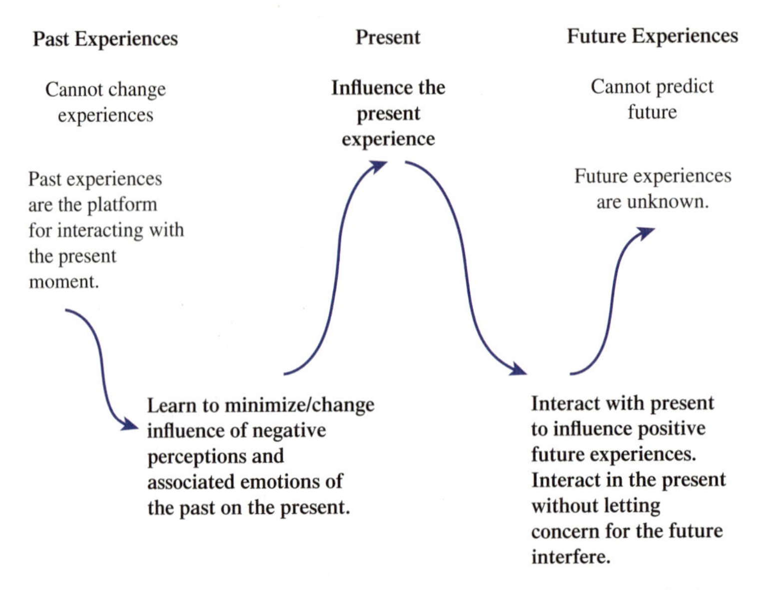

One way to learn how to separate past experiences and the attached emotions from the present is to examine your self-dialogue. Take time to complete the following self-dialogue matrix. Record what you say to yourself regarding negative past experiences and emotions that creep up in a situation that causes nervousness or anxiety. Rehearse the positive dialogue.

Negative Dialogue Regarding Past	Emotions from Negative Dialogue	Positive Dialogue to Replace Negative Dialogue Regarding Past	Emotions from the New Dialogue

Write dialogue that reminds you that the past does not have to completely overwhelm the present situation, dialogue that reminds you of your strengths and keeps you focused on the present moment and not worrying about the future, even just a few hours into the future.

It is likely that you will need to use this strategy quite often in different situations during your college years.

Another strategy is to find other students and form a study group that meets once a week. A study group keeps you focused on the task and limits the opportunities to drift off into negative self-dialogue about the past or future. It also provides support as each member takes turns with challenges he or she might have. Study groups are also fun.

Emotions and Ability to Focus in the Present

Remember discussing the working memory required by the brain to learn? Studies repeatedly have associated anxiety with diminished working memory ability. Yes, life happens while in college, but successful students learn how to set aside interfering emotions in order to study.

It may not make sense when you first think about it, but the more you fight emotions, the stronger the emotions become. The more you worry about being anxious, the more mental time you spend on being anxious and the anxiety takes over. The reverse is also true. Perhaps you were accepted into the club of your first choice or someone said yes when you asked him/her out for a date. You are excited and can't concentrate on anything. However, many students allow anxiety to overcome them. When you learn to interact with anxiety and accept that it is present and do not judge yourself, the less anxiety you will experience. *The first step is to learn that emotions do not define who we are.* A popular *mindfulness* analogy for learning to accept emotions without judgement is illustrated in the following figure.

Emotions are not good or bad, and they do not define who you are. The main steps to interacting with emotions are (1) acknowledge them, (2) think about why you are having them in the present moment, (3) create a self-dialogue that helps you to prevent the emotion from overwhelming you in the present situation, and (4) design a plan to complete the task at hand.

Strategies for Learning How to Interact with Emotions

1. Find friends who are positive and nonjudgmental. They should also be good listeners. In turn, you can be the same type of friend for them. They will be the type of students to invite to go to the library and study, even if you are not in the same courses.
2. Meditation is a key intervention for examining and learning how to be okay with emotions. You can download online meditation guides that last from a few minutes to an hour or more. Some colleges even offer meditation sessions and yoga classes. If not, find some other friends and start a yoga or meditation group.
3. In addition, select a study strategy to use during study time. Active study strategies require your attention, and thus there is less time to let the mind drift.
4. Think of a creative way to symbolize setting the emotion aside for the moment. For example, take a piece of paper and write the emotion on it. Wad up the paper and throw it in the waste basket.

Life keeps going even though you are in college. When emotions begin to overwhelm you on a regular basis, it is important to seek assistance. Most colleges have a counseling center, which is a good place to begin. It is just important to admit that you cannot manage your emotions and to find help. Many students experience this. The ones who persist are the ones who seek assistance. Remember, there are solutions for the challenges you face in college.

Self-efficacy and Hanging in There When Statistics Gets Challenging

"Self-efficacy is the extent to which people believe they are capable of performing specific behaviors in order to attain certain goals" (chirr.nlm.nih.gov/self-efficacy.php). For example, answer the following question.

Question	Disagree Strongly	Disagree	Agree	Agree Strongly
I am confident that I can earn a B in this statistics course.				

If you answer "agree" or "agree strongly," you have a higher self-efficacy score, which means you are more likely to find solutions when the statistics curriculum is challenging and follow through with learning because you believe you can. If your self-efficacy score is lower, the opposite is likely to happen, and you will either lower your expectations and/or give up.

Self-efficacy is also very specific regarding different tasks. For instance, you may be confident about understanding the concepts in each chapter, but when it comes to the *Explorations in Statistical Thinking* sections, your self-efficacy might be lower. Sometimes certain concepts are easy to grasp while others are difficult, and you think you will never understand them. In this case, self-efficacy depends on the difficulty of the task. When you face situations like these specific tasks and you are discouraged, what do you do?

1. Identify what task is challenging you.
2. Do not procrastinate—it will take more time to accomplish the task.
3. If emotions like frustration, anxiety, anger, or depression emerge, use the strategies mentioned to manage the emotions.
4. Find sources to help you: online sources, classmates, instructor, teaching assistant, tutor.
5. Learn to preview future tasks, so you can prepare ahead of time if they are difficult.

Locus of Control

When you approach academic challenges with strategies from this chapter, you are considered to have internal locus of control. Locus of control is the "tendency of people to believe that control resides internally within them, or externally, with others or the situation" (changingminds.org/explanations/preferences/locus_control.htm). A student with internal locus of control will look at the task at hand and design a strategy to complete it. If the outcome is not as expected, the student does not blame others or external situations for the disappointment. Rather, the student examines the process that was used and revises it when the next task must be completed. For instance, if a student does not perform like he wanted on the first statistics test, he can either blame circumstances or others for the low performance, or he can look at what actions he took to study for the test. He might see that he did not allow enough time to review, or maybe recognize that he did not study enough on a daily basis during the previous weeks. A student with internal locus of control will readjust the study plan for the next portion of the statistics course.

SECTION 0.4 Putting It All Together

Every concept in this chapter relates to persistence when an academic course, like a statistics, gets tough. If you take all these suggestions seriously and use them, you increase the likelihood of persisting and reaching the desired academic goals. It requires action.

It is time to return to the scenario of Steven. Reread the scenario, while also thinking about what you have just learned. Here is the continuation of the scenario.

> *You are a junior in college and one of Steven's new friends. Steven returns from shooting hoops with some friends. You were one of his friends shooting hoops. You notice that Steven chills the rest of the night alone in his dorm room instead of joining in on a video game. At breakfast, you see him and sit down to eat with him. He shares what he went through yesterday. What is your advice? Use strategies mentioned in this chapter to complete the following chart. Assess Steven's challenges. Then select strategies for studying and for managing emotions like stress, anxiety, low self-esteem. Describe how these strategies could help Steven.*

Steven's Challenges	Strategy	How Will This Strategy Help Steven with the Challenge?
1.		
2.		
3.		
4.		

Steven's situation is not unique. Throughout college, students face similar challenges that not only threaten course studies but also the drive to persevere in order to reach desired goals. The following exercise models setting behavioral goals to reach more long-term goals like earning a bachelor's degree in your desired field of study. At this time, however, the exercise will focus on what you should do to successfully perform in the statistics course. Of course, you can do this for all your other courses.

Take Action!

It is time to set behavioral goals. If you focus on how you study and maintain a healthy attitude, you will reach the tangible, end-of-the-course goals you desire.

Complete the following exercises. As you progress in the statistics course, revise them if they are not working.

Objective	Strategies
Improve listening	1. 2. How do these strategies help the brain process information?
Record information in class	1. 2. How do these strategies help the brain process information?
Rework class notes into study tools	1. 2. How do these strategies help the brain process information?
Use your textbook to study	1. 2. How do these strategies help the brain process information?
Completing homework	1. 2. How do these strategies help the brain process information?
Design review strategies for statistic tests	1. 2. How do these strategies help the brain process information?
Coping with frustrations	1. 2.
Managing emotions when they interfere with studying	1. 2.

(*Continued*)

Objective	Strategies
What will you do if you can't manage emotions and academic studies?	1. 2. Remember, there are times when we cannot solve our own issues and must seek assistance.

Even though this chapter focused on how to study and maintain positive attitudes during a statistics course, each of these strategies transfers into the rest of your college experience. Remember, choose what type of college experience you want. Make plans, revise them, learn from mistakes, and celebrate each success!

1 Getting Ready for Statistics

Statistics is different from mathematics, but you do need some basic math knowledge and skills in order to be successful in your introductory statistics course. This chapter reviews several basic topics that will help you to prepare for statistics.

SECTION 1.1 Numbers and the Number Line—A Quick Review

You encounter numbers in many different everyday situations. Numbers allow you to count and to measure various quantities that might be of interest. A big part of statistics involves summarizing and analyzing numerical data, so a quick review of numbers and simple operations on numbers seems like a good place to start the review of the math you need to know in order to be successful in your introductory statistics class.

In your previous math classes, you were introduced to the following terms and definitions. If this all seems familiar to you, you can look quickly at these terms and move on. But if it has been a while since you last studied math, take a few minutes now to review them.

Term	Explanation or Definition
Whole numbers	These are the numbers 0, 1, 2, 3, ... They are the numbers that are used for counting.
Positive number	A positive number is a number that is greater than 0. Positive numbers can be written using the "+" symbol, but often this symbol is omitted. For example, +4 is usually just written as 4. Both +4 and 4 represent a number that is 4 units above 0.
Negative number	A negative number is a number that is less than 0. Negative numbers are written using the "−" symbol. For example, −8 represents a number that is 8 units below 0, and is read as "negative 8" or "minus 8."
Decimal number	Decimal numbers allow you to represent numbers that are not whole numbers. The number 12.345 is an example of a decimal number that is between 12 and 13. Decimal numbers include a decimal point ("."). Each digit in a decimal number has a place value, depending on its location relative to the decimal point. Decimal numbers can be positive, negative, or 0.

The four basic arithmetic operations that are used in calculations are addition (+), subtraction (−), multiplication (×), and division (÷). Although these symbols for the operators are common, there are several other common ways to denote multiplication and division.

Ways multiplication might be represented: There are at least four different ways to represent multiplication. For example, 5 multiplied by 3 might be written in any of the following ways:

$$5 \times 4$$
$$5 \cdot 4$$
$$5 * 4$$
$$(5)(4)$$

Ways division might be represented: There are at least two different ways to represent division. For example, 25 divided by 3 might be written in either of the following ways:

$$25 \div 3$$
$$25/3$$

You will probably be using a calculator to handle any arithmetic that you might need to do in your statistics class, so we won't review these four basic operations here. But it is worth reminding you of some things to remember:

- ✔ Adding a negative number is like subtraction. For example, $12 + (-8)$ is the same a $12 - 8$.
- ✔ Subtracting a negative number is like addition. For example, $12 - (-8)$ is the same as $12 + 8$.
- ✔ Multiplying two negative numbers results in a positive number. For example, $(-5)(-6) = 30$.
- ✔ Multiplying a positive number by a negative number results in a negative number. For example, $7 \times (-2) = -14$.
- ✔ Dividing a negative number by a negative number results in a positive number. For example, $-36 \div -4 = 9$.
- ✔ Dividing a positive number by a negative number (or dividing a negative number by a positive number) results in a negative number. For example, $60 \div -10 = -6$.

Check It Out!

Calculate the following and then check your answers.

a. $-25 \times (-3)$

b. $13 + (-5)$

c. $32 - (-57)$

d. $3 \times (-4)$

e. -6×6

f. $28 \div (-7)$

g. $-72 \div 3$

Answers

a. 75

b. 8

c. 89

d. -12

e. -36

f. -4

g. -24

Relative Position and the Number Line

A number line is a way to represent numbers visually on a scale represented by a horizontal line with values marked off at equal intervals. One example of a number line is shown here.

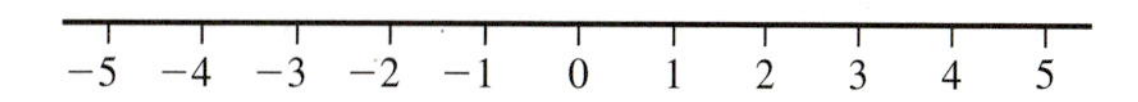

The relative position of two numbers can be determined by where they fall on the number line. Notice that positive numbers are to the right of 0 and negative numbers are to the left of 0. Also notice that numbers increase as you move to the right and decrease as you move to the left along the number line. It is important to remember this, because it is easy to get confused when trying to decide when one number is greater or less than another, especially when working with negative numbers.

For example, -4 is less than 2, and -5 is less than -3.

Check It Out!

For each of the following pairs of numbers, indicate which number is greater and then check your answers.

a. 9 and 4

b. 1 and -7

c. -4 and 1

d. -8 and -9

e. 6 and 95

f. 32 and -8

g. -31 and 13

h. -9 and 0

Answers

a. 9

b. 1

c. 1

d. -8

e. 95

f. 32

g. 13

h. 0

Plotting Points on the Number Line

It is common to represent numbers as points drawn on a number line. For example, the number 4 is represented as:

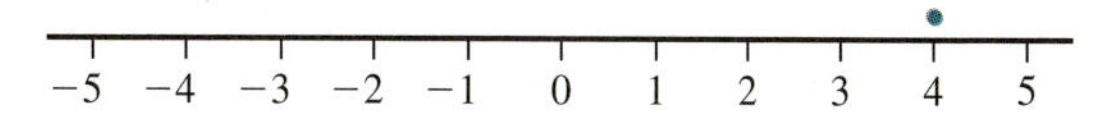

You can also add the number 3.2 to this number line:

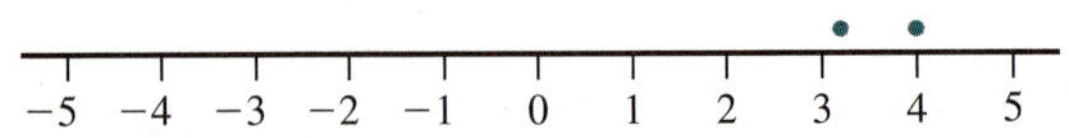

When you get to the chapter on creating graphical displays of data in your statistics course, you will need to be able to plot numbers along a number line in order to create a type of graph called a *dotplot*.

Check It Out!

For each of the following sets of numbers, plot the numbers along a number line and then check your answers.

2, −2, −1, and 3

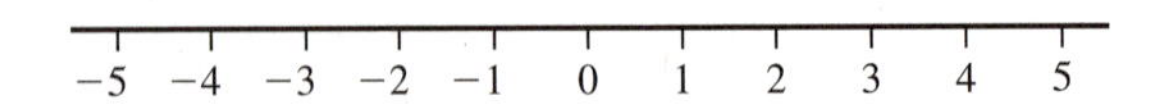

−4.9, 2.1, 4.1 and −1.3

−5 −4 −3 −2 −1 0 1 2 3 4 5

−2.2, −2.5, 3.6, and 3

−5 −4 −3 −2 −1 0 1 2 3 4 5

−0.5, −1.5, 0.6, and 0

−5 −4 −3 −2 −1 0 1 2 3 4 5

Answers

2, −2, −1, and 3

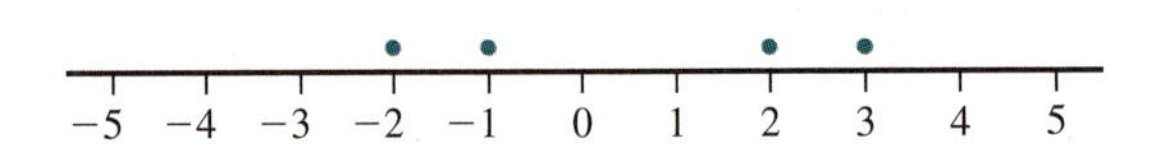

−4.9, 2.1, 4.1 and −1.3

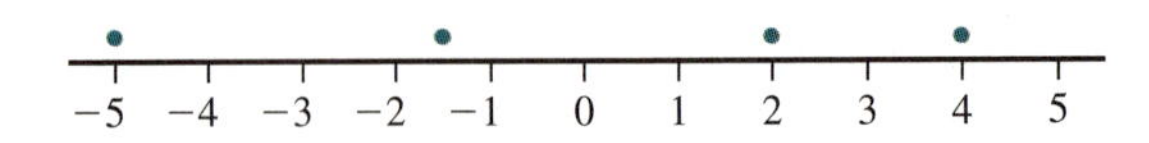

−2.2, −2.5, 3.6, and 3

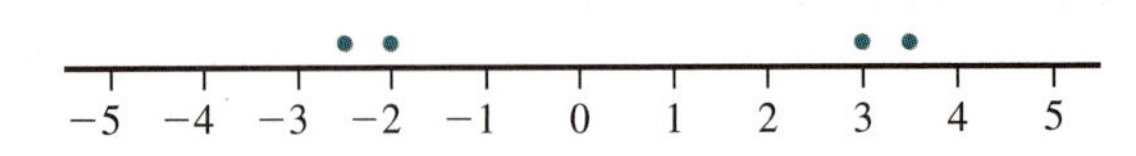

−0.5, −1.5, 0.6, and 0

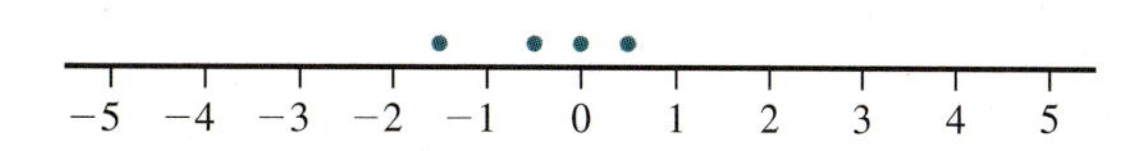

SECTION 1.1 EXERCISES

Calculate the values of the following expressions.

1.1 $68 \div 34$
1.2 $25 \times (-9)$
1.3 $-5 - (-9)$
1.4 -13×-16
1.5 $-8 \div 4$
1.6 $-6 - 92$
1.7 $9 \div 3$
1.8 $5 \div 1$
1.9 $-14 - 3$
1.10 $-4 + -3$

Compare the numbers, and identify which is greater.

1.11 6 and 0
1.12 −4 and 7
1.13 47 and 12
1.14 1 and −8
1.15 −64 and −84
1.16 4 and 7
1.17 −73 and 1
1.18 14 and −3
1.19 36 and −14
1.20 3 and 4

For each of the following sets of numbers, plot the numbers along a number line.

1.21 4, −2.7, −4.7, −1, and 2

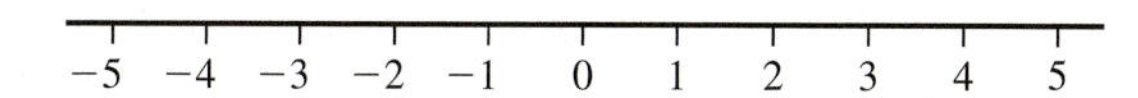

1.22 4, −3, −1, and 4

−5 −4 −3 −2 −1 0 1 2 3 4 5

1.23 −1, 4.3, 6.1, 4.3, 4, and −9

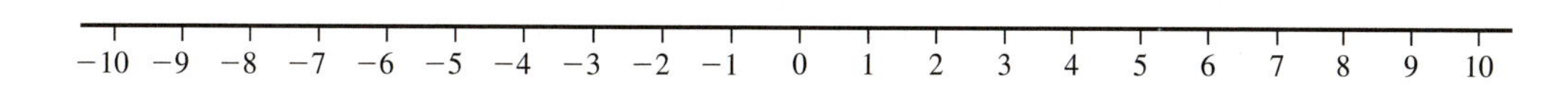

1.24 8, 9.9, 1.3, 8, and −5.5

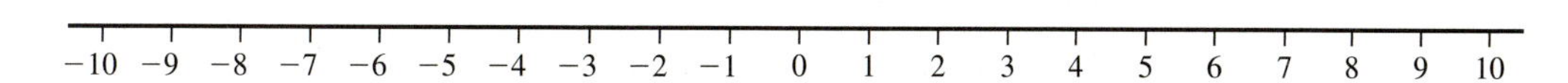

1.25 6, −83.8, −56.4, −54

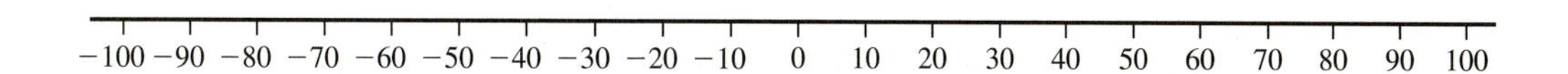

SECTION 1.2 Rounding Decimal Numbers

In your statistics course, you will encounter many situations where you will need to round decimal numbers. For example, you may want to round 24.8976724 to three decimal places. This means that you want only three digits after the decimal place.

To round a number to a specified number of decimal places, you look at the digit that is in the specified position and the first digit that follows it. For example, if you are rounding to three decimal places, you would look at the third digit after the decimal place and the first digit to the right of it, which would be the fourth digit after the decimal place.

The following rules are then used to round the number:

- ✔ If the first digit to the right of the place you are rounding to is 0, 1, 2, 3, or 4, then you drop that digit and all of the digits that follow it.
- ✔ If the first digit to the right of the place you are rounding to is 5, 6, 7, 8 or 9, then you drop that digit and all of the digits that follow it, and then add 1 to the last remaining digit.

For example, to round 24.8976724 to three decimal places, you focus on the digit in the third decimal place (which is 7) and the first digit to the right of it (which is 6).

The first digit to the right of 7 is 6, which is 5 or greater, so drop it and the rest of the digits that follow and add 1 to the previous digit.

Rounded to three decimal places: 24.898

The number 114.03246 rounded to two decimal places is 114.03, because the digit in the third decimal place is less than 5.

Sometimes instead of saying "round to two decimal places" you might be asked to "round to the nearest hundredth." This is just another way of saying the same thing, because the second decimal place is the hundredths place. Here are some other examples of alternate wording that you might encounter:

Round to one decimal place	**is the same as**	Round to the nearest tenth
Round to two decimal places	**is the same as**	Round to the nearest hundredth
Round to three decimal places	**is the same as**	Round to the nearest thousandth

Check It Out!

Round each of the following numbers to the specified number of decimal places, and then check your answers.

a. 4.1783, to two decimal places

b. −3.4316902614, to five decimal places

c. −60.4547, to one decimal place

d. 261.460454700, to three decimal places

e. 2541.270990, to three decimal places

Answers

a. 4.18

b. −3.43169

c. −60.5

d. 261.460

e. 2541.271

SECTION 1.2 EXERCISES

Round the following numbers.

1.26 −31.3175711, to five decimal places

1.27 1220.2387521, to six decimal places

1.28 −69.9674624824, to six decimal places

1.29 −60.329, to one decimal place

1.30 −1.95, to one decimal place

1.31 −2.2904, to one decimal place

1.32 9366.908369, to four decimal places

1.33 −7.49585786, to eight decimal places

1.34 −336.1352, to one decimal place

1.35 8893.4314975776, to two decimal places

SECTION 1.3 Ordering Decimal Numbers

Most people don't have trouble deciding which of two numbers is greater if the numbers are whole numbers, such as 46 and 123. But with decimal numbers it can sometimes be a bit more difficult, especially if the numbers you are trying to order are decimal numbers between 0 and 1. As you study statistics, there are many places where you may have to arrange a list of numbers in order from least to greatest, or compare two decimal numbers to decide which is greater. For this reason, it is worth taking a bit of time to review how to order decimal numbers.

Ordering Positive Decimal Numbers

The easiest way to decide which of two positive decimal numbers is greater is to compare them digit by digit. Start by writing the two numbers you want to order so that they have the same number of digits before the decimal place and the same number of digits after the decimal place. You can do this by adding 0's to the beginning or end of a number as needed—this doesn't change the value of the number. For example, if you want to compare 0.02416 and 0.0237, you would start by adding a 0 to the end of the second number so that both numbers have one digit before the decimal place and five digits after the decimal place.

First number: 0.02416
Second number: 0.02370

Once you have done this, you compare the numbers digit by digit, starting with the left-most digit. As soon as you find a place where the digits differ, the number that has the greater value in this position is the greater number. For the example above:

First number: 0.02416
Second number: 0.02370

This is the first place where the digits differ. Because 4 is greater than 3, the first number is greater than the second number.

Check It Out!

Identify which positive decimal number in the following pairs is greater, and then check your answers.

a. 12.20 or 12.30

b. 0.875216 or 0.875219

c. 9.67462 or 9.6744

d. 824.60329 or 824.60231

e. 95.22 or 95.92

Answers

a. 12.30

b. 0.875219

c. 9.67462

d. 824.60329

e. 95.92

Ordering Negative Decimal Numbers

Ordering negative numbers is a bit different, and some people find it odd that −3 is greater than −5. It helps to keep the number line in mind and to remember that numbers to the left on the number line are less than numbers to the right. Also remember that any positive number is greater than a negative number. This means that 3 is greater than −50, simply because 3 is positive and it is greater than any negative number.

The process is similar to the one used to decide which of two positive decimal numbers is greater, but the rule for deciding is reversed when you are working with negative numbers. Start by writing the two numbers you want to order so that they have the same number of digits before the decimal place and the same number of digits after the decimal place. You can do this by adding 0's to the beginning or end of a number as needed—this doesn't change the value of the number. For example, if you want to order −63.0 and −603.0, you would start by adding a 0 to the beginning of the first number so that both numbers have three digits before the decimal place and one digit after the decimal place.

First number: −063.0
Second number: −603.0

Once you have done this, you compare the numbers digit by digit starting with the leftmost digit. As soon as you find a place where the digits differ, the number that has the SMALLER value in this position is the larger number. For the example above:

First number: −063.0
Second number: −603.0

This is the first place where the digits differ. Because 0 is SMALLER than 6, the first number is greater than the second number.

If you want to compare −10.02416 and −10.0237, you would start by adding a 0 to the end of the second number so that both numbers have two digits before the decimal place and five digits after the decimal place.

First number: −10.02416
Second number: −10.02370

Once you have done this, you compare the numbers digit by digit starting with the leftmost digit. As soon as you find a place where the digits differ, the number that has the SMALLER value in this position is the larger number. For the example above

First number: −10.02416
Second number: −10.02370

This is the first place where the digits differ. Because 3 is SMALLER than 4, the second number is greater than the first number.

Check It Out!

Identify which negative decimal number in the following pairs is greater, and then check your answers.

a. −90.436 or −90.46

b. −0.9083 or −9.063

c. −97.49 or −96.45

d. −85.786336 or −85.7813

e. −35.28893 or −35.28493

Answers

a. −90.436

b. −0.9083

c. −96.45

d. −85.7813

e. −35.28493

SECTION 1.3 Exercises

Compare the numbers and identify which is greater.

1.36 −3.14975 or −3.74975

1.37 −763.3134 or −763.3634

1.38 −97.2 or −97.4

1.39 −411.2732 or −44.2832

1.40 −97.99842 or −97.99892

1.41 1.168025 or 1.168225

1.42 −2.6628751 or −2.6629751

1.43 4.572933 or 4.57297

1.44 8808.33434 or 8808.33464

1.45 −9.50625 or −9.576

SECTION 1.4 Getting to Know Your Calculator—Order of Operations, Powers of Numbers, Square Roots, and Scientific Notation

A scientific or graphing calculator will be an important tool in your statistics course. Even if your class will use statistical computer software, there will still be times when you will need to do basic calculations or to evaluate simple expressions using your calculator.

Take a look at your calculator and see if you can locate the following keys (most calculators have these keys, but your particular calculator might not have all of them).

Key	What It Is Used For
+	Adding two numbers
−	Subtracting one number from another
× or *	Multiplying two numbers
/ or ÷	Dividing one number by another
=	Completing a calculation
+/− or (−)	Changing a number from positive to negative (or negative to positive); used to enter a negative number
x^2	Calculating the square of a number
y^x	Calculating the power of a number
$\sqrt{\ }$	Calculating the square root of a number
( )	Grouping parts of an expression that should be evaluated first

Each of these calculator functions will be considered in more detail in the discussion that follows. Have your calculator handy so that you can follow along.

Simple Arithmetic Expressions with One Operator—The +, −, × or *, / or ÷, and = keys

You are probably familiar with these keys on your calculator—they are used to evaluate simple arithmetic expressions (things like $17 + 43$ or $29 \div 2$ or 58×4). Depending on your calculator, division may be represented as ÷ or /. Multiplication might be represented by × or by *.

The "=" is used to signal that you want to calculate the value of an arithmetic expression that you have entered.

Check It Out!

Use your calculator to evaluate each of the following arithmetic expressions, and then check your answers.

a. $4 + 8.19$

b. 5×37.9

c. $3 - 3.26$

d. $29.4/3$

e. $2 - 3.23$

Answers

a. 12.19

b. 189.5

c. −0.26

d. 9.8

e. −1.23

Simple Arithmetic Expressions with More Than One Operator—Grouping and Order of Operations

Some arithmetic expressions involve more than a single operator. For example, consider the following expressions:

$$3 - 4 - 6 - 8$$
$$3 \times 6 + 54 - 18$$
$$3 \times 8 \div 6$$

To understand how you evaluate expression like these, you need to understand two things: grouping and order of operations.

Grouping

Parentheses are used in expressions to group some of the operations in the expression and indicate that what is inside the parentheses should be evaluated first before evaluating the rest of the expression.

For example, consider the following two expressions:

Expression	Value
$(3 \times 5) + 2$	17
$3 \times (5 + 2)$	21

In the first expression, the parentheses tell you to evaluate 3×5 first, and then add 2. This would result in $15 + 2 = 17$. In the second expression, the parentheses tell you to evaluate $5 + 2$ first, and then multiply that number by 3. This would result in $3 \times 7 = 21$. The placement of the parentheses makes a difference!

Important Note: Sometimes parentheses are used to indicate multiplication. For example, (8)(7) represents 8×7. If you see an expression where two sets of parentheses are next to each other with no operator between them, you should read this as multiplication.

For example, in the expression $(8 + 3)(4)$, you would evaluate what is in the first pair of parentheses first to get (11)(4). Now the two pairs of parentheses next to each other indicate multiplication.

$$(8 + 3)(4) = (11)(4) = 44$$

It is common to see parentheses used to indicate multiplication in statistics.

Check It Out!

Use your calculator to evaluate each of the following expressions, and then check your answers.

a. $13 + (68 + 45)$

b. $(9.8 + 7.2) + (5 \div 7)$

c. $(85 - 68) \times 2$

d. $(8.2 \times 0.7) - 8$

e. $1.9 + (2.3 \times 3.3)$

Answers

a. 126

b. 17.71 (rounded)

c. 34

d. −2.26

e. 9.49

Order of Operations

Consider the following arithmetic expression:

$$3 \times 5 + 40 - 6 \times (8 - 5)$$

How do you evaluate an expression like this? It looks complicated, but there is an agreed upon way to go about it. Just follow these steps in the order given.

1. **Evaluate what is inside parentheses first.** If there are parentheses inside of parentheses, such as in the expression $2 \times (3 + (4 \div 2))$, start with the innermost set of parentheses. For example, to evaluate the expression $2 \times (3 + (4 \div 2))$ you would start with the innermost set of parentheses and evaluate $4 \div 2$ first. This would result in $2 \times (3 + 2) = 2 \times 5 = 10$.

2. **Perform any multiplications and divisions, working from left to right.** For example, if you are evaluating $8 \div 4 \times 2 + 6$, you would first divide 8 by 4 to get $2 \times 2 + 6$. You would then calculate 2×2 to get $4 + 6 = 10$.

3. **Perform any additions and subtractions, working from left to right.** For example, if you are evaluating $8 + 5 + 2 - 6$, you would work from left to right to obtain $13 + 2 - 6 = 15 - 6 = 9$.

These steps specify what is called the "order of operations." There is one more step that will be added when we look at more complicated expressions, but you don't need to worry about that yet.

Check It Out!

Use your calculator to evaluate each of the following expressions, and then check your answers.

a. $6 + (2 - 7) - (9 \times 2) + (5 - 2)$

b. $2 \times (4 - 6)$

c. $(9 \div 3) - (14 \div 2) + (6 - 4)$

d. $(7 + 3) \div (2 + 3) - 2$

e. $(2 \times 4) - 3$

Answers

a. -14

b. -4

c. -2

d. 0

e. 5

Working with Negative Numbers—The +/− key

If your calculator has a +/− key, you use it to change a positive number to a negative number. This is what allows you to enter a negative number. For example, to enter the number −26, you would enter 26 and then press the +/− key to change 26 to −26.

To evaluate the expression $15 + (-26)$, you would

enter 15
press the + key
enter 26
press the +/− key
press the =

If you do this, you should get −11.

Check It Out!

Use your calculator to evaluate each of the following expressions, and then check your answers.

a. $9 \div (5 - (4 - 94))$

b. $(5 - 5) + 81$

c. $4 - ((-4) \div (-8))$

d. $((-82) \div 6) - (-92)$

e. $((-8) \times (-2)) \times (69 \div (-8))$

Answers

a. 0.95 (rounded)

b. 81

c. 3.5

d. 78.3 (rounded)

e. -138

Powers and Square Roots—The x^2, y^x, and $\sqrt{}$ keys

Raising a number to a power is a way to indicate multiplying a number by itself some number of times. For example, 7^3 is read as 7 to the third power, and it is shorthand for $7 \times 7 \times 7$. The **exponent** is the small number written as a superscript (for example, in 7^3, the exponent is 3), and it specifies how many times the number is to be multiplied by itself. Some other examples are shown in the table below.

Expression	Means
6^5	$6 \times 6 \times 6 \times 6 \times 6$
5^6	$5 \times 5 \times 5 \times 5 \times 5 \times 5$
4^2	4×4

To raise a number to a power using your calculator, look for a y^x key. To use this key, you enter the number you want to raise to a power, then press the y^x key, then enter the exponent, and then press the = key.

When you multiply a number by itself one time, this is called squaring the number, and this can be expressed using an exponent of 2. For example, 7 squared means 7×7 and could be written 7^2. Squaring numbers comes up a lot in statistics, so you want to make sure you are familiar with what this means.

Some calculators have a special x^2 key that can be used to calculate the square of a number, but if your calculator doesn't have this key, you can always use the y^x key.

Check It Out!

Use your calculator to evaluate each of the following expressions, and then check your answers.

a. $(-6)^4$

b. $(-1)^5$

c. 14^6

d. 5^3

e. 7^6

Answers

a. 1296

b. -1

c. 7,529,536

d. 125

e. 117,649

More on Exponents—Some Special Cases

The discussion above is really about how to raise a number to a power when the power is a positive integer that is greater than 1. But there are some other cases that should be mentioned. Sometimes you will encounter an expression that has a number raised to a power of 0 or a power of 1. These are defined in the following box.

Any number raised to the power 1 is equal to that number. For example, $8^1 = 8$.

Any number (except 0) raised to the power 0 is defined to be equal to 1. For example, $6^0 = 1$.

0 raised to a power of 0 is undefined.

It is possible to raise a number to a power that is a negative integer. You probably won't encounter this in your statistics course, but just in case you were wondering, negative powers are defined in the following box.

A number raised to a negative power (a negative exponent) is the reciprocal of the number raised to the same positive power. For example, $5^{-4} = \frac{1}{5^4}$ written as a fraction.

Check It Out!

Use your calculator to evaluate each of the following expressions, and then check your answers.

a. 9^1

b. 14^1

c. 4^{-1}

d. 3^{-1}

e. 74^0

Answers

a. 9

b. 14

c. $\frac{1}{4} = 0.25$

d. $\frac{1}{3} = 0.33$ (rounded)

e. 1

Square Roots

The square root of a number is represented using the square root symbol, $\sqrt{\ }$. For example, $\sqrt{16}$ represents the square root of the number 16. When you calculate the square root of a number, you are finding a value that can be squared to obtain that number. For example, if you want to know the value of $\sqrt{16}$, you are looking for a number that can be squared (multiplied by itself) to get 16. Because $4 \times 4 = 16$, the square root of 16 is 4.

It is also possible to square -4 to get 16, because -4 multiplied by -4 equals 16. This means that technically both $+4$ and -4 are square roots of 16. In statistics, any time you need to find a square root, you will want the positive square root, so you don't really need to worry about the negative square root.

Your calculator should be able to calculate square roots for you. Look for the $\sqrt{\ }$ key. To calculate a square root of a number, just enter that number and then press the $\sqrt{\ }$ key.

Check It Out!

Use your calculator to evaluate each of the following expressions, and then check your answers.

a. $\sqrt{8}$

b. $\sqrt{6.2}$

c. $\sqrt{63}$

d. $\sqrt{6}$

e. $\sqrt{0.3}$

Answers

a. 2.83

b. 2.49

c. 7.94

d. 2.45

e. 0.55

Where Do Powers and Square Roots Fit in the Order of Operations?

If you want to evaluate an expression that includes powers and or square roots, you would evaluate them before other operators (multiplication, division, addition, or subtraction. So a more complete set of steps specifying order of operations would be:

1. Evaluate what is inside parentheses first.
2. Evaluate any powers and square roots, working from left to right.
3. Perform any multiplications and divisions, working from left to right.
4. Perform any additions and subtractions, working from left to right.

For example, to evaluate $2^4 - 7$ you would evaluate the power first and then do the subtraction: $2^4 - 7 = 2 \times 2 \times 2 \times 2 - 7 = 9$. To evaluate $(2 - 7)^4$, you would evaluate the subtraction first (because of the parentheses) and then raise that number to the fourth power: $(2 - 7)^4 = (-5)^4 = -5 \times -5 \times -5 \times -5 = 625$.

Check It Out!

Use your calculator to evaluate each of the following expressions, and then check your answers.

a. $\sqrt{((59 - 9) \div (45 + 1))}$

b. $\sqrt{((8 \times 9) \times 9)}$

c. $9 + 3^6$

d. 9^4

e. $\sqrt{97} + 72$

Answers

a. 1.04

b. 25.5

c. 738

d. 6,561

e. 81.85

One Last Calculator "Need to Know"—Scientific Notation

Some calculators use scientific notation to represent very big or very small numbers that can't be expressed using the number of digits available on the calculator display. If that happens, you might see something like 3.768E-8 or 1.277E9. With this notation, the "E" tells you that you need to move the decimal point in the number that precedes the E (which will be a number between 1 and 10). The number that follows the E tells you how many places that you need to move the decimal point. If the number following the E is negative, you move the decimal point that many places to the left. If the number following the E is positive, you move the decimal point that many places to the right.

For example, if you see 3.768E-8, you would start with 3.768 and then move the decimal point eight places to the left, filling in with 0's as needed. This would give you 0.00000003768. If you see 1.277E9, you would start with 1.277 and move the decimal point nine places to the right. This would give you 1277000000.

You may not see this very often, but if you do see it, now you know what it means!

SECTION 1.4 EXERCISES

Evaluate the following expressions.

1.46 -25×6

1.47 $(2 \times 6^3)^2$

1.48 $\sqrt{4} + (-6)$

1.49 $\sqrt{7^3 \div 94}$

1.50 $(39 - 29) \div \sqrt{4}$

1.51 $-(\sqrt{66} \times 4) \times 9$

1.52 $6 - (64 + (-6))$

1.53 $3^3 - 4 - 19$

1.54 $(\sqrt{99} - 4 - \sqrt{57})^2$

1.55 $(8 \div 85) + (-72)$

1.56 $\sqrt{3} \div (77 + 9) \div (-76)$

1.57 $4^2 - \sqrt{5} + 14^2 \times (-5)$

1.58 $(8 + \sqrt{9} + 3^4 \times 7)^3$

1.59 $-(7 \div 61 \div 2 \times 6)$

1.60 $((-4) - 8 - 24)^2 \times 4$

1.61 $70 \div (4 - 0)^2 \times 88 \times 9$

1.62 $(\sqrt{94} - 5^2) \times (5 - (-65)) \div (-6)$

1.63 $(5 + 83 + \sqrt{9} + 55) \div 4$

1.64 $(\sqrt{65} \div 6)^3 - \sqrt{4} \div \sqrt{28} + (-3)$

1.65 $16 \div 8^2 \times 4 \div 7^3 \times (-1)$

2 Creating Graphical Displays—The Math You Need to Know

SECTION 2.1 Review—Rounding Decimal Numbers, Plotting Points on the Number Line

Several of the topics covered in Chapter 1 will be important as you learn to create and interpret graphical displays of data in your statistics course. Here is a brief review of two of them, rounding decimal numbers and plotting points on a number line.

Rounding Decimal Numbers

When creating graphical displays, you may need to round decimal numbers that are between 0 and 1. Remember that to round decimal numbers, you look at the digit that is in the specified position and the one digit that follows it. The following rules are then used to round the number:

- ✔ If the digit that follows the place you are rounding to is 0, 1, 2, 3, or 4, you drop that digit and all of the digits that follow it.
- ✔ If the digit that follows the place you are rounding to is 5, 6, 7, 8, or 9, you drop that digit and all of the digits that follow it, and then add 1 to the last remaining digit.

For example, to round 0.36924 to three decimal places, you would look at the digit in the fourth decimal place. In this example, that digit is 2. Because this digit is less than 5, to round to three decimal places you just drop the 2 and all digits that follow, resulting in 0.369.

To round 0.732876 to three decimal places, you would look at the digit in the fourth decimal place. This digit is 8, and because 8 is greater than 5, you would drop this digit and all that follow and then add 1 to the digit in the third decimal place. This results in 0.733.

Plotting Points on the Number Line

Recall that a number line is a horizontal line with a scale that consists of numbers marked off at equal intervals. Numerical values are represented on a number line by drawing a dot above the number line in the position that corresponds to that numerical value.

For example, the numerical values 8, 12, 6, and 8 are represented on the number line shown here. Notice that if some of the numbers you are representing on the number line are the same, they are stacked vertically above that value.

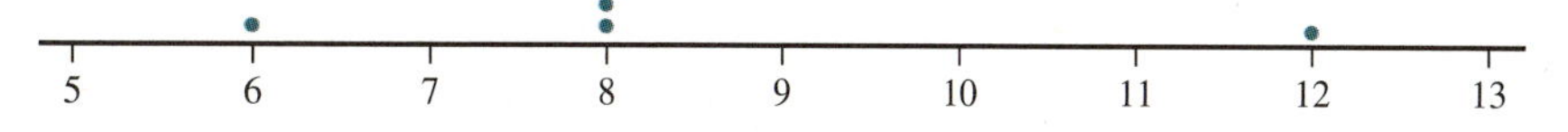

SECTION 2.1 EXERCISES

Round the following numbers.

2.1 0.8856, to two decimal places

2.2 0.3307, to one decimal place

2.3 Plot the numbers using the given number line.
$-3.8, 5, 0.9, -3.7, -0.1$

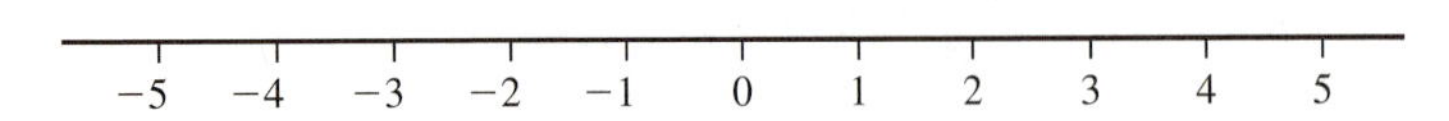

2.4 Plot the numbers using the given number line.
$-2.5, -4.0, -1.4, -1, 2$

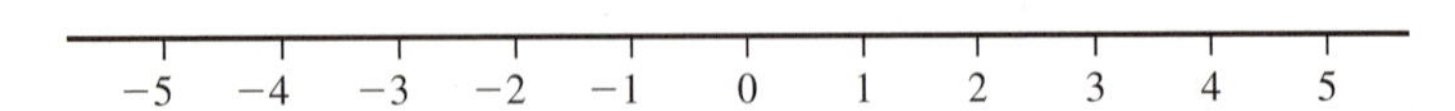

SECTION 2.2 Selecting an Appropriate Numerical Scale

In statistics, you will be creating graphical displays of data. Some of the displays you will encounter (dotplots and histograms, for example) require you to decide on an appropriate numerical scale for your graph.

Let's start with selecting an appropriate scale for a dotplot. A dotplot is constructed by plotting a set of numerical values on a number line. You would start by drawing a horizontal line:

The next step is to mark off an appropriate numerical scale. To do this, start by finding the smallest number and the largest number that you need to represent. When you add a scale to your graph, you need to make sure that the smallest number and the largest number will have a place in the graph. So, when you choose the scale for the graph, you want to start with a value that is below the smallest number you want to represent. You also want the scale to end with a value that is above the largest number that you want to represent.

For example, if the smallest value in the data set is 17 and the largest value is 54, you might choose to start your scale at 15 and end it at 55. You could then add additional scale values between these two endpoints, making sure to keep the spacing between values equal. For this example, you might choose to mark the scale in increments of 5 (using 5, 10, 15, 20, … , 50, 55). This would result in

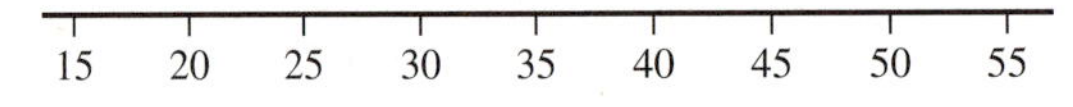

Another reasonable choice is

As you choose a scale, keep the following in mind.

- ✔ Make sure the scale will accommodate all the values that you want to represent.
- ✔ Make sure that the distances between the values that label the scale are equal.
- ✔ Choose scale labels that will make it easy to plot the data values. For example, it may be easier to locate data values if the labels are multiples of 10 or 5 than if they are multiples of 7.
- ✔ There is no one "right" way to construct a scale (but there are some incorrect ways!).

Let's look at an example. Suppose that you have the following data on the time (in minutes) that it took each of 10 people to solve a puzzle.

7 3 14 6 9 22 4 9 11 10

The smallest number in this data set is 3 and the largest is 22. One reasonable choice for a scale for a dotplot would be to start the scale at 0 and to mark off the scale in increments of 5, ending at 25. You should also add a label to the number line that indicates what the scale represents. In this example, you could label the number line "Time to solve puzzle (in minutes)" as shown.

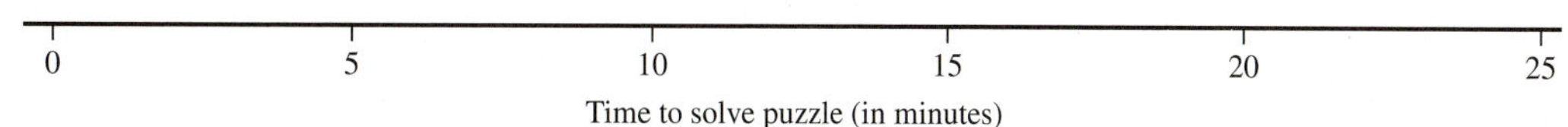

Once you have chosen a scale, you can plot each of the data values using this numerical scale. Notice that the dots are stacked vertically if there are repeated values in the data set.

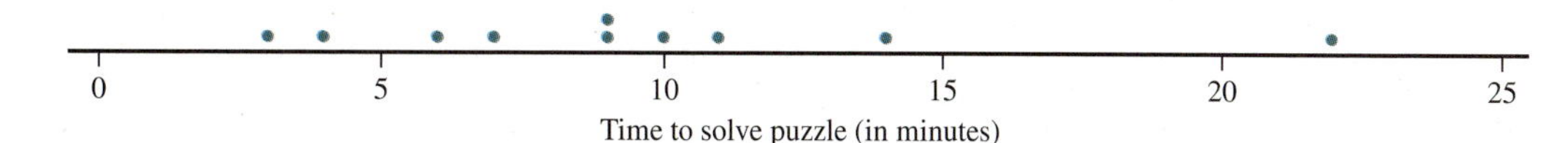

Check It Out!

a. Consider these data: 5, 6, 6, 9, 1, 2, 0, 0, 1, and 7.
Identify at least one problem with the following dotplot:

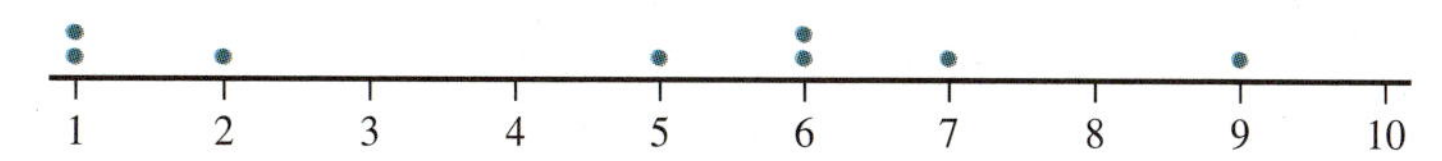

b. Consider these data: 7, 8, 6, 0, 5, 0, 8, 8, 6, and 9.
Identify at least one problem with the following dotplot:

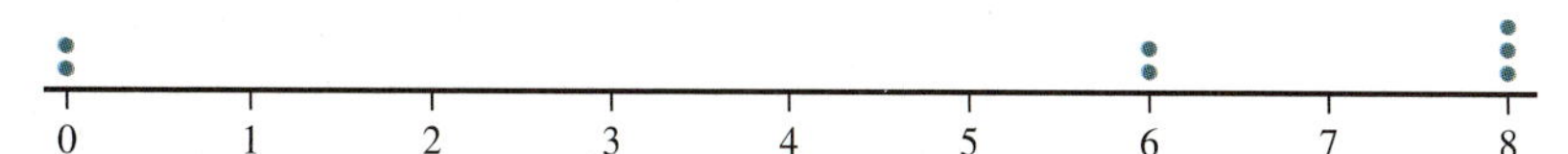

c. Consider these data: 87, 7, 9, 24, 74, 9, 8, 9, 3, and 9.
Identify at least one problem with the following dotplot:

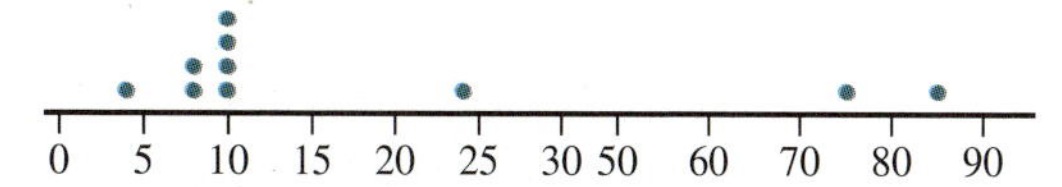

d. Consider these data: 0.6, 8, 4, 1, 4, 0.3, 0.5, 7, 0.5, and 0.7.
Identify at least one problem with the following dotplot:

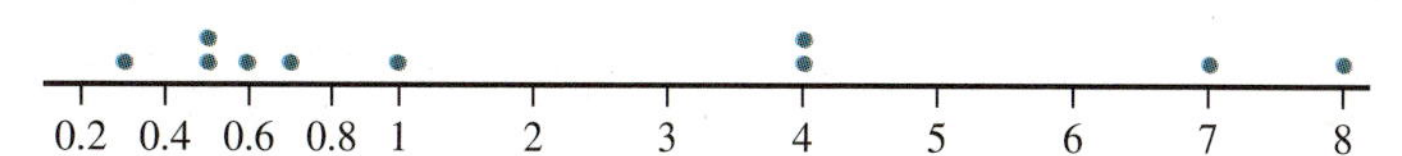

e. Construct an appropriate dotplot for each of the following data sets:
   i. 0.7, 0.8, 2.3, 8.5, 7.2, 3.5, 4.8, 9.6, 7.2, and 1.7
   ii. 9.5, 7.0, 2.5, 9.6, 9.6, 7.4, 4.2, 4.4, 4.7, and 7.1
   iii. 97, 71, 29, 35, 41, 70, 93, 66, 84, and 47
   iv. 12, 76, 78, 81, 25, 23, 99, 71, 86, and 68

Answers

a. The two 0 values are missing, because 0 was not included on the number line.

b. The odd-numbered data values (5, 7, and 9) are missing from the dotplot.

c. The spacing on the number line is different for values greater than 50 and values less than 50.

d. The spacing on the number line is different for values between 0 and 1 and for values between 1 and 10.

e.

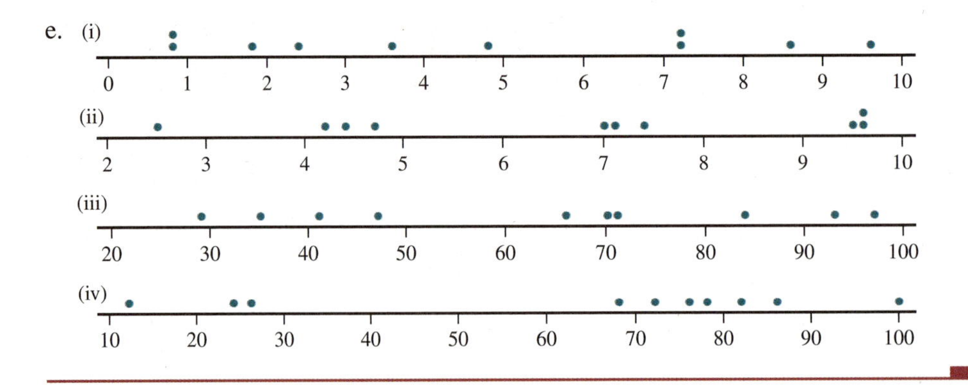

Graphical Displays with Two Axes

The numerical scales that we have used so far have been represented by a horizontal number line. Some of the graphical displays that you will work with in your statistics course have both a horizontal scale and a vertical scale. The term used to describe each of these scales is "axis."

In graphs that use both a vertical scale and a horizontal scale, the axes are used to represent different quantities that you want to graph, and you will need to choose an appropriate scale for each axis. You use the same process that was described previously to add an appropriate scale to each of the axes.

For example, when you construct a graph called a histogram, you might use the vertical axis to represent relative frequency. We will talk more about relative frequency in Section 2.6, but for now what you need to know is that relative frequencies are numbers that are between 0 and 1. To choose an appropriate vertical scale to represent relative frequency, you always start the scale at 0 and then label it at equal intervals. If the largest relative frequency that you need to represent is 0.36, you might use one of the scales shown here.

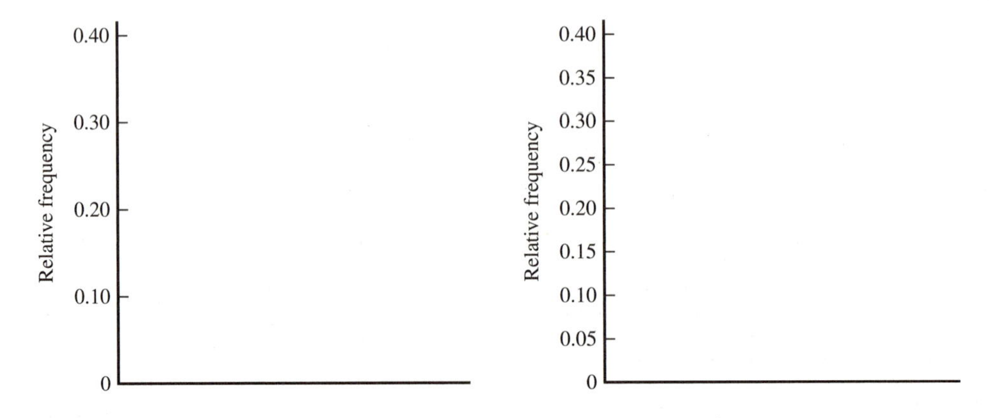

Check It Out!

a. Suppose that the largest relative frequency that you will have to represent is 0.23. Add a scale to the vertical axis.

b. Suppose that the largest relative frequency that you will have to represent is 0.70. Add a scale to the vertical axis.

Answers

a. Answers vary, but one reasonable labeling would be 0, 0.05, 0.10, 0.15, 0.20, and 0.25.

b. Answers vary, but one reasonable labeling would be 0, 0.1, 0.2, 0.3, 0.4, 0.5, 0.6, 0.7, and 0.8.

You will also see graphical displays that use two axes when you cover scatterplots. Adding scales to the axes in this type of graph will be discussed in Section 2.5.

SECTION 2.2 EXERCISES

2.5 Consider these data: 4, 64, 70, 61, 31, 56, 91, 89, 66, and 73.

Identify at least one problem with the following dotplot:

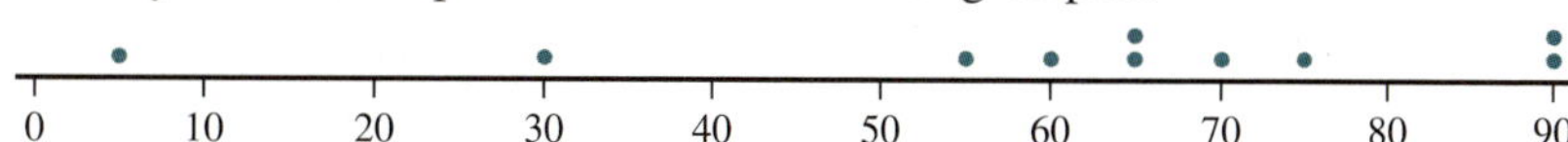

2.6 Consider these data: 8, 2, 4, 1, 7, 7, 2, 7, 6, and 0.

Identify at least one problem with the following dotplot:

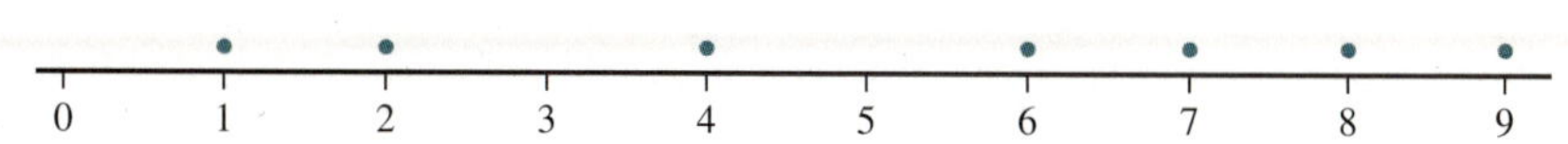

2.7 Consider these data: 10, 10, 6, 14, 15, 5, 4, 14, and 9.

Identify at least one problem with the following dotplot:

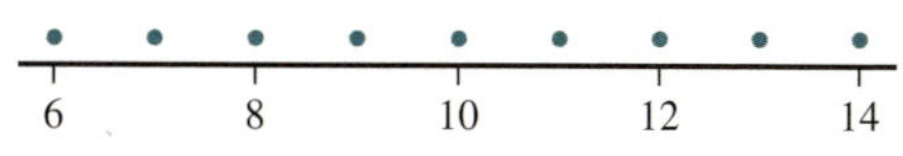

2.8 Consider these data: 0.7, 0.2, 0.1, 1.7, 3.1, 1.6, and 1.1.

Identify at least one problem with the following dotplot:

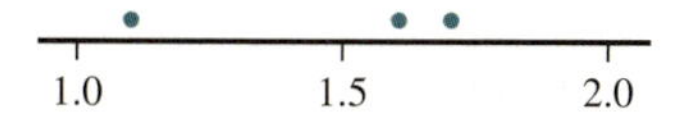

2.9 Consider these data: 0.9, 1.4, 0.8, −0.3, −0.4, 2.7, −0.4, −0.6, and −2.3.

Identify at least one problem with the following dotplot:

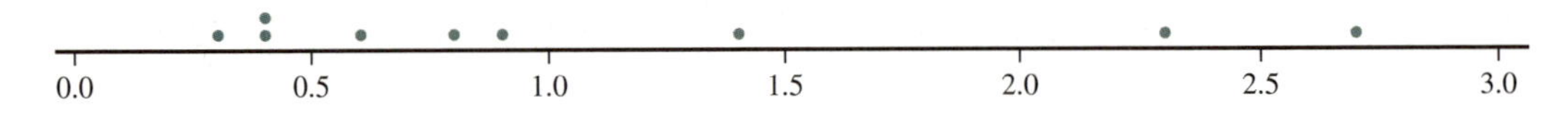

2.10 Consider these data: 0, 15, 15, 26, 30, 40, 50, 60, 70, and 90.

Identify at least one problem with the following dotplot:

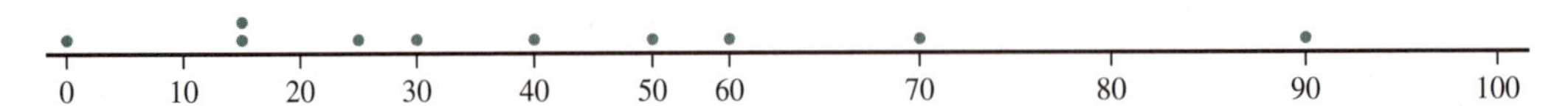

2.11 Samples of 10 students selected from each homeroom at a middle school were asked if they would prefer healthier food choices to be offered at lunch. The proportion of students responding "yes" was recorded for each classroom. Construct an appropriate dotplot for the resulting data set:

0.6, 0.5, 0.6, 0.4, 0.5, 0.5, 0.5, 0.4, 0.5, 0.7, 0.6, and 0.3

2.12 Construct an appropriate dotplot for the following data set:

0.3, 0.8, 0.0, 1.0, 0.7, 0.9, 0.1, and 0.9

2.13 Construct an appropriate dotplot for the following data set:

27, 90, 69, 31, 9, 2, 85, and 88

2.14 Times (in minutes) required to complete an exam in a math class were recorded for nine students. Construct an appropriate dotplot for the resulting data:

34, 35, 46, 24, 74, 72, 64, 68, and 53

2.15 Suppose you think that the largest relative frequency that you will have to represent is 0.31. Add a scale to the vertical axis.

2.16 Suppose that the largest relative frequency that you will have to represent is 0.53. Add a scale to the vertical axis.

SECTION 2.3 Intervals and Interval Widths

In your statistics course, you will work with intervals when you construct graphical displays called histograms and later in the course when you learn about confidence intervals. An interval is a range of values along the number line.

Intervals are represented by two numbers, which are called the endpoints of the interval. The interval itself is made up of all the numbers between the two endpoints. Some intervals include one or both endpoint values.

The usual way to represent an interval is to list the two endpoints inside parentheses and separated by a comma. The smaller endpoint is listed first. For example, (3.0, 6.5) represents the interval that consists of all the numbers between 3.0 and 6.5. To indicate that you want to include the endpoint value in the interval, you can use a square bracket instead of a parenthesis. Here are a few examples:

Interval	Description of Interval
(2, 9)	All of the numbers between 2 and 9
[2, 9)	2, and all of the numbers between 2 and 9
[6, 14]	6, 14, and all of the numbers between 6 and 14
(−2, 2)	All of the numbers between −2 and 2

Check It Out!

Describe the following intervals in words and check your answers.

a. [4, 7]

b. [−99, 1)

c. (−85, −8]

For each of the intervals described below, give the interval notation that matches the description. Check your answers.

d. −7, −2, and all of the numbers between −7 and −2

e. 18, and all of the numbers between 2 and 18

f. 8, and all of the numbers between 8 and 9

Answers

a. 4, 7, and all of the numbers between 4 and 7

b. −99, and all of the numbers between −99 and 1

c. −8, and all of the numbers between −85 and −8

d. [−7, −2]

e. (2, 18]

f. [8, 9)

Sometimes intervals are also described using words and the mathematical symbol "less than," $<$. When working with frequency distributions and histograms, you may see intervals described in the form:

lower endpoint to $<$ upper endpoint

This notation represents an interval that includes the lower endpoint but not the upper endpoint. For example, the interval [2, 4) could be written as 2 to < 4. Both represent the interval that includes 2 and all the numbers between 2 and 4, but does not include 4.

Check It Out!

Describe the following intervals in words, and check against the given answers.

a. -1 to < 9

b. 5 to < 25

c. 0 to < 60

For each of the following intervals described in words, write the interval using the "<" notation. Check to make sure you answer matches the given solution.

d. −7, and all of the numbers between −7 and −6

e. 73, and all of the numbers between 73 and 75

f. −2, and all of the numbers between −2 and 10

Answers

a. −1, and all of the numbers between −1 and 9

b. 5, and all of the numbers between 5 and 25

c. 0, and all of the numbers between 0 and 60

d. -7 to < -6

e. 73 to < 75

f. -2 to < 10

Interval Width

When working with intervals used to construct frequency tables and histograms (and later in the course when you work with confidence intervals), you may need to calculate the width of an interval.

The **width of an interval** is the distance between the lower endpoint and the upper endpoint, which is calculated by subtracting the lower endpoint from the upper endpoint.

$$\text{Interval width} = \text{upper endpoint} - \text{lower endpoint}$$

Note: Interval width doesn't depend on whether the endpoints are included in the interval.

Here are some examples of finding interval widths:

Interval	Interval Width
(4, 7)	$7 - 4 = 3$
(4, 7]	$7 - 4 = 3$
15 to < 20	$20 - 15 = 5$
6 to < 8	$8 - 6 = 2$
1.5 to < 2.0	$2.0 - 1.5 = 0.5$

Check It Out!

Calculate the width of each interval, checking your answers as you go.

a. (7, 9]

b. [−75, −56)

c. $-13 < -5$

d. 6 to < 8

e. -9.3 to < 2.0

Answers

a. $9 - 7 = 2$

b. $-56 - (-75) = 19$

c. $-5 - (-13) = 8$

d. $8 - 6 = 2$

e. $2.0 - (-9.3) = 11.3$

SECTION 2.3 EXERCISES

2.17 Describe the interval (7.5, 9.0] in words.

2.18 Describe the interval [−4.6, 9.4) in words.

2.19 Use interval notation to represent the following interval: All of the numbers between 9 and 82, and 82.

2.20 Use interval notation to represent the following interval: −4.8, 22.7, and all of the numbers between −4.8 and 22.7.

2.21 Calculate the width of the interval (−7.8, 8.0].

2.22 Calculate the width of the interval (6, 26).

2.23 Calculate the width of the interval (35, 51).

2.24 Place the following four intervals in order, from narrowest to widest:

(3.9, 8.0]
(3.0, 8.9]
(−3.9, 8.0]
(0.8, 3.9]

SECTION 2.4 Proportions, Decimal Numbers, and Percentages

Proportions and Decimal Numbers

In statistics, a **proportion** is a number between 0 and 1 (including 0 and 1, which are also possible values for a proportion) that represents a part of a whole group. For example, you might be interested in the proportion of females in the group of people attending a concert. If there were 586 people at the concert and 212 of them are female, the proportion of females at the concert is the fraction $\frac{212}{586}$. This fraction represents the "female" part of the whole group of 586 people. You could say that the proportion of women at the concert is $\frac{212}{586}$, but it is also common to represent a proportion as a decimal number.

To convert $\frac{212}{586}$ to a decimal, you divide 212 by 586:

$$\frac{212}{586} = 0.361775$$

Rounding this decimal to three decimal places would result in a proportion of 0.362.

Suppose that you were interested in the proportion of students at a particular college who live within 1 mile of the campus. If there are 6000 students at this college and 2417 of them live within 1 mile of campus, the proportion of students at this college who live within 1 mile of campus is

$$\frac{2417}{6000} = 0.402833$$

Rounded to three decimal places, this proportion is 0.403.

Check It Out!

a. Convert the fractions to decimals, rounded to three places. Check your answers as you go.

i. $\frac{5}{7}$

ii. $\frac{18}{66}$

iii. $\frac{213}{948}$

b. A college track coach needs to plan recruiting trips, and realizes that 64 of the 88 athletes on the current team will return next year. Rounded to three decimal places, what proportion of the current team athletes will return next year?

Answers

a. i. 0.714

ii. 0.273

iii. 0.225

b. $\frac{64}{88} = 0.727$

Proportions and Percentages

Suppose that you are told that the proportion of students at a particular college who use Facebook is 0.834. What does this mean? The easiest way to interpret a proportion is to convert it to a **percentage**. In fact, there will be many situations in your statistics course that will require you to convert proportions to percentages and to convert percentages to proportions.

To convert a proportion to a percentage:

1. If the proportion is not already expressed as a decimal number, convert the fraction to a decimal number.
2. Multiply the proportion by 100 and add a percent symbol (%).

To convert a percentage to a proportion:

1. Drop the % symbol.
2. Divide the percentage by 100 to get a decimal number.

If the proportion of students at a particular college who have a Facebook account is 0.834, you can convert this proportion to a percentage by multiplying by 100 and adding the % symbol:

$$0.834 \text{ as a percentage is } (0.834)100\% = 83.4\%$$

This means that 83.4% of the students at the college have a Facebook account.

Suppose that you are told that 64.3% of the students at the college are registered to vote. If you want to express this as a proportion, you would drop the % symbol and divide by 100:

$$64.3\% = \frac{64.3}{100} = 0.643$$

Check It Out!

a. Convert the proportions to percentages. Check your answers.

i. 0.75

ii. 0.694

iii. 0.732

iv. 0.7

b. During flu season, the proportion of employees who missed at least 1 day of work during a single week at a local company was 0.123. What percentage of employees does this represent?

c. Convert the percentages to proportions. Check your answers along the way.

i. 46%
ii. 6.9%
iii. 15.6%
iv. 94%
v. 88.5%

d. A newspaper reported that 88.5% of community residents surveyed support keeping the local public swimming pool open for an extra hour each evening. What proportion of residents support keeping the pool open for an extra hour each evening?

Answers

a. i. 75%
ii. 69.4%
iii. 73.2%
iv. 70%

b. 0.123 as a percentage is $(0.123)100\% = 12.3\%$.

c. i. 0.46
ii. 0.069
iii. 0.156
iv. 0.94
v. 0.885

d. $88.5\% = \frac{88.5}{100} = 0.885$.

SECTION 2.4 EXERCISES

2.25 Convert the fraction $\frac{4}{12}$ to a decimal, rounded to three-places.

2.26 Convert the fraction $\frac{63}{85}$ to a decimal, rounded to three places.

2.27 Convert the fraction $\frac{3}{9}$ to a decimal, rounded to three places.

2.28 Convert the fraction $\frac{49}{298}$ to a decimal, rounded to three places.

2.29 Convert 0.474 to a percentage.

2.30 Convert 0.005 to a percentage.

2.31 Convert 0.91 to a percentage.

2.32 Convert 0.335 to a percentage.

2.33 Convert the percentage 0.1% to proportion.

2.34 Convert the percentage 58.2% to proportion.

2.35 Convert the percentage 4.3% to proportion.

2.36 Convert the percentage 36.4% to proportion.

SECTION 2.5 Plotting Points in Two Dimensions

Sometimes you will need to create a display of a data set that consists of pairs of measurements. For example, you might have data on both the time spent playing video or computer games (in hours per day) and reaction time (in seconds). Each of the characteristics that is measured (such time spent playing video or computer games and reaction time) is called a **variable**. It is common to represent variables using letters such as x and y. For example, you could define the variables as follows:

$$x = \text{time spent playing video or computer games}$$

and

$$y = \text{reaction time}$$

If both variables were measured for each of 10 ninth-grade boys, you would have an x value and a y value for each boy.

Suppose that one boy plays video or computer games 2 hours per day and has a reaction time of 0.6 seconds. For that boy, $x = 2$ and $y = 0.6$.

A data set that consists of values for each of two variables is called a **bivariate data set**. It is common to represent bivariate data using ordered pair notation. An ordered pair is two numbers enclosed in parentheses and separated by a comma. Examples of ordered pairs are (2, 0.6) and (1, 0.7). The first number in each pair is the value of x and the second number is the value of y. With x and y representing the variables defined earlier, the ordered pair (2, 0.6) represents a boy who plays video or computer games 2 hours per day and has a reaction time of 0.6 seconds. The pair (1, 0.7) represents a boy who plays video or computer games 1 hour per day and has a reaction time of 0.7 seconds.

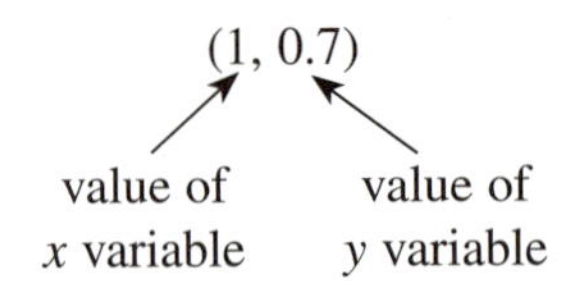

When you have bivariate data, a question you will explore in your statistics course is whether there is a relationship between the two variables that define the data set. For example, you might wonder if reaction time is related to the amount of time spent playing video or computer games. Creating a graphical display of the data is a good place to start.

A **scatterplot** is a graph of bivariate data, and it involves plotting the data using a pair of number lines that cross, as shown here. These two number lines are called axes. The horizontal axis is called the x axis and the vertical axis is called the y axis.

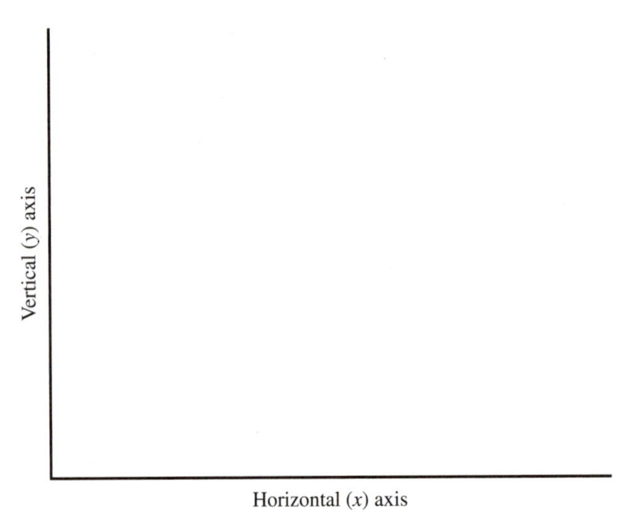

To plot data in a scatterplot, you start by drawing the axes and then add a scale to each axis. Earlier you saw how to choose an appropriate scale when you have one axis. The same strategy applies here, but you must choose a scale for each of the two axes.

Start by looking at the x values in the data set and considering the smallest and largest values to choose an appropriate scale for the x axis. Then consider the smallest and largest values for the y variable and choose an appropriate scale for the y axis.

> Note: In your past math courses, when you plotted points in two dimensions, you may have learned that the axes for the graph should cross at the point (0, 0). However, when you are graphing bivariate data, sometimes the range of the x values, the range of the y values, or both, may be far away from 0. If this is the case, you do not need to have axes that cross at (0, 0) when constructing a scatterplot.

Once you have added a scale to each axis, you graph the (x, y) pairs in the data set by representing each pair as a dot based on the corresponding x and y values. For example, to represent the pair (4, 15.5), you would move across to 4 on the x axis and then move up to 15.5 on the y axis.

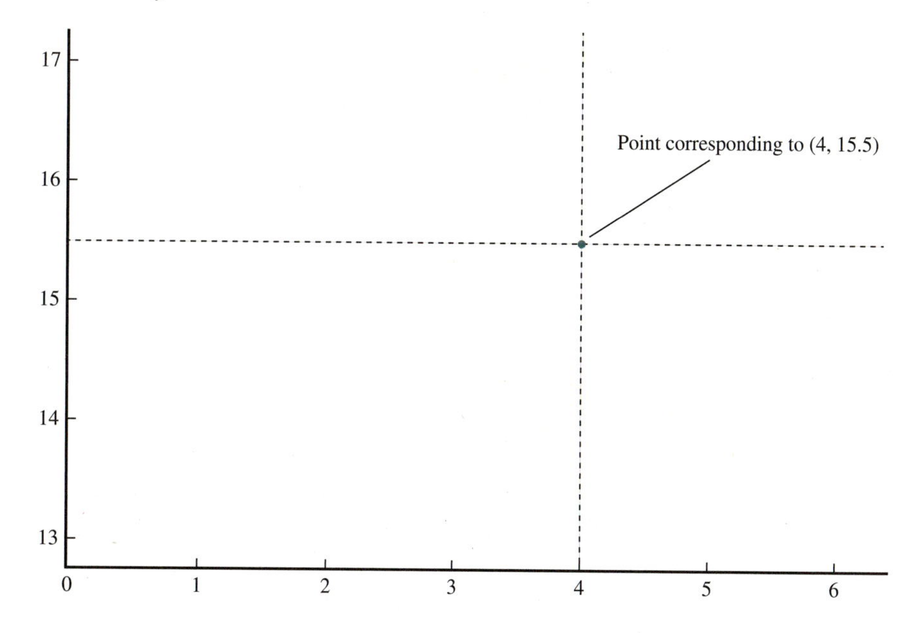

Check It Out!

Plot the specified point on the axes provided. Check your answers along the way.

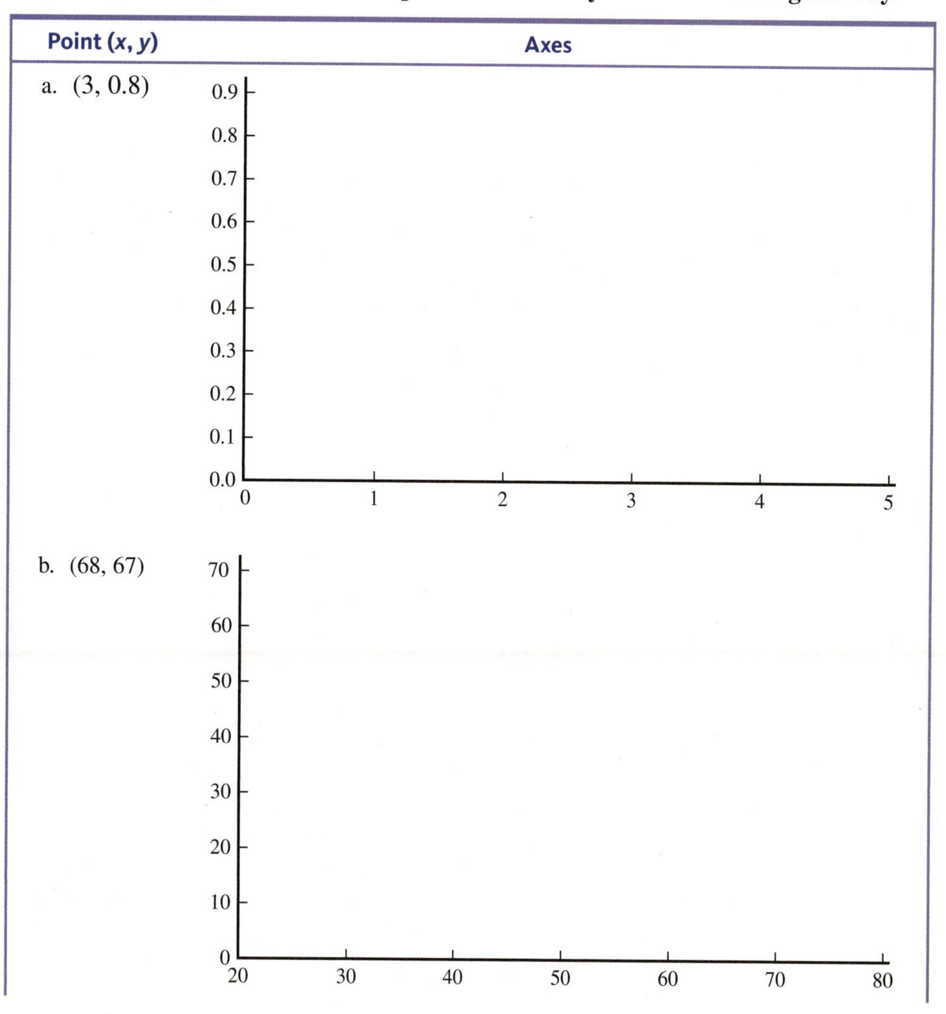

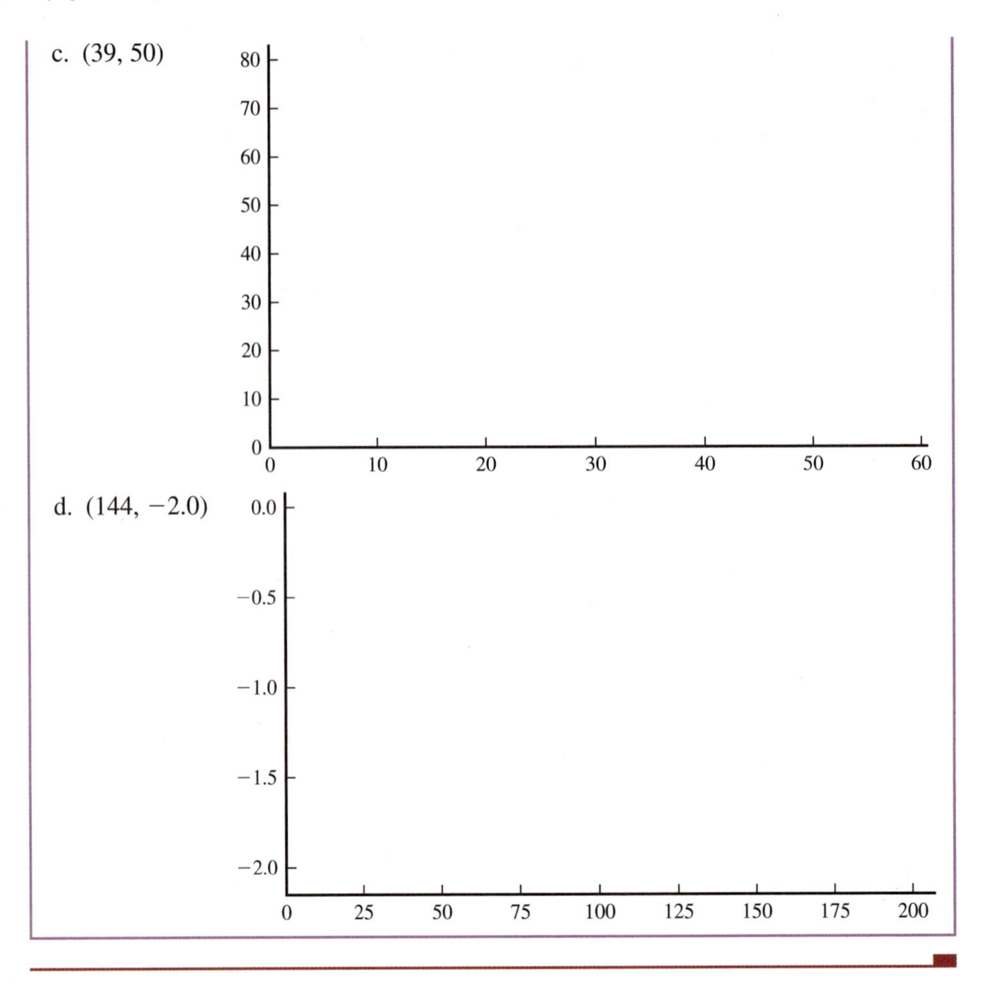

c. (39, 50)
80
70
60
50
40
30
20
10
0
0
10
20
30
40
50
60
d. (144, −2.0)
0.0
−0.5
−1.0
−1.5
−2.0
0
25
50
75
100
125
150
175
200

Answers

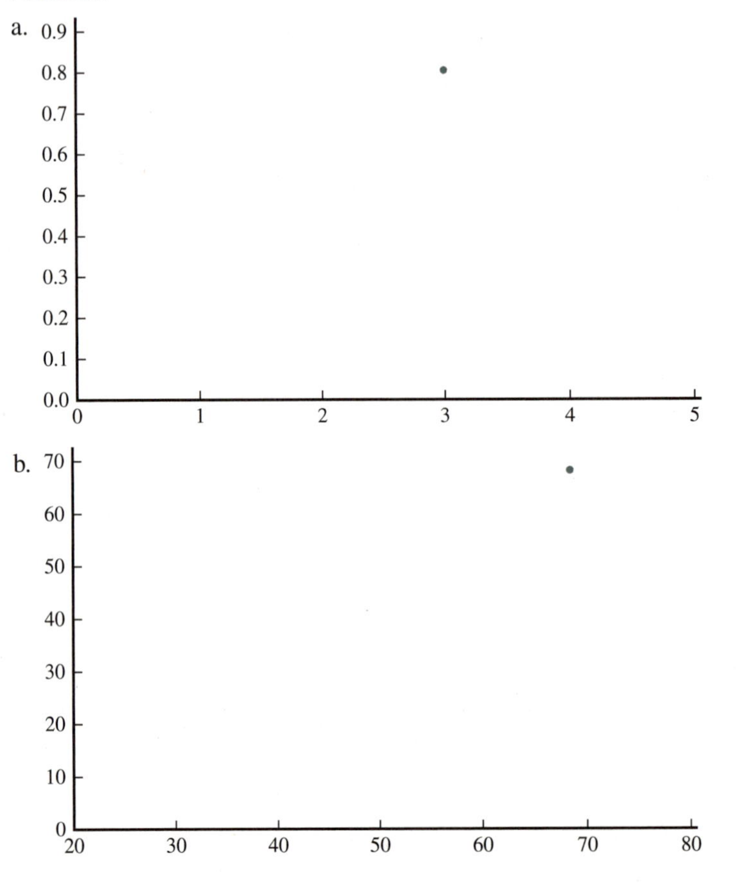

a.
0.9
0.8
0.7
0.6
0.5
0.4
0.3
0.2
0.1
0.0
0
1
2
3
4
5
b.
70
60
50
40
30
20
10
0
20
30
40
50
60
70
80

c.

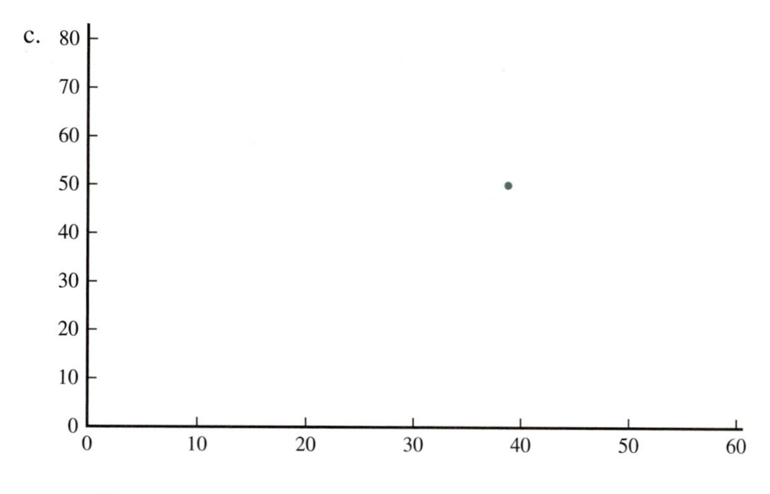

d.

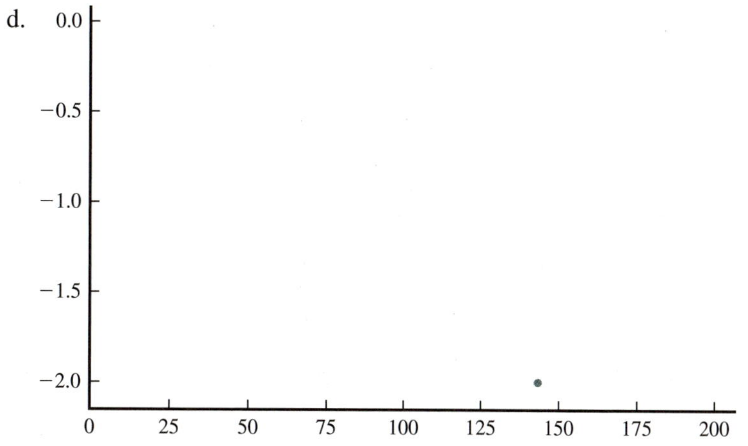

Let's look at an example. Data on

$$x = \text{time spent playing video or computer games (hours per day)}$$

and

$$y = \text{reaction time (seconds)}$$

for 10 ninth-grade students are given in the table below.

Student	Time Spent Playing Video or Computer Games (x)	Reaction Time (y)
1	2	0.6
2	1	0.7
3	0	0.9
4	1	1.1
5	3	0.5
6	1	0.9
7	2	0.8
8	4	0.4
9	0	1.0
10	3	0.4

Start by drawing a pair of axes and label each axis with the appropriate variable name.

Next, choose an appropriate scale for the x axis. The x values in the data set range from 0 to 4, so you might use a scale that goes from 0 to 5 in increments of 1.

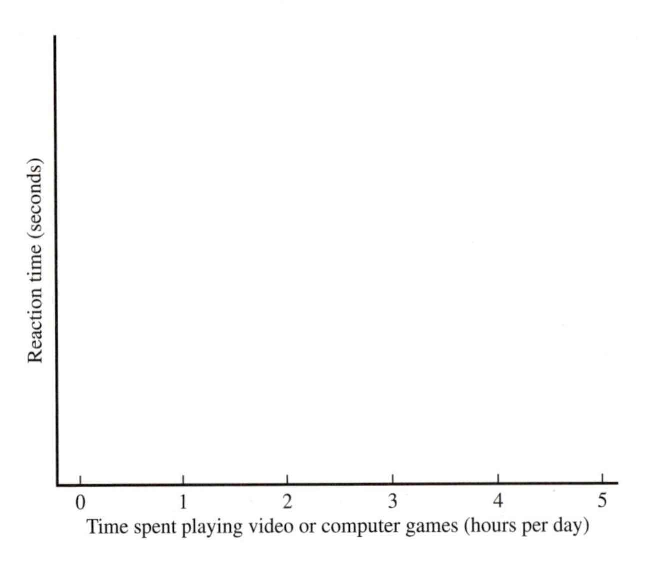

The y values range from 0.4 to 1.1, so a reasonable choice for the y axis scale would be to start at 0.4 and go up to 1.2 in increments of 0.1.

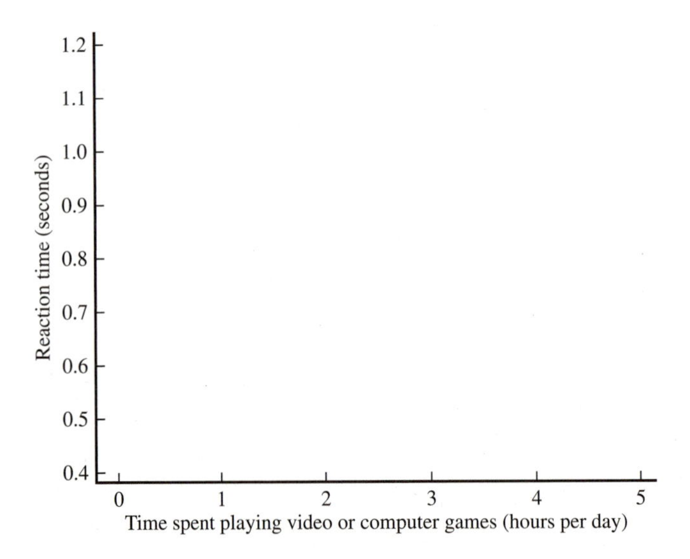

Now you are ready to add points to the scatterplot to represent the data pairs. The first point is (2, 0.6), so you would add a dot to represent that pair as shown.

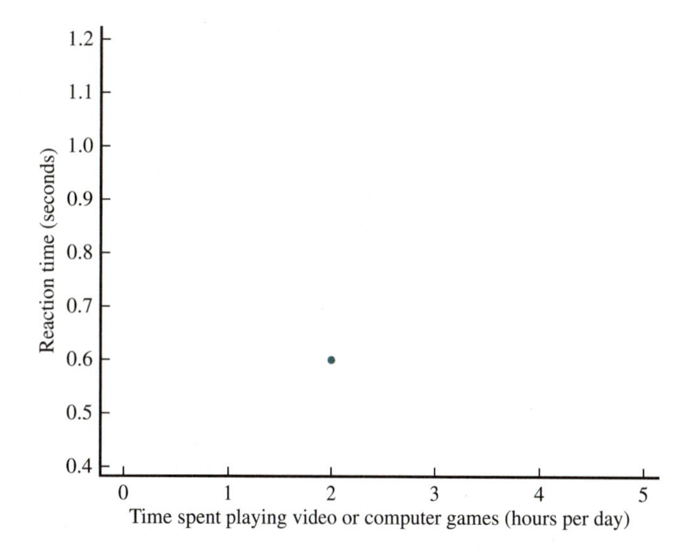

The scatterplot is completed by adding points for the other nine data pairs. The resulting scatterplot is shown here.

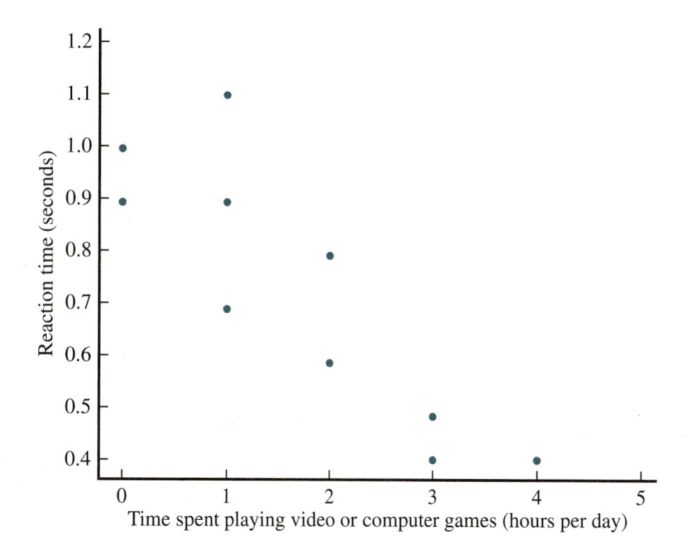

Check It Out!

a. Construct a scatterplot for the following bivariate data set. Be sure to include axis labels. Check your answer.

x = Height (inches), y = Weight (pounds) for a sample of male athletes

Pair	x	y
1	69	160
2	72	185
3	78	205
4	74	175
5	76	185
6	76	185
7	77	185

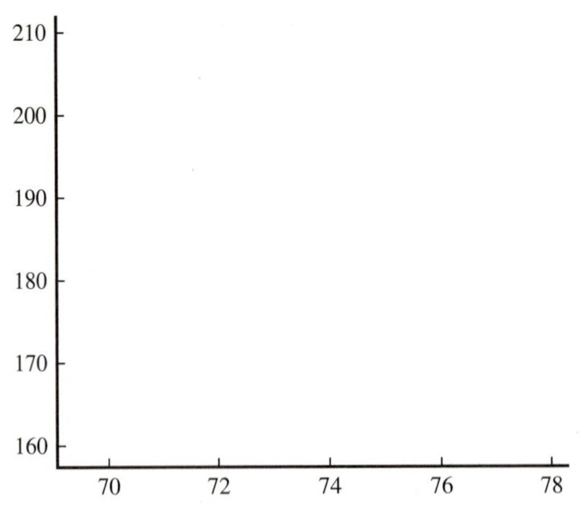

b. Construct a scatterplot for the following bivariate data set. Be sure to include axis labels. Check your answer.

x = Average weekly high temperature (degrees F), y = Number of ice cream cones sold at a local shop in a week

Week	x	y
1	76	579
2	88	525
3	85	615
4	92	680

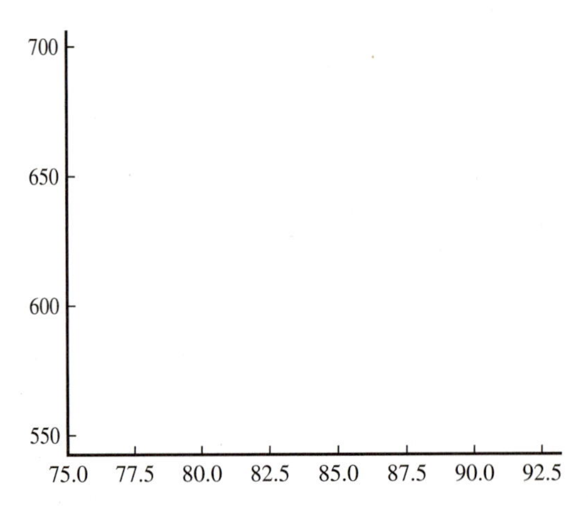

c. Construct a scatterplot for the following bivariate data set. Be sure to include axis labels and scales on the axes. Check your answers as you go.

x = Elevation (m), y = Air temperature (degrees C) during a mountain hike

Pair	x	y
1	100	25.7
2	120	25.6
3	140	23.8
4	160	22.5
5	180	21.3
6	200	20.1
7	220	19.9
8	240	19.8
9	260	20.1
10	280	19.8

Answers

a.

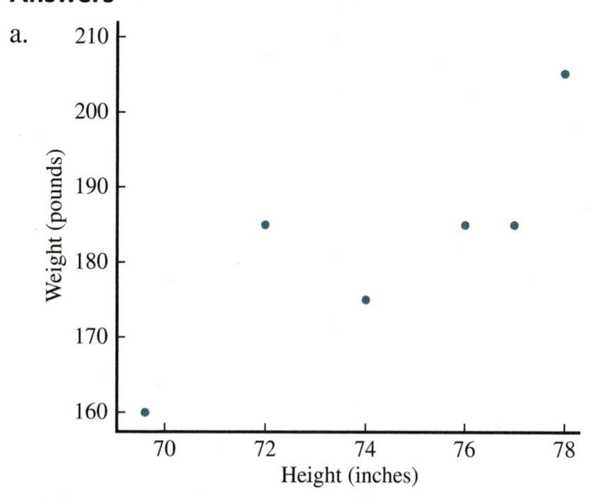

b.

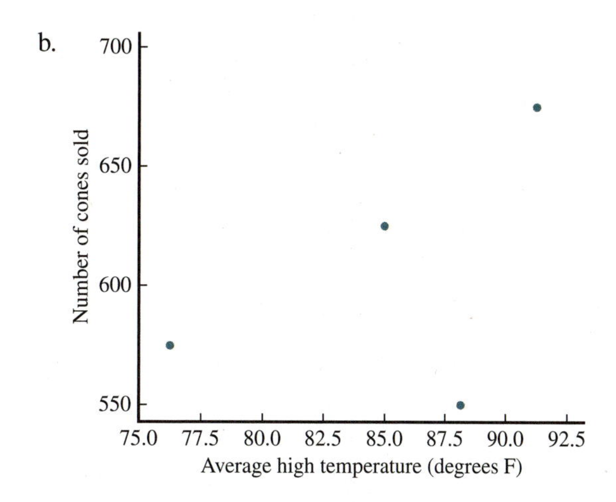

c.

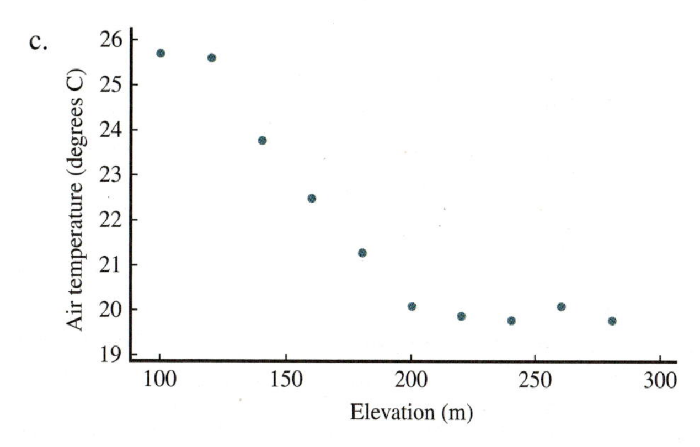

Once you have constructed a scatterplot, you want to look at it to assess whether there appears to be a pattern in the plot or whether the points appear to be scattered at random. This provides insight into a possible relationship between the two variables. If there is a pattern, then it means that there may be a relationship between the two variables and that they tend to change together in a predictable way.

For example, in the scatterplot of the data on time spent playing video or computer games and reaction time, there is a pattern in the plot, with reaction time tending to decrease as the time spent playing video or computer games increases. This means that boys who spent more time playing video or computer games tended to have quicker reaction times.

If the points in a scatterplot appear to be scattered at random with no pattern, then there may be no relationship between the two variables. For example, suppose that the scatterplot of reaction time versus time spent playing video and computer games was as shown here:

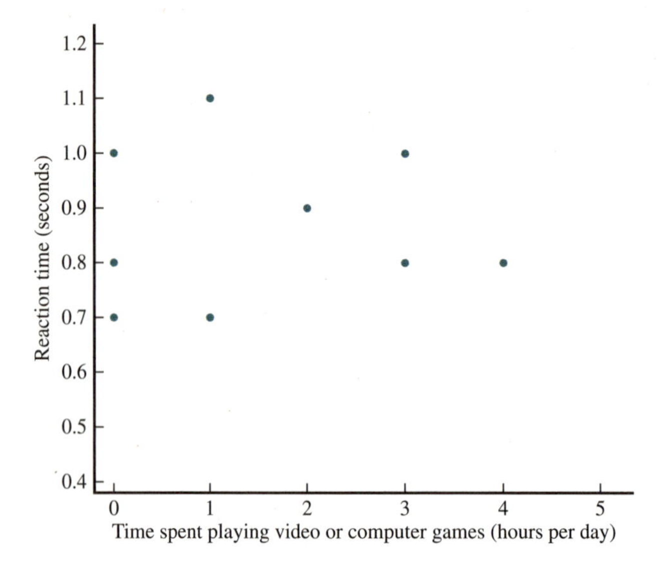

There does not appear to be any pattern in this scatterplot, and this would lead you to conclude that there may be no relationship between reaction time and time spent playing video or computer games.

Check It Out!

Describe the pattern—if any—that you detect in each of the following scatterplots. Check your answers as you go.

a.

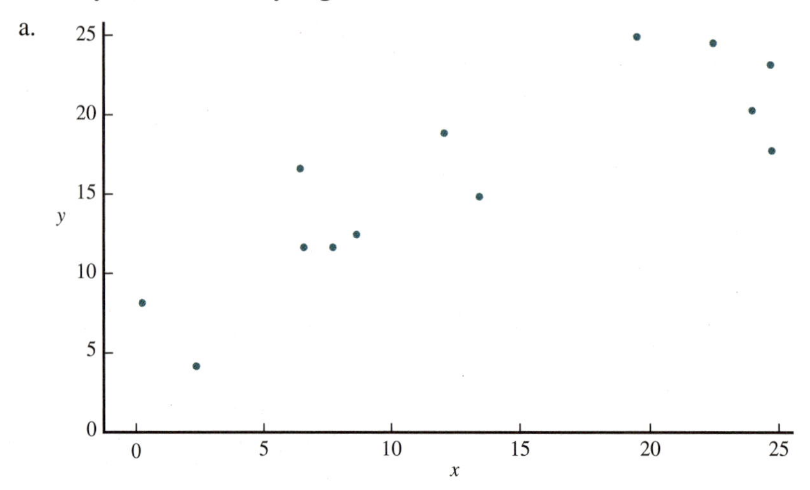

b.

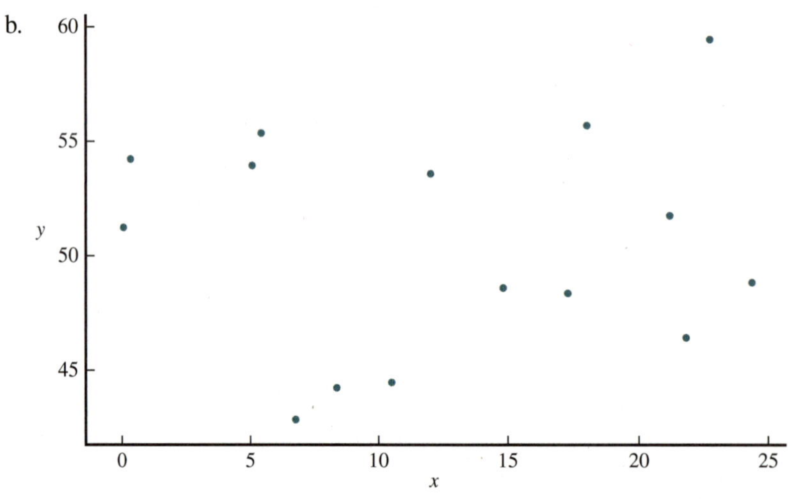

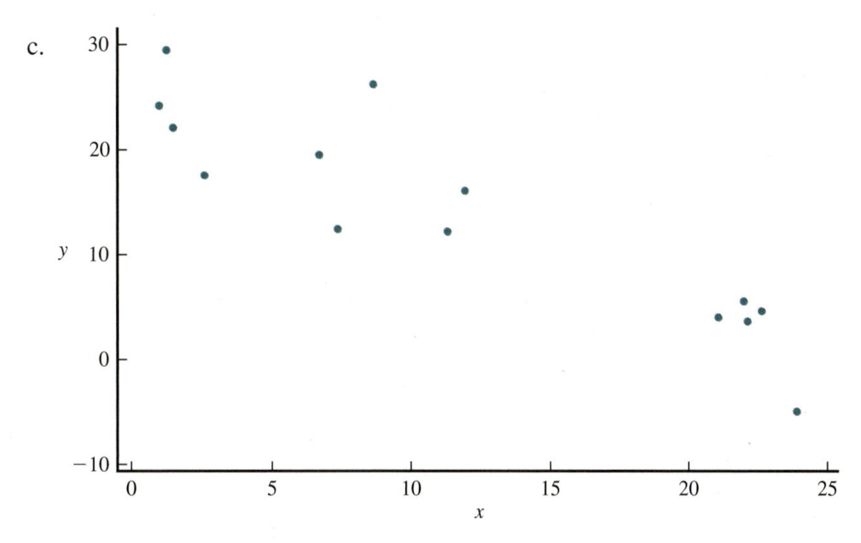

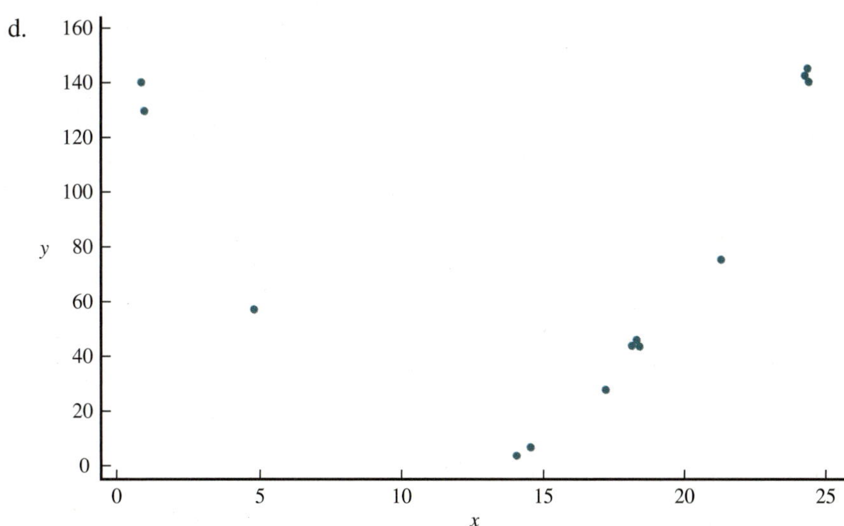

Answers

a. As x increases, y also tends to increase.

b. There is no pattern—the points fall in a random scatter.

c. As x increases, y tends to decrease.

d. As x increases from 0 to about 10, y tends to decrease, but after about 10, as x increases, y also increases.

Linear Patterns in Scatterplots

A common pattern in scatterplots is a linear pattern. You say that there is a linear pattern when the points in the plot appear to be scattered around some straight line. Here are some examples of scatterplots that show linear patterns:

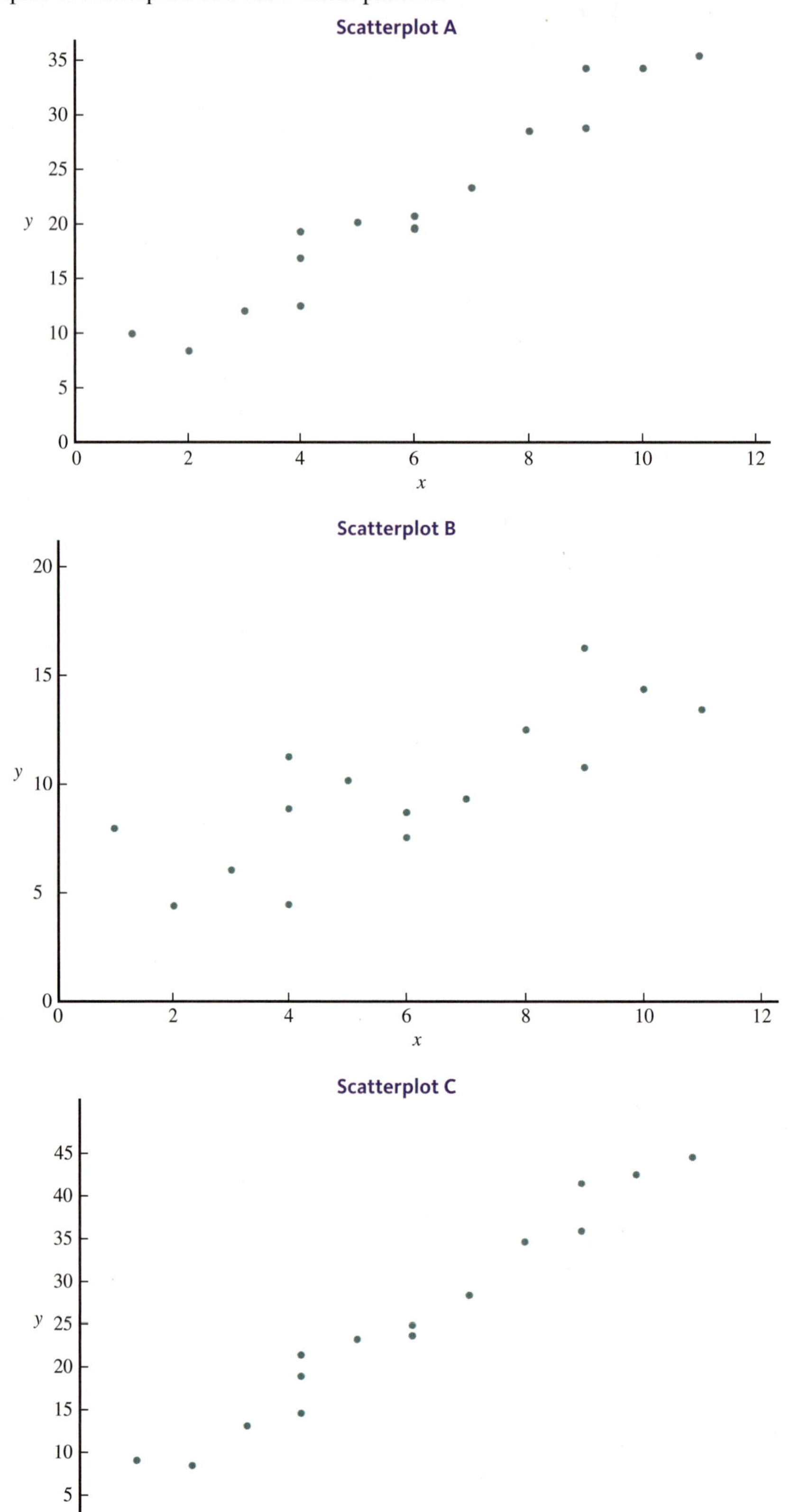

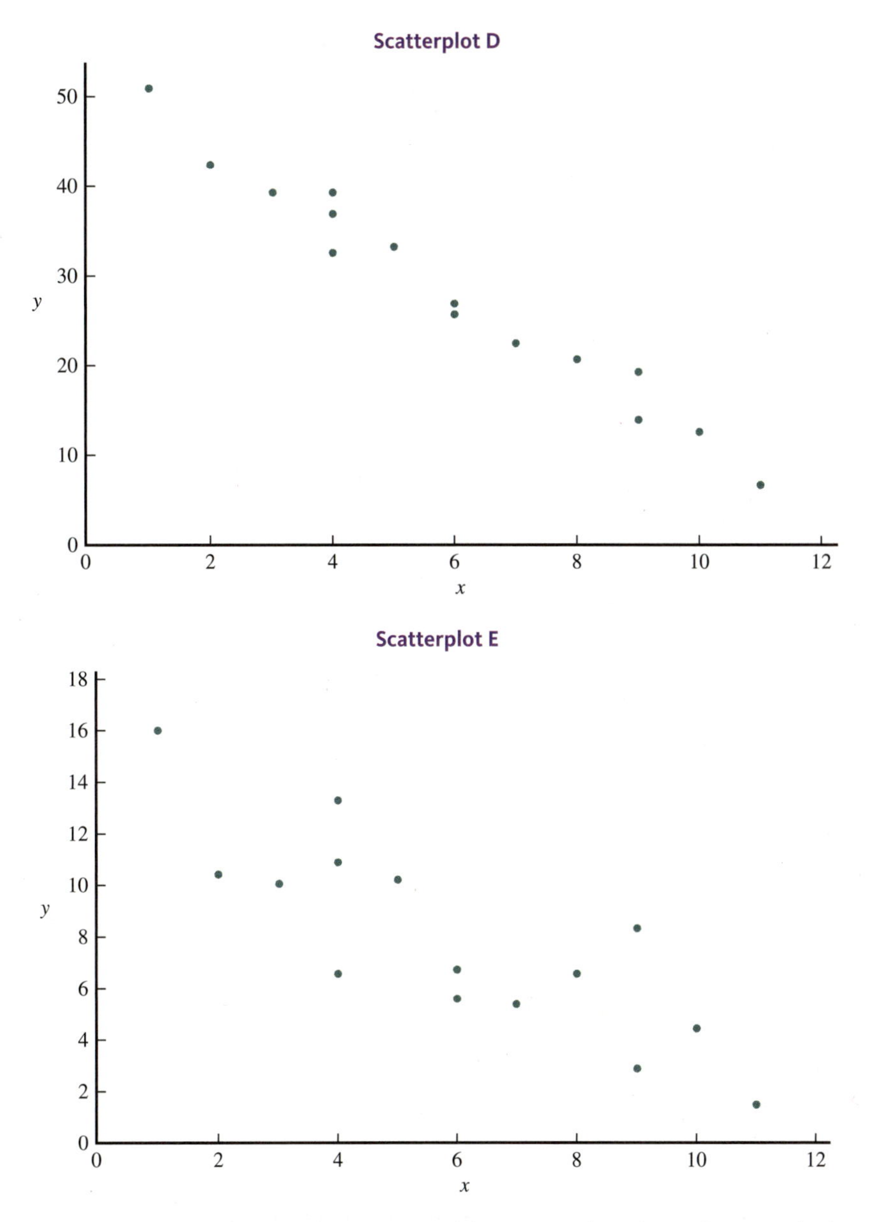

Linear patterns are described as either positive or negative, depending on whether the pattern slopes upward or downward. If the pattern slopes upward, then the y value tends to increase as x increases, and the pattern is described as a positive linear pattern. Scatterplots A, B, and C all show positive linear patterns. If the pattern slopes downward, the y value tends to decrease and x increases, and the pattern is described as negative. Scatterplots D and E above show negative linear patterns.

Check It Out!

For each of the scatterplots, identify whether there is a positive linear pattern, a negative linear pattern, or no linear pattern. Check your answers as you go.

a.

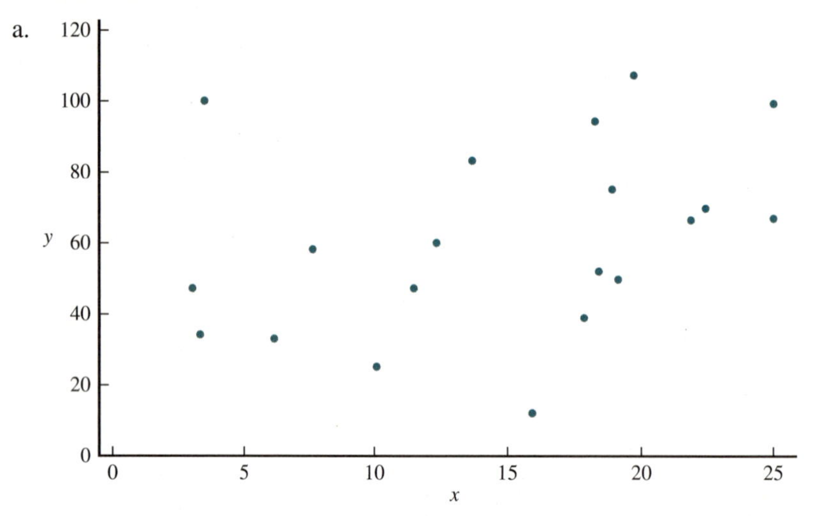

b.

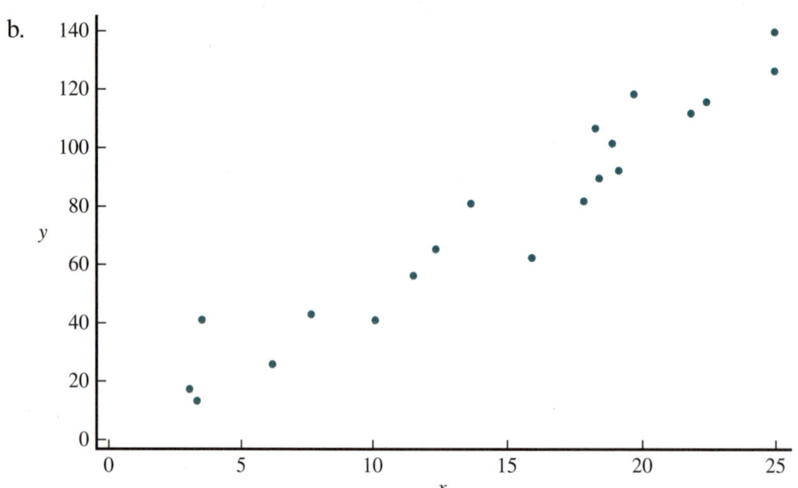

c.

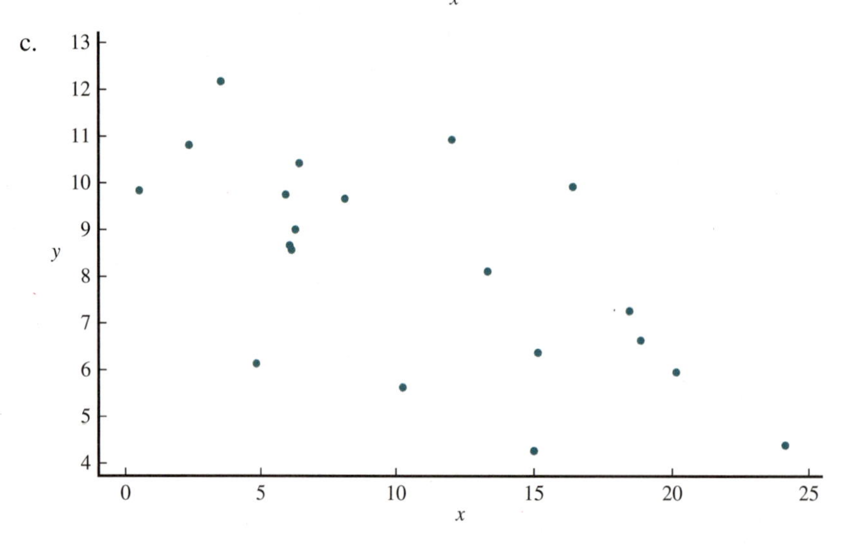

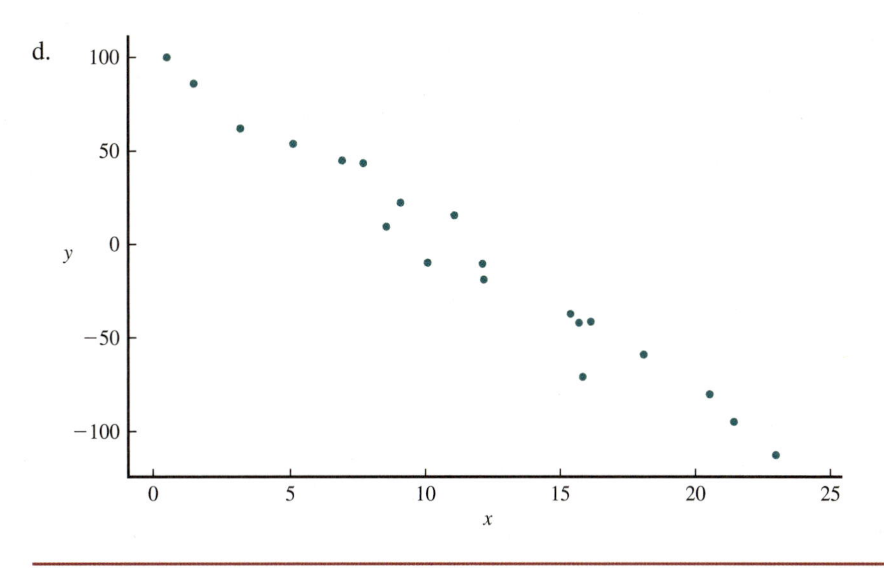

Answers

a. No linear pattern

b. Positive linear pattern

c. Negative linear pattern

d. Negative linear pattern

SECTION 2.5 EXERCISES

2.37 Plot the specified point on the axes provided.

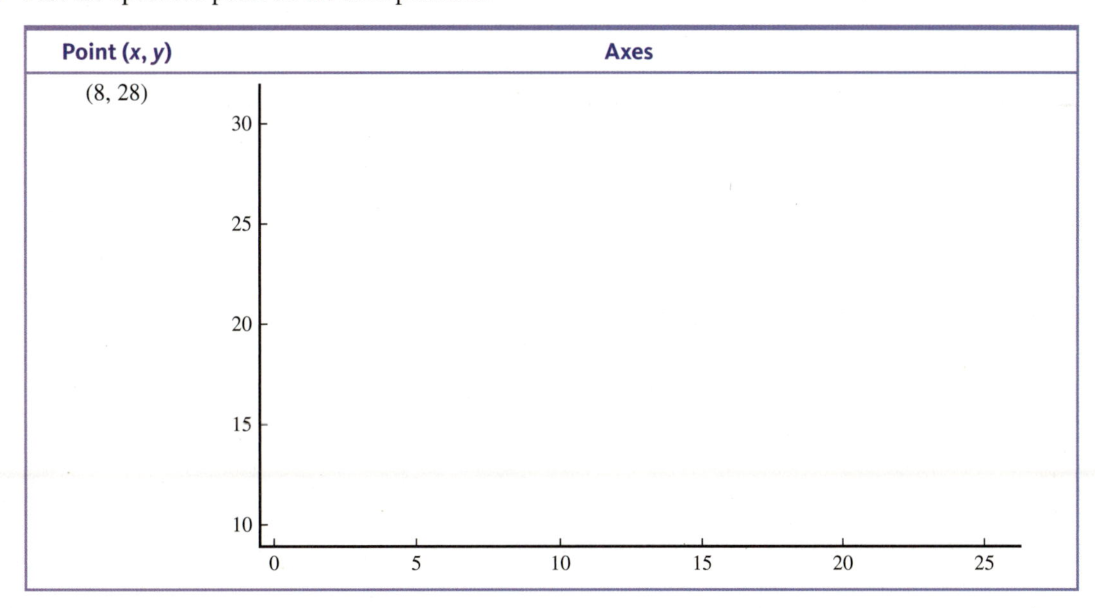

Point (x, y)	Axes
(8, 28)	

2.38 Plot the specified point on the axes provided.

Point (x, y)	Axes
(87, 18)	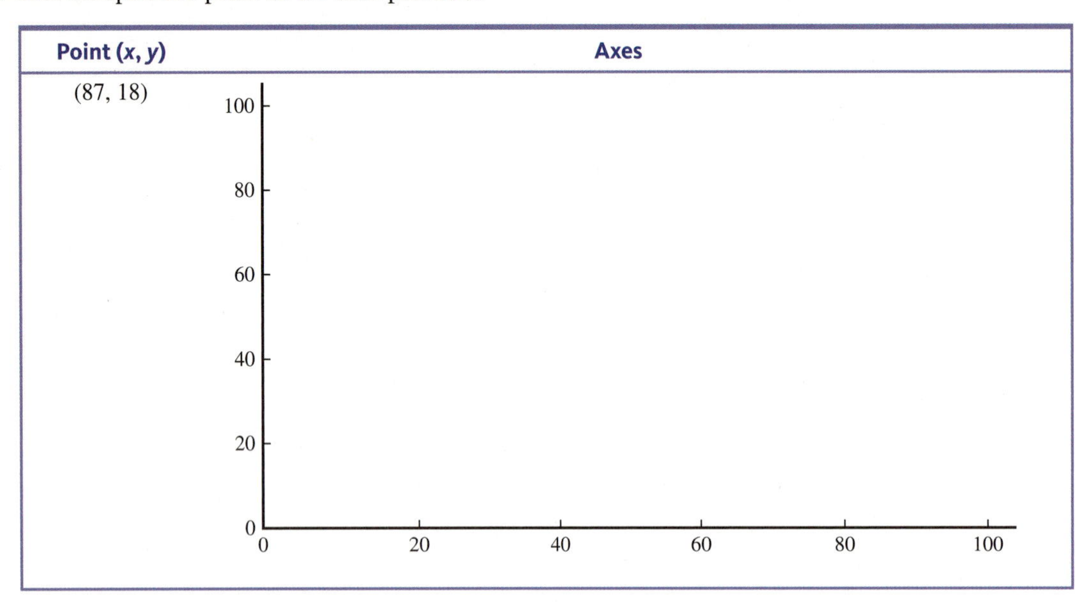

2.39 Use the given axes to construct a scatterplot of the given data on x = Price per pound (dollars) for fresh peaches and y = Number of peaches sold at a sample of local farmers' markets. Be sure to label the axes.

Market	x	y
1	1.87	45
2	1.71	61
3	1.73	55
4	1.45	74
5	1.36	71

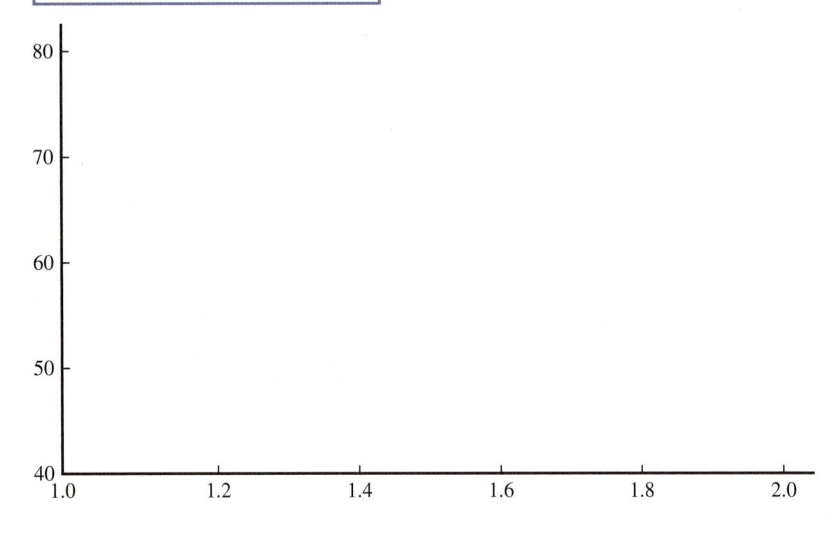

2.40 Construct a scatterplot for the given data on x = Age (years), y = Final exam score (percent) for the students in an evening class. Choose appropriate scales for the axes, and remember to include axis labels.

Student	x	y	Student	x	y
1	18	82	7	32	71
2	19	80	8	33	67
3	21	82	9	41	66
4	28	68	10	44	77
5	31	80	11	46	84
6	31	89	12	47	87

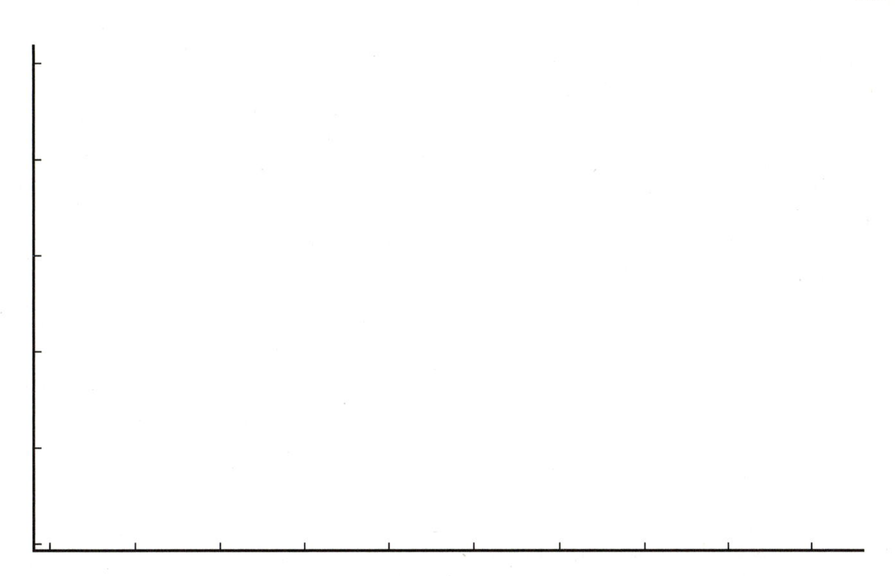

2.41 Describe the pattern—if any—in the following scatterplot.

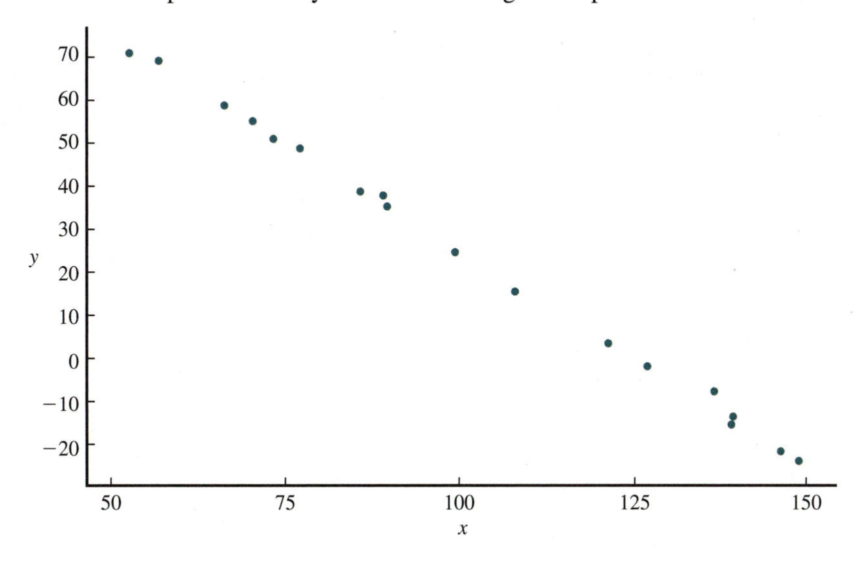

2.42 Describe the pattern—if any—in the following scatterplot.

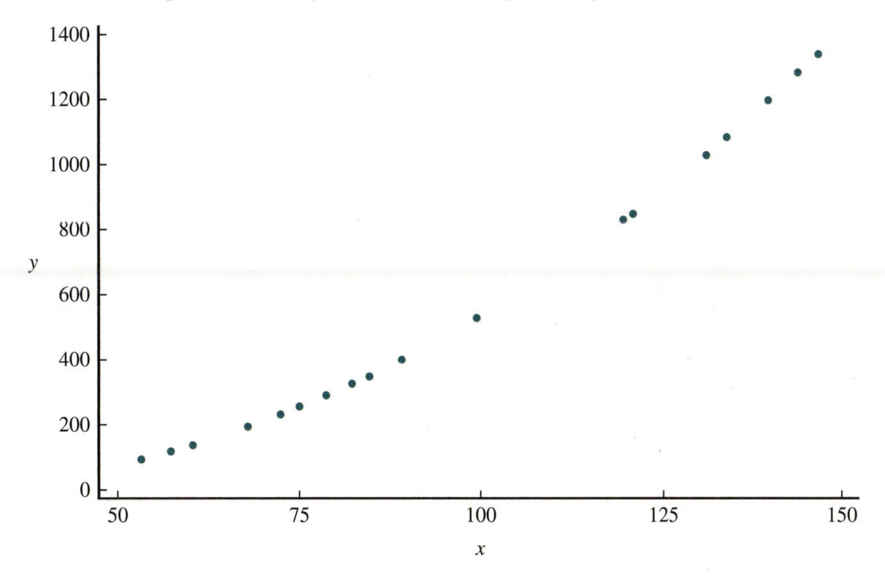

2.43 Describe the pattern—if any—in the following scatterplot.

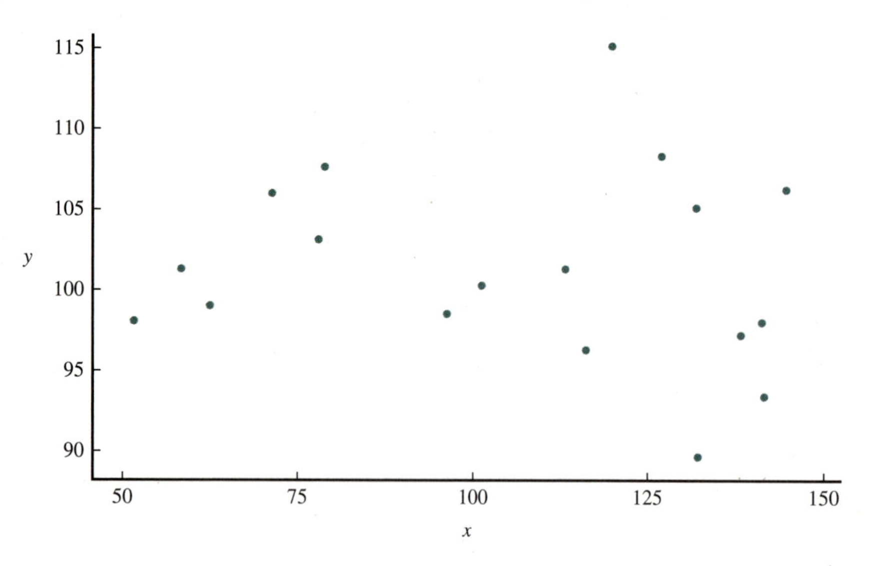

2.44 Describe the pattern—if any—in the following scatterplot.

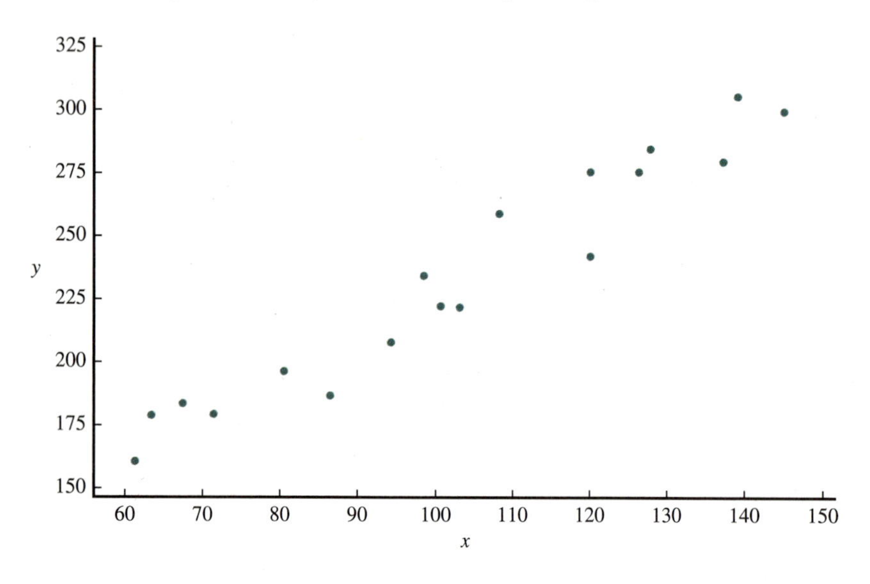

2.45 Identify whether there is a positive linear pattern, a negative linear pattern, or no pattern in the following scatterplot.

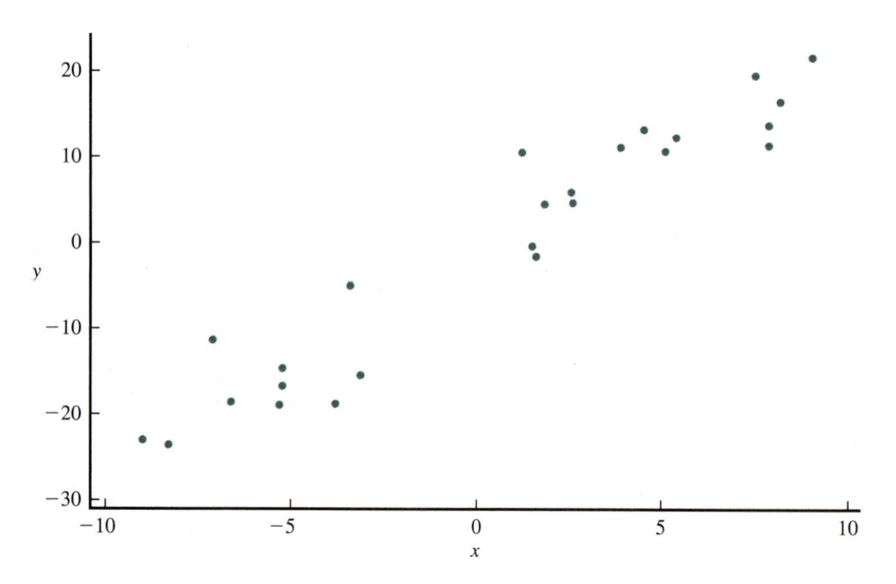

2.46 Identify whether there is a positive linear pattern, a negative linear pattern, or no pattern in the following scatterplot.

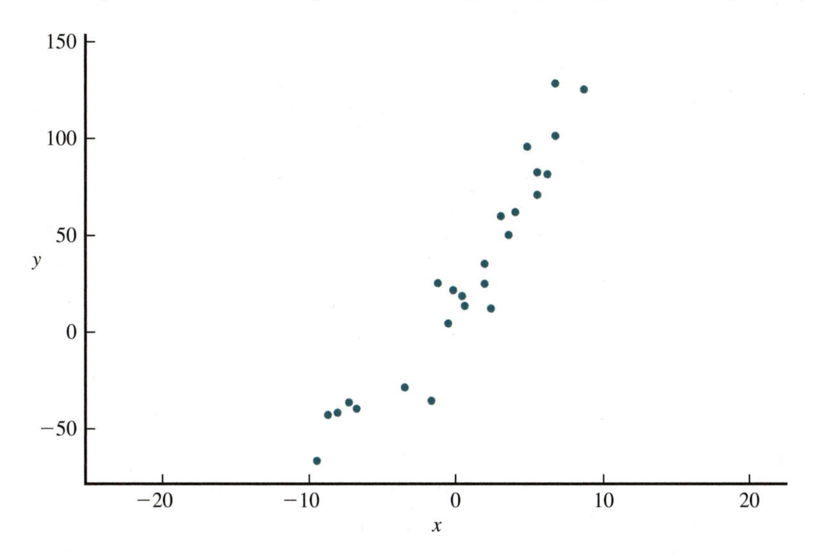

2.47 Identify whether there is a positive linear pattern, a negative linear pattern, or no pattern in the following scatterplot.

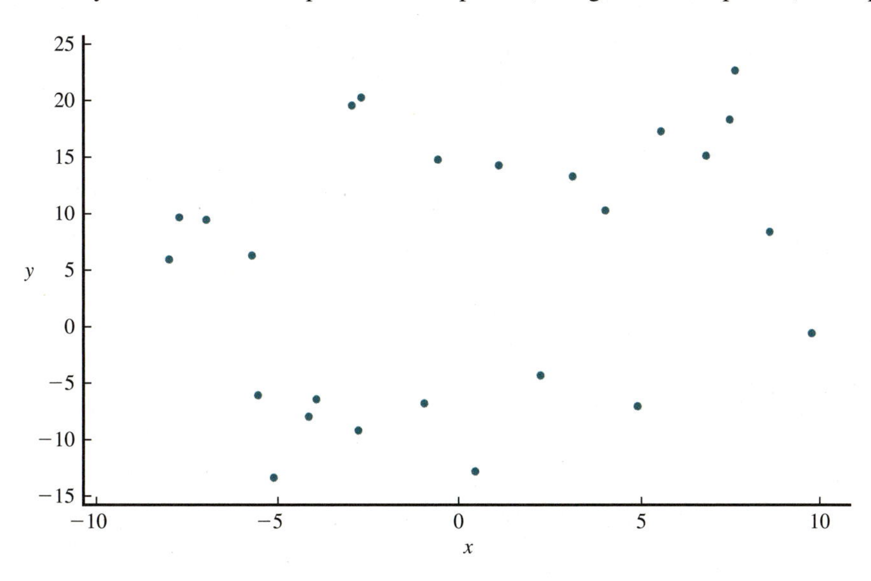

2.48 Identify whether there is a positive linear pattern, a negative linear pattern, or no pattern in the following scatterplot.

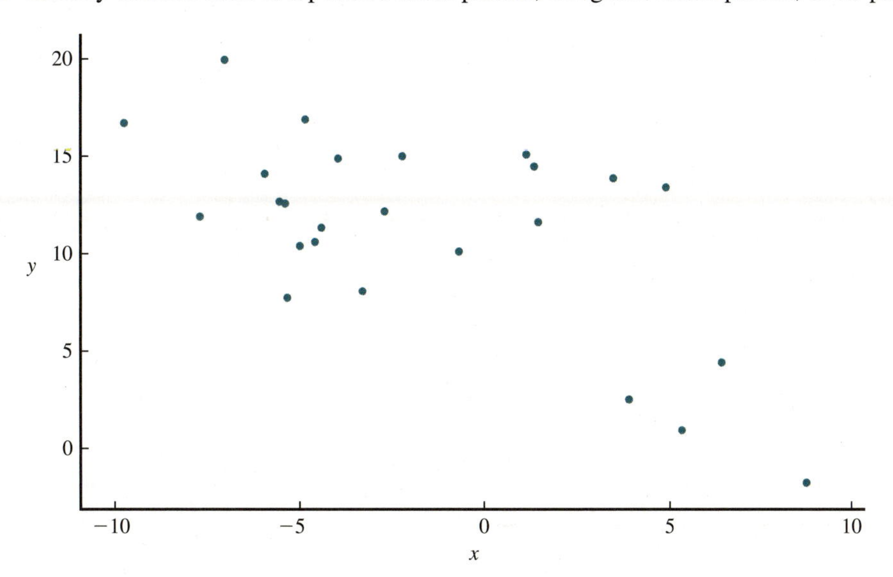

SECTION 2.6 Evaluating Expressions

When constructing graphical displays, you often must calculate relative frequencies, interval widths, and sometimes densities. Let's look at each of these.

Relative Frequency

Relative frequency is defined as the number of observations (the frequency) for a particular category or falling in a particular interval, divided by the total number of observations.

$$\text{relative frequency} = \frac{\text{frequency}}{\text{total number of observations}}$$

For example, suppose you collected data by asking 50 students at your college what they would choose for a pet if they could have one: dog, cat, bird, or no pet. If 21 of these people responded "dog," the relative frequency for the category "dog" would be

$$\text{relative frequency} = \frac{\text{frequency}}{\text{total number of observations}} = \frac{21}{50} = 0.42$$

If you had asked each of these 50 people how much he or she spends each month on transportation, this would result in numerical data. Suppose these data were classified into intervals of \$0 to < \$50, \$50 to < \$100, \$100 to < \$150, and \$150 to < \$200. If 17 reported transportation costs that fell in the interval \$100 to < \$150, the relative frequency for the interval \$100 to < \$150 is

$$\text{relative frequency} = \frac{\text{frequency}}{\text{total number of observations}} = \frac{17}{50} = 0.34$$

Check It Out!

a. Calculate the values of relative frequency, expressed as a decimal rounded to three decimal places. Check your answers along the way.

Frequency	Total Number of Observations	Relative Frequency
38	75	
49	81	
157	318	
470	1,246	
124	613	

b. When 489 fans attending a minor league baseball game were asked if they planned to attend another game this season, 278 said "yes." Calculate the relative frequency of fans who said "yes," as a decimal rounded to three decimal places.

Answers

a.

Frequency	Total Number of Observations	Answer
38	75	0.507
49	81	0.605
157	318	0.494
470	1,246	0.377
124	613	0.202

b. relative frequency $= \dfrac{\text{frequency}}{\text{total number of observations}} = \dfrac{278}{489} = 0.569$

Interval Width

Recall from Section 2.3 that you calculate the width of an interval by subtracting the lower endpoint of the interval from the upper endpoint of the interval.

interval width = upper endpoint − lower endpoint

For example, the width of the interval (19, 24) is

$$\text{Interval width} = \text{upper endpoint} - \text{lower endpoint} = 24 - 19 = 5$$

and the width of the interval 100 to < 125 is

$$\text{Interval width} = \text{upper endpoint} - \text{lower endpoint} = 125 - 100 = 25$$

Check It Out!

Calculate the width for each interval using the lower and upper endpoints. Check your answers along the way.

Lower Endpoint	Upper Endpoint	Interval Width
6	9	
78	115	
30	77	
3.32	9.9	
6.3	39	

Answers

Lower Endpoint	Upper Endpoint	Answer
6	9	3
78	115	37
30	77	47
3.32	9.9	6.58
6.3	39	32.7

Density

When you are constructing graphs based on data that have been summarized in a table that uses unequal width intervals, you will need to calculate a density for each interval. For example, suppose that data on distance from home to campus for 75 students who attend a particular college was summarized, resulting in the table on the next page.

Notice that the intervals used in the table are not all the same width. To construct a graph of these data, you need to calculate a density for each interval.

A **density** is calculated by dividing the relative frequency for an interval by the corresponding interval width.

$$\text{density} = \frac{\text{relative frequency}}{\text{interval width}}$$

Distance from Home to Campus (in miles)	Frequency
0 to < 1	4
1 to < 5	17
5 to < 10	31
10 to < 20	10
20 to < 50	9
50 to < 100	4

For example, to calculate the density for the interval 5 to < 10 miles for the data summarized in the table, first calculate the relative frequency and the interval width for that interval:

$$\text{relative frequency} = \frac{\text{frequency}}{\text{total number of observations}} = \frac{31}{75} = 0.413$$

$$\text{Interval width} = \text{upper endpoint} - \text{lower endpoint} = 10 - 5 = 5$$

Now you can calculate the density for the interval 5 to < 10:

$$\text{density} = \frac{\text{relative frequency}}{\text{interval width}} = \frac{0.413}{5} = 0.083$$

Check It Out!

Calculate densities for the intervals in the table. Check your answers as you go.

Interval	Frequency	Density
0 to < 1	4	
1 to < 5	17	
5 to < 10	31	0.083
10 to < 20	10	
20 to < 50	9	
50 to < 100	4	

Answers

Interval	Frequency	Answer
0 to < 1	4	0.053
1 to < 5	17	0.057
5 to < 10	31	0.083
10 to < 20	10	0.013
20 to < 50	9	0.040
50 to < 100	4	0.001

SECTION 2.6 EXERCISES

2.49 Calculate the value of the relative frequency, expressed as a decimal rounded to three decimal places.

Frequency	Total Number of Observations
46	75

2.50 Calculate the value of the relative frequency, expressed as a decimal rounded to three decimal places.

Frequency	Total Number of Observations
10	83

2.51 On the first day of class, a professor asked how many students had previous experience using a spreadsheet program. Out of 54 students in the class, 7 had previous experience with a spreadsheet program. Calculate the value of the relative frequency of students with spreadsheet experience, expressed as a decimal rounded to three decimal places.

2.52 Out of 40 campers at a youth summer camp, 13 chose Arts and Crafts as their preferred activity for the day. Calculate the value of the relative frequency of campers who chose Arts and Crafts, expressed as a decimal rounded to three decimal places.

2.53 Calculate the width of the interval with the following lower and upper endpoints.

Lower Endpoint	Upper Endpoint
5.1	5.5

2.54 Calculate the width of the interval with the following lower and upper endpoints.

Lower Endpoint	Upper Endpoint
73	195

2.55 Calculate the width of the interval with the following lower and upper endpoints.

Lower Endpoint	Upper Endpoint
54	78

2.56 Calculate the width of the interval with the following lower and upper endpoints.

Lower Endpoint	Upper Endpoint
7.8	28.8

2.57 Calculate the densities for each of the intervals and frequencies given in the table.

Interval	Frequency
0 to $<$ 1	11
1 to $<$ 10	36
10 to $<$ 50	8
50 to $<$ 100	52

3 Measures of Center and Variability—The Math You Need to Know

SECTION 3.1 Review—Ordering Decimal Numbers, Square Roots, Distance between Two Points

Several of the topics covered earlier will also be important as you learn to how numerical summaries of center and variability are used to learn about data. This section provides a quick review of those topics.

Ordering Decimal Numbers

Several of the measures you will encounter as you learn how to calculate and interpret numerical summaries of data (such as the median and quartiles) will require you to put a set of numbers in order from smallest to largest.

Remember that the easiest way to decide which of two positive decimal numbers is greater is to compare them digit by digit. It may help to write the numbers you want to put in order so that they have the same number of digits before the decimal place and the same number of digits after the decimal place. To order two positive numbers, compare the numbers digit by digit, moving from left to right. As soon as you find a position where the digits are different, the one that has the greater digit in this position is the greater of the two numbers.

It gets a little trickier if you need to order two negative numbers, because the rule for deciding which one is greater is reversed. You still compare the two numbers digit by digit from left to right, but when you find a position where the digits are different, the one that has the greater digit in this position is the *lesser* of the two numbers. For example, -12 is greater than -25. It helps to keep the number line in mind and remember that as you move to the right anywhere on the number line, the numbers increase.

Also, remember that any positive number is greater than any negative number, so if the data you are trying to order includes both positive and negative numbers, it might be easiest to put the positive numbers in list that is in order from least to greatest first, and then put the negative numbers in a list that is in order from least to greatest. You can then combine the two lists to form one ordered list that contains all the numbers in the data set.

Let's look at an example. Suppose 12 people play a game. Each person who plays the game is asked a series of 20 questions. A point is added to the person's score for each question that is answered correctly and a point is subtracted from the score for each question that is answered incorrectly. For example, a person who answers all 20 questions correctly would have a score of 20, a person who answers 16 questions correctly would have a score of $16 - 4 = 12$, a person who answers 10 questions correctly would have a score of $10 - 10 = 0$, and a person who answers 5 correctly would have a score of $5 - 15 = -10$. Suppose the scores for the 12 people were as follows:

$$-4 \quad 2 \quad 10 \quad -6 \quad -2 \quad 8 \quad 4 \quad 10 \quad -4 \quad 4 \quad 4 \quad -8$$

To put these scores in order, you can start by putting the positive numbers in order:

$$2 \quad 4 \quad 4 \quad 4 \quad 8 \quad 10 \quad 10$$

Then put the negative numbers in order:

$$-8 \quad -6 \quad -4 \quad -4 \quad -2$$

Then the complete list of ordered scores is

$$-8 \quad -6 \quad -4 \quad -4 \quad -2 \quad 2 \quad 4 \quad 4 \quad 4 \quad 8 \quad 10 \quad 10$$

Check It Out!

Put the lists of numbers in order from least to greatest. Check that your answers are correct as you go.

List 1: −3, −72, 9, −7, 4, −26, −53

List 2: −0.016, −0.539, 0.888, −0.486, 0.502, −0.781, −0.793, 0.975

List 3: −74, −2.4, −74.9, 11, 38, −59

List 4: 5.28, 7.41, −3.7, −82.9, 46.152, 9.241, −89.040, −43.2, 2.974

Answers

List 1: −72, −53, −26, −7, −3, 4, 9

List 2: −0.793, −0.781, −0.539, −0.486, −0.016, 0.502, 0.888, 0.975

List 3: −74.9, −74, −59, −2.4, 11, 38

List 4: −89.040, −82.9, −43.2, −3.7, 2.974, 5.28, 7.41, 9.241, 46.152

Square Roots

Remember that the square root of a positive number is represented using the square root symbol $\sqrt{\ }$. When you calculate the square root of a positive number, you are finding values that can be squared to obtain that number. For example, if you want to know the value of $\sqrt{16}$, you are looking for numbers that can be squared (multiplied by themselves) to get 16. Because $4 \times 4 = 16$ and $(-4) \times (-4) = 16$, the square roots of 16 are 4 and −4. These are sometimes written together as ±4. In your statistics course, you will only be interested in the positive square root.

For some numbers, like 4 and 9, you may be able to determine the square roots of the number without using your calculator, but for other positive numbers, you will need your calculator. This is certainly the case when you are working with positive numbers that have a square root that is not a whole number, or if you want to find the square root of a positive decimal number. Use your calculator to verify the values of the following positive square roots, which have been rounded to three decimal places:

Number	Square Root
7	2.646
3.8	1.949
26.45	5.143
114.987	10.723
3425.4	58.527

Check It Out!

Find the positive square root of the following numbers. Check your answers as you go. Note that the answers are rounded to three decimal places.

a. 594.6

b. 7.90

c. 895.94

d. 3.677

Answers

a. 24.384

b. 2.811

c. 29.932

d. 1.918

Distance between Two Points

The distance between two numbers (points on the number line) is calculated by subtracting the smaller number from the larger number. Sometimes you will also need to calculate the point that is half way between two numbers (called the midpoint). To calculate the midpoint when you are given two numbers, you simply add the two numbers and then divide the sum by 2:

$$\text{midpoint between two numbers} = \frac{\text{sum of the two numbers}}{2}$$

For example, the midpoint between 8 and 41 is $\frac{8 + 41}{2} = \frac{49}{2} = 24.5$. The midpoint between 105 and 136 is $\frac{105 + 136}{2} = \frac{241}{2} = 120.5$.

Check It Out!

Find the midpoints for the given pairs of numbers. Check your answers as you go.

a. 60 and 1

b. 7 and 47

c. 973 and 898

d. 15.6 and 7.8

Answers

a. 30.5

b. 27

c. 935.5

d. 11.7

SECTION 3.1 EXERCISES

For Exercises 3.1–3.4, put the lists of numbers in order from least to greatest.

3.1 49, 35, 49, −24, −28, 40, −62, −56, −69

3.2 5.24, −4.77, 2.68, 6.99, 5.5, −0.1, −0.52, 3.63

3.3 −51.4, 57.9, −3.1, 3.91, 66.8

3.4 −0.968, 0.043, 0.395, −0.766, −0.271, −0.543, −0.255, 0.668

For Exercises 3.5–3.8, find the positive square root of the number given.

3.5 514.84

3.6 452.4

3.7 2.783

3.8 3.28

For Exercises 3.9–3.12, find the midpoint of the given pair of numbers.

3.9 59 and 53

3.10 5 and 9

3.11 98 and 0

3.12 3.5 and 87.3

SECTION 3.2 Variables and Algebraic Expressions

In Chapter 1, you worked with arithmetic expressions, which consist of numbers and arithmetic operators (such as +, −, and $\sqrt{\ }$). As you continue your work in statistics, you will also work with algebraic expressions. An algebraic expression is one that consists of numbers, operators, and **variables**. A variable is a letter (for example, x or y) that is used to represent a number that might vary. Variables are used in expressions to stand in for a number.

For example, consider the algebraic expression $6x + 4$. Remember that x represents a number. When a number is written next to a variable with no operator between them (like $6x$ in the expression $6x + 4$), it means the number and the number represented by the variable should be multiplied together. If the value of x is 3, the value of the expression $6x + 4$ is $6x + 4 = (6)(3) + 4 = 18 + 4 = 22$. If the value of x is 2.6, the value of the expression $6x + 4$ is $6x + 4 = (6)(2.6) + 4 = 15.6 + 4 = 19.6$.

Check It Out!

Evaluate the algebraic expressions. Check your answers as you go.

Expression 1: $8x \div 4$, for $x = 1$

Expression 2: $5y - (-7)$, for $y = 3$

Expression 3: $6x \times (-4)$, for $x = 9.4$

Answers

Expression 1: 2

Expression 2: 22

Expression 3: −225.6

In statistics, variables usually represent the value of some characteristic (such as wing length for birds or weight for oranges) that is measured when data are collected. For example, in a study of the relationship between the amount of time spent playing online games and reaction time, the following variables might be defined:

$$x = \text{time spent playing online games (in hours per day)}$$

and

$$y = \text{reaction time (in seconds)}$$

Check It Out!

For each of the studies described, define two variables of interest. Check your answers as you go.

Study 1: Shoppers at a grocery store are asked to rate the appeal of the packaging for a new brand of organic yogurt on a scale from 1 = Least Appealing to 10 = Most Appealing. Some of the shoppers are shown packaging that is Red and other shoppers are shown packaging that is Green.

Study 2: A town is considering whether to build a new outdoor skate park, and a survey is sent to a random sample of households in the town. The survey asks whether there are children under 18 living in the household, and whether the person responding to the survey is in favor of building a new outdoor skate park.

Answers

Study 1:

x = Color of the packaging (either Red or Green)
y = Rating given by a shopper (a number from 1 to 10)

Study 2:

x = Children under 18 years old living in the household (Yes or No)
y = Favor building a new skate park (Yes or No)

Suppose that the following variable was measured for each student in a sample of ten 5th-graders:

x = time required to complete a math homework assignment (in minutes)

This results in a data set that consists of ten observed values for the variable x. Those data values might be

12 17 9 18 16 18 20 11 9 25 18

Each of these numbers represents a value for x. Sometimes it is also helpful to have a way of denoting the individual values in a data set—for example, a way to represent the first value of x, the second value of x, and so on. This is done by using subscripts (a number written next to and a little below a variable). For example, x_1 is used to represent the first value of x in a data set, x_2 represents the second value of x in a data set, and x_7 represents the seventh value of x in a data set. Notice that x_1 represents the first value in a data set, not necessarily the smallest value.

For the data set shown above,

$$x_1 = 12 \quad x_2 = 17 \quad x_7 = 20$$

Check It Out!

For each of the data sets, find the specified data values. Check your answers as you go.

Data Set 1: −49, 98, −5, 0, −27, 11, 85, 87, −9, 22
Find x_1 and x_6

Data Set 2: 71, 8, 8, 4, 9, 1
Find x_5, x_4, and x_6

Data Set 3: 5, 0, 9, 2, 56, −7, −7, −3, 26, 7
Find x_2, x_5, x_4, and x_{10}

Answers

Data Set 1: $x_1 = -49$, and $x_6 = 11$

Data Set 2: $x_5 = 9$, $x_4 = 4$, and $x_6 = 1$

Data Set 3: $x_2 = 0$, $x_5 = 56$, $x_4 = 2$, and $x_{10} = 7$

SECTION 3.2 EXERCISES

For Exercises 3.13–3.16, evaluate the algebraic expressions.

3.13 $5x \div 10$, for $x = 7$

3.14 $-77y - (-5)$, for $y = 6$

3.15 $-3u \times 4$, for $u = 13$

3.16 $-35v + (-71)$, for $v = -8$

For the studies described in Exercises 3.17–3.20, define two variables of interest.

3.17 An instructor is curious to know whether scores (out of 100 points) on the first test in a math course can be used to predict the scores (also out of 100 points) for the same students on the final exam.

3.18 A local polling organization sends a representative to houses in a town. At each house, the representative asks an adult about his or her political affiliation (Democrat, Republication, or Other) and also about his or her support for raising the local income tax to help to pay for repairing streets (Favor or Oppose).

3.19 Researchers think that patients taking a new medication might have a greater chance of developing hives (a type of skin rash) as a side effect of the medication. The researchers record whether patients in a study took the new medication or an older, existing medication, and whether they did or did not develop hives.

3.20 A soccer coach records how many goals the team scored in games played on their home field, and how many goals the team scored in games played away from their home field.

For each of the data sets in Exercises 3.21–3.24, find the specified data values.

3.21 Data Set: 38, 75, 49, 81, 15, 73, 18, 47, −1, 24
Data values: x_9, x_2, x_{10}, x_7, and x_3

3.22 Data Set: −6, −1, 3, 6, 1, 5, −5, 8, 4
Data values: x_1, x_7, x_8, x_9, and x_5

3.23 Data Set: 8.4, −6.3, −9.6, −9.9, 8.1, 5.9, 8.2, −6.2, −8.6
Data values: x_3 and x_2

3.24 Data Set: 67, 97, 81, −15, 30, 77, 33, 29, 90
Data values: x_6, x_4, x_2, and x_3

SECTION 3.3 Summation Notation

Calculating many of the summary measures used to describe data sets involves adding up a set of numbers. For example, suppose that students in a statistics class can take a quiz multiple times and that you have data on the number of attempts it took eight students to pass the quiz. The variable of interest in this setting is

$$x = \text{number of attempts}$$

and the eight data values would be denoted by x_1, x_2, x_3, x_4, x_5, x_6, x_7, and x_8. You might want to calculate the sum $x_1 + x_2 + x_3 + x_4 + x_5 + x_6 + x_7 + x_8$.

Summation notation is a shorthand notation to represent addition. The summation symbol, $\sum$, is the Greek letter capital sigma, and is used to represent addition. When you see it in an expression or a formula it is just telling you to add up the values of whatever follows this symbol. For example, if you see $\sum x$, this means you should add up all the values of the variable x in a data set. For the variable x = number of attempts described above, $\sum x = x_1 + x_2 + x_3 + x_4 + x_5 + x_6 + x_7 + x_8$. If the eight data values were

1 5 2 1 3 1 1 4

then $\sum x = 1 + 5 + 2 + 1 + 3 + 1 + 1 + 4 = 18$.

Here are a few more examples of the use of summation notation. Use your calculator to verify each of these results, and pay attention to the differences between $\sum x$, $\sum x^2$, and $(\sum x)^2$.

Data Set: 3, 9, 14, 6, 4		
If You See This...	**It Means Do This...**	**The Value for This Data Set Is...**
$\sum x$	Add up all the values in the data set.	$\sum x = 3 + 9 + 14 + 6 + 4 = 36$
$\sum x^2$	Square each value in the data set and then add up the squared values.	$\sum x^2 = 3^2 + 9^2 + 14^2 + 6^2 + 4^2$ $= 9 + 81 + 196 + 36 + 16$ $= 338$
$\left(\sum x\right)^2$	Add up all the values in the data set and then square the sum.	$\left(\sum x\right)^2 = (3 + 9 + 14 + 6 + 4)^2 = 36^2 = 1296$

Data Set: 2.1, 1.3, 4.2, 3.9, 1.8, 0.5		
If You See This...	**It Means Do This...**	**The Value for This Data Set Is...**
$\sum x$	Add up all the values in the data set.	$\sum x = 2.1 + 1.3 + 4.2 + 3.9 + 1.8 + 0.5 = 13.8$
$\sum x^2$	Square each value in the data set and then add up the squared values.	$\sum x^2 = 2.1^2 + 1.3^2 + 4.2^2 + 3.9^2 + 1.8^2 + 0.5^2$ $= 4.41 + 1.69 + 17.64 + 15.21 + 3.24 + 0.25$ $= 42.44$
$\left(\sum x\right)^2$	Add up all the values in the data set and then square the sum.	$\left(\sum x\right)^2 = (2.1 + 1.3 + 4.2 + 3.9 + 1.8 + 0.5)^2$ $= 13.8^2 = 190.44$

Check It Out!

For each data set, calculate $\sum x$, $\sum x^2$, and $(\sum x)^2$. Check your answers as you go.

Data Set 1: −49, 98, −5, 0, −27, 11, 85, 87, −9, 22
Data Set 2: 71, 8, 8, 4, 9, 1
Data Set 3: 5, 0, 9, 2, 56, −7, −7, −3, 26, 7

Answers

Data Set 1: $\sum x = 213$, $\sum x^2 = 28{,}239$, and $(\sum x)^2 = 45{,}369$

Data Set 2: $\sum x = 101$, $\sum x^2 = 5267$, and $(\sum x)^2 = 10{,}201$

Data Set 3: $\sum x = 88$, $\sum x^2 = 4078$, and $(\sum x)^2 = 7744$

SECTION 3.3 EXERCISES

For each data set in Exercises 3.25–3.28, calculate $\sum x$, $\sum x^2$, and $(\sum x)^2$.

3.25 38, 75, 49, 81, 15, 73, 18, 47, −1, 24
3.26 −6, −1, 3, 6, 1, 5, −5, 8, 4
3.27 8.4, −6.3, −9.6, −9.9, 8.1, 5.9, 8.2, −6.2, −8.6
3.28 67, 97, 81, −15, 30, 77, 33, 29, 90

SECTION 3.4 Deviations from the Mean, Squared Deviations, Sum of Squared Deviations

In your statistics course, you will use measures of center and measures of variability to describe the values in a data set. A measure of center describes a data set with a single number that gives you a sense of what a typical value is for the data set. When you learned about measures of center, you saw that the mean of a data set (denoted by $\bar{x}$ for the mean of a sample and by μ for the mean of an entire population) is the arithmetic average of the values in the data set.

For example, suppose that each person in a sample of six teenagers was asked how many text messages he or she sent in a typical day, resulting in the following data set:

16 5 0 12 28 14

To find the sample mean for this data set, you would add up all of data values and then divide by the number of observations, which is 6:

$$\bar{x} = \frac{16 + 5 + 0 + 12 + 28 + 14}{6} = \frac{75}{6} = 12.5 \text{ text messages}$$

Check It Out!

For each data set, calculate the sample mean, which is denoted by $\bar{x}$. Check your answers as you go.

Data Set 1: 38, 75, 49, 81, 15, 73, 18, 47, −1, 24
Data Set 2: −6, −1, 3, 6, 1, 5, −5, 8, 4
Data Set 3: 8.4, −6.3, −9.6, −9.9, 8.1, 5.9, 8.2, −6.2, −8.6
Data Set 4: 67, 97, 81, −15, 30, 77, 33, 29, 90

Answers

Data Set 1: $\bar{x} = 41.9$
Data Set 2: $\bar{x} = 1.67$
Data Set 3: $\bar{x} = -1.11$
Data Set 4: $\bar{x} = 54.3$

Measures of variability are used to describe how much the values in a data set tend to differ from one another. A common measure of variability is the standard deviation. To understand how the standard deviation measures variability, you need to understand **deviations from the mean**. There is a deviation from the mean for each value in a data set and it is calculated by subtracting the mean from that value.

For example, suppose that each person in a sample of six teenagers was asked how many text messages he or she sent in a typical day, resulting in the following data set:

16 5 0 12 28 14

To find the mean for this data set, you would add up all of data values and then divide by the number of observations, which is 6:

$$\bar{x} = \frac{16 + 5 + 0 + 12 + 28 + 14}{6} = \frac{75}{6} = 12.5$$

The deviation from the mean for the first value in the data set is $16 - 12.5 = 3.5$. The deviation from the mean for the second value is $5 - 12.5 = -7.5$.

Notice that a deviation from the sample mean might be positive or it might be negative. It will be negative for data values that are less than the mean and it will be positive for data values that are greater than the mean.

Check It Out!

a. For the text message data given above, calculate the deviations for the third, fourth, fifth, and sixth values in the data set to complete the following table.

Data Value	Deviation from the Mean
16	3.5
5	−7.5
0	
12	
28	
14	

b. Add up the deviations, and show that they sum to zero.

Answers

a.

Data Value	Deviation from the Mean
16	3.5
5	−7.5
0	−12.5
12	−0.5
28	15.5
14	1.5

b. $\sum(x - \bar{x}) = 3.5 + (-7.5) + (-12.5) + (-0.5) + 15.5 + 1.5 = 0$

When you calculate the standard deviation, you work with the sum of the squared deviations from the mean. The table below explains the expressions you will encounter:

Expression	What It Means
$(x - \bar{x})$	This represents a single deviation from the mean. There is one deviation from the mean corresponding to each value in the data set. You calculate the deviation from the mean by subtracting the mean from the value in the data set.
$(x - \bar{x})^2$	This represents a squared deviation from the mean. There is one squared deviation from the mean corresponding to each value in the data set. You calculate a squared deviation from the mean by first calculating the deviation from the mean and then squaring that number.
$\sum(x - \bar{x})^2$	This represents the sum of the squared deviations from the mean. You calculate the sum or the squared deviations from the mean by adding up (remember the $\sum$ means add up what follows) the squared deviations from the mean for all the values in the data set.

Look back at the data set consisting of the number of text messages sent for a sample of six teenagers. The deviations from the mean were

$$3.5 \quad -7.5 \quad -12.5 \quad -0.5 \quad 15.5 \quad 1.5$$

The squared deviation from the mean corresponding to the first data value is $(3.5)^2 = 12.25$. The squared deviation corresponding to the second data value is $(-7.5)^2 = 56.25$.

The complete set of squared deviations is

$$12.25 \quad 56.25 \quad 156.25 \quad 0.25 \quad 240.25 \quad 2.25$$

You can now calculate the sum of the squared deviations:

$$\sum(x - \bar{x})^2 = 12.25 + 56.25 + 156.25 + 0.25 + 240.25 + 2.25 = 467.50$$

The sum of the squared deviations from the mean is used in the calculation of the standard deviation, which is a measure of the variability in the data set. Why does this make sense? Look at the text message data and the corresponding squared deviations from the mean.

Data Value	Deviation from the Mean	Squared Deviation from the Mean
16	3.5	12.25
5	−7.5	56.25
0	−12.5	156.25
12	−0.5	0.25
28	15.5	240.25
14	1.5	2.25

Keep in mind that the mean for this data set is 12.5. One of the data values was 12, which is very close to the mean. This data value has a deviation from the mean that is close to 0, and this results in a small squared deviation from the mean. The farther away a data value is from the mean, the greater the squared deviation from the mean. If there is a lot of variability in a data set, the data values are quite spread out and some will be far from the mean. If there is not as much variability in the data set, the values in the data set tend to be similar and this means that they will tend to be close to the mean. The more the data values spread out from the center of the data set, the greater the sum of squared deviations will be. This is why the sum of squared deviations provides information about variability in a data set.

Check It Out!

For each data set, calculate the sum of the squared deviations, which is denoted by: $\sum(x - \bar{x})^2$. Check your answers as you go.

Data Set 1: 38, 75, 49, 81, 15, 73, 18, 47, −1, 24
Data Set 2: −6, −1, 3, 6, 1, 5, −5, 8, 4
Data Set 3: 8.4, −6.3, −9.6, −9.9, 8.1, 5.9, 8.2, −6.2, −8.6
Data Set 4: 67, 97, 81, −15, 30, 77, 33, 29, 90

Answers

Data Set 1: $\sum(x - \bar{x})^2 = 7138.9$
Data Set 2: $\sum(x - \bar{x})^2 = 188$
Data Set 3: $\sum(x - \bar{x})^2 = 569.37$
Data Set 4: $\sum(x - \bar{x})^2 = 10{,}974$

SECTION 3.4 EXERCISES

For the data sets in Exercises 3.29–3.32, calculate $\bar{x}$ and the sum of the squared deviations, $\sum(x - \bar{x})^2$.

3.29 38, 75, 49, 81, 15, 73, 18, 47, −1, 24
3.30 −6, −1, 3, 6, 1, 5, −5, 8, 4
3.31 8.4, −6.3, −9.6, −9.9, 8.1, 5.9, 8.2, −6.2, −8.6
3.32 67, 97, 81, −15, 30, 77, 33, 29, 90

SECTION 3.5 Evaluating Expressions

This section provides a review of the formulas that are used to calculate measures of center and variability and looks at how you evaluate the algebraic expressions that appear in the formulas.

Sample Mean

$$\bar{x} = \frac{\sum x}{n}$$

In this formula,

$\sum x$ represents the sum of all the x values in the data set

n represents the total number of observations in the data set

To calculate the sample mean, you first calculate the sum of the x values and then divide that sum by the total number of observations in the data set.

For example, suppose that the total number of calories in the food ordered for lunch by five customers at a fast food restaurant are

540 860 490 1250 720

To calculate the mean of this sample, first calculate the sum of the x values:

$$\sum x = 540 + 860 + 490 + 1250 + 720 = 3860$$

Then divide this sum by n. For this example, $n = 5$.

$$\bar{x} = \frac{\sum x}{n} = \frac{3860}{5} = 772.0 \text{ calories}$$

Sample Variance

$$s^2 = \frac{\sum (x - \bar{x})^2}{n - 1}$$

In this formula,

$\sum (x - \bar{x})^2$ represents the sum of the squared deviations from the mean

n represents the total number of observations in the data set

To calculate the sample variance, you first calculate the deviations and the squared deviations. Then you add up the squared deviations. Finally, you divide that sum by $n - 1$.

For example, suppose that the total number of calories in the food ordered for lunch by five customers at a fast food restaurant are

540 860 490 1250 720

The sample mean is 772 calories (see above).

The table below shows the calculation of the deviations from the mean and the squared deviations for this data set.

Data Value	Deviation from the Mean	Squared Deviation from the Mean
540	$540 - 772 = -232$	$(-232)^2 = 53{,}824$
860	$860 - 772 = 88$	$(88)^2 = 7{,}744$
490	$490 - 772 = -282$	$(-282)^2 = 79{,}524$
1250	$1250 - 772 = 478$	$(478)^2 = 228{,}484$
720	$720 - 772 = -52$	$(-52)^2 = 2{,}704$

Now you can calculate the sum of the squared deviations:

$$\begin{aligned}\sum (x - \bar{x})^2 &= (-232)^2 + (88)^2 + (-282)^2 + (478)^2 + (-52)^2 \\ &= 53{,}824 + 7744 + 79{,}524 + 228{,}484 + 2704 \\ &= 372{,}280\end{aligned}$$

Next calculate the value of the sample variance by dividing by $n - 1$. Since $n = 5$ for this example, $n - 1 = 4$.

$$s^2 = \frac{\sum (x - \bar{x})^2}{n - 1} = \frac{372{,}280}{4} = 93{,}070 \text{ calories}^2$$

Sample Standard Deviation

$$s = \sqrt{s^2} = \sqrt{\frac{\sum (x - \bar{x})^2}{n - 1}}$$

To calculate the sample standard deviation, you first calculate the sample variance and then calculate the positive square root of that number.

For example, suppose that the total number of calories in the food ordered for lunch by five customers at a fast food restaurant are

540 860 490 1250 720

The sample variance is 93,070 calories^2 (see above).

The sample standard deviation is the square root of the variance, so

$$s = \sqrt{s^2} = \sqrt{93{,}070} = 305.074 \text{ calories (rounded to three decimal places)}$$

Check It Out!

For each data set, calculate the sample variance, which is denoted by s^2, and the sample standard deviation, which is denoted by s. Check your answers as you go.

Data Set 1: 38, 75, 49, 81, 15, 73, 18, 47, −1, 24
Data Set 2: −6, −1, 3, 6, 1, 5, −5, 8, 4
Data Set 3: 8.4, −6.3, −9.6, −9.9, 8.1, 5.9, 8.2, −6.2, −8.6
Data Set 4: 67, 97, 81, −15, 30, 77, 33, 29, 90

Answers

Data Set 1: $s^2 = 793.211$, and $s = 28.16$
Data Set 2: $s^2 = 23.500$, and $s = 4.85$
Data Set 3: $s^2 = 71.171$, and $s = 8.44$
Data Set 4: $s^2 = 1371.75$, and $s = 37.04$

z-Scores

A *z-score* is defined to be a value minus the corresponding mean, and then divided by the corresponding standard deviation. This is what the definition looks like in a formula:

$$z = \frac{x - \text{mean}}{\text{standard deviation}}$$

Sometimes *z-scores* are called "standard scores," because this process is known as "standardizing." To calculate a *z-score*, you must know the mean and standard deviation of the data set. Then the *z-score* corresponding to a particular value, x, is calculated by subtracting the mean from x and then dividing by the standard deviation.

For example, suppose that the total number of calories in the food ordered for lunch by five customers at a fast food restaurant are

540   860   490   1250   720

The sample mean and standard deviation are $\bar{x} = 772$ calories and $s = 305.074$ calories (see above).

To calculate the *z-score* corresponding to the data value $x = 540$, you would first subtract the mean from 540 and then divide that number by the standard deviation:

$$z = \frac{x - \text{mean}}{\text{standard deviation}} = \frac{540 - 772}{305.074} = \frac{-232}{305.074} = -0.760$$

(rounded to three decimal places)

Check It Out!

For each value, mean, and standard deviation, calculate the value's *z-score*. Check your answers as you go.

x value	Mean	Standard Deviation
54	49	20
6.8	5.6	0.7
64	67	8
10.3	20.6	5.7
47	63	7.8

Answers

$z = 0.25$

$z = 1.71$

$z = -0.38$

$z = -1.81$

$z = -2.05$

SECTION 3.5 EXERCISES

For each data set in Exercises 3.33–3.36, calculate the sample mean $\bar{x}$, the sample variance, s^2, and the sample standard deviation, s.

3.33 6.1, 3.6, 5.6, 1.4, 6.7, 6.5, 5.9, 1.2, −3.8, 6.2, 4.5

3.34 38, 74, 64, 72, 77, 62, 64, 92

3.35 6, 7, 2, −2, −7

3.36 −6, 1, 3, 1, 3, 4, 7, 3, 0, 3, 8, 9

For each data set in Exercises 3.37–3.40, calculate the *z-score* for the given x-value.

3.37 $x = 25$, using $\bar{x} = 41.9$ and $s = 28.16$

3.38 $x = -2.5$, using $\bar{x} = -2.5$ and $s = 23.18$

3.39 $x = 0$, using $\bar{x} = -1.11$ and $s = 8.44$

3.40 $x = 100$, using $\bar{x} = 115.1$ and $s = 14.5$

4 Describing Bivariate Numerical Data—The Math You Need to Know

SECTION 4.1 Review—Variables, Scatterplots, Linear and Nonlinear Patterns, *z-Scores*

Variables

In statistics, variables usually represent the value of some characteristic (such as percentage of the people that have a college degree for U.S. states or height for trees in a forest) that is measured when data are collected. For example, in a study of the relationship between cooking time and the number of unpopped kernels in bags of microwave popcorn, the following variables might be defined:

$$x = \text{cooking time (in seconds)}$$

and

$$y = \text{number of unpopped kernels}$$

Sometimes you want to describe the relationship between two variables because you would like to use the value of one of the variables to predict the value of the other variable. For example, you may want to describe the relationship between cooking time and the number of unpopped kernels so that you could predict how many unpopped kernels will remain if you cook a bag of microwave popcorn for 75 seconds or the number if you cook a bag of microwave popcorn for 100 seconds. When this is the case, the variable that you would like to predict is called the **dependent variable** and the variable that you will use to make the prediction is called the **independent variable**.

Note:

The dependent variable is sometimes also called the **response variable**.

The independent variable is sometimes also called the **predictor variable**.

In statistics, when you distinguish between an independent variable and a dependent variable, it is usual to use the letter x to represent the independent variable and the letter y to represent the dependent variable.

Note:

x is used to represent the independent variable.

y is used to represent the dependent variable.

Notice that in the previous example, x represents cooking time and y represents the number of unpopped kernels. This is because you want to predict number of unpopped kernels based on cooking time. This makes the number of unpopped kernels the dependent (y) variable and cooking time the independent (x) variable.

Check It Out!

For each of the following pairs of variables, identify which is likely to represent the independent (x) variable and which represents the dependent (y) variable. Check your answers as you go.

Variables	Independent (x)	Dependent (y)
Variable 1: Amount of money spent by colleges per student (in dollars) Variable 2: College four-year graduation rate (percentage)		
Variable 1: Quality rating for new vacuum cleaners Variable 2: Price for new vacuum cleaners		
Variable 1: Size of houses for sale (square feet) Variable 2: Sale price of houses (in thousands of dollars)		

Answers

Variables	Answer
Variable 1: Amount of money spent by colleges per student (in dollars) Variable 2: College four-year graduation rate (percentage)	x = Variable 1: Amount spent per student y = Variable 2: Graduation rate
Variable 1: Quality rating for new vacuum cleaners Variable 2: Price for new vacuum cleaners	x = Variable 2: Price y = Variable 1: Quality rating
Variable 1: Size of houses for sale (square feet) Variable 2: Sale price of houses (in thousands of dollars)	x = Variable 1: Size y = Variable 2: Sale price

Scatterplots

In Chapter 2, you constructed scatterplots, which are graphs of bivariate data. Bivariate numerical data consist of values for each of two variables, and are represented by pairs of numbers.

For example, consider the data in the accompanying table, gives the 2015 unemployment rate for each of the eight states in the South Atlantic region of the United States (from the report "Regional and State Unemployment—2015," Bureau of Labor Statistics, bls.gov/news.release/pdf/srgune.pdf). Also given in the table is the percentage of people with no remaining natural teeth (from the report "Worst Dental Health: States," bloomberg.com/graphics/best-and-worst/#worst-dental-health-states).

State	Percentage with No Natural Teeth	2015 Unemployment Rate (%)
Delaware	16.4	4.9
Florida	13.3	5.4
Georgia	21.0	5.9
Maryland	13.6	5.2
North Carolina	21.5	5.7
South Carolina	21.6	6.0
Virginia	15.0	4.4
West Virginia	36.0	6.7

The authors of the article "More Americans Turning to Dentures to Get, Keep Jobs" (*USA TODAY*, August 10, 2015) were interested in whether there was a relationship between the unemployment rate and the percentage of the population with no natural teeth. The article includes the following statement:

Some areas that top the nation for unemployment also fare worst in a health measure that can keep people from getting jobs—missing teeth.

Suppose you were interested in seeing if the unemployment rate could be predicted based on the percentage of the population with no natural teeth. Then the percentage of people with no natural teeth would be the independent variable and the unemployment rate would be the dependent variable. You could define the two variables of interest as shown here:

$$x = \text{percentage of people with no natural teeth}$$

$$y = \text{2015 unemployment rate (in percent)}$$

To investigate whether there is a relationship between these two variables, you could start by making a scatterplot of the data.

Recall that a scatterplot is a graph of bivariate data that involves plotting the data using a pair of number lines that cross. These two number lines are called axes, and the horizontal axis is called the x axis and the vertical axis is called the y axis. You start by drawing the axes and then add a label and a scale to each axis. Once you have added a scale to each axis, you graph the (x, y) pairs in the data set by representing each pair as a dot based on the corresponding x and y values. For example, to represent the data for Delaware, (16.4, 4.9), you would move across to 16.4 on the x axis and then move up to 4.9 on the y axis.

Here is a scatterplot of the data for all eight states, with the data point for Delaware indicated:

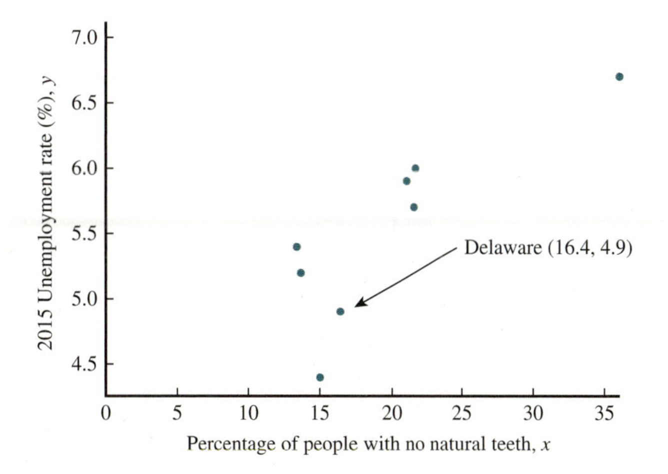

Notice that based on the scatterplot, it appears that states with a higher percentage of people with no natural teeth do tend to have higher unemployment rates.

Check It Out!

a. The human resources office at a company keeps track of how y = Salary (in thousands of dollars per year) for office workers employed by the company is related to x = Years since graduating from college. The accompanying table contains data for a sample of employees:

Employee Number	Years Since College Graduation	Salary
1	2	32.0
2	4	33.0
3	7	41.0
4	17	53.5
5	19	59.0
6	36	79.0

Construct a scatterplot representing the dependent (y) variable plotted against the independent (x) variable. Comment on how the dependent variable (y) changes as the independent variable (x) changes. Check your answer.

b. Which of the following two scatterplots (A or B) represents two variables that are not related? Explain your reasoning, then check your answer.

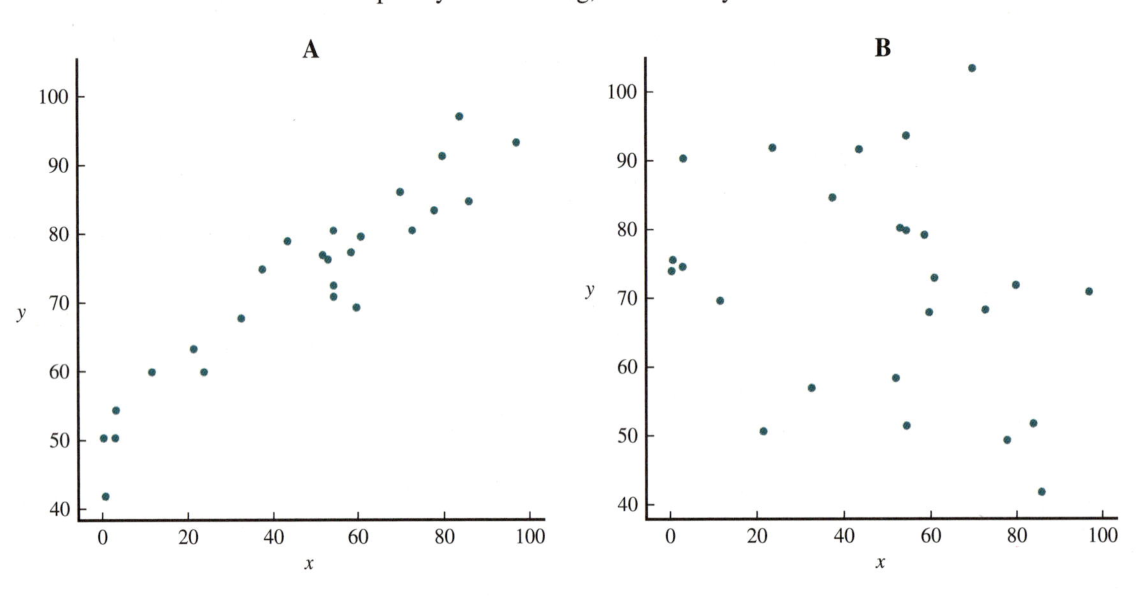

c. Match the scatterplots to the descriptions. Check your answers.

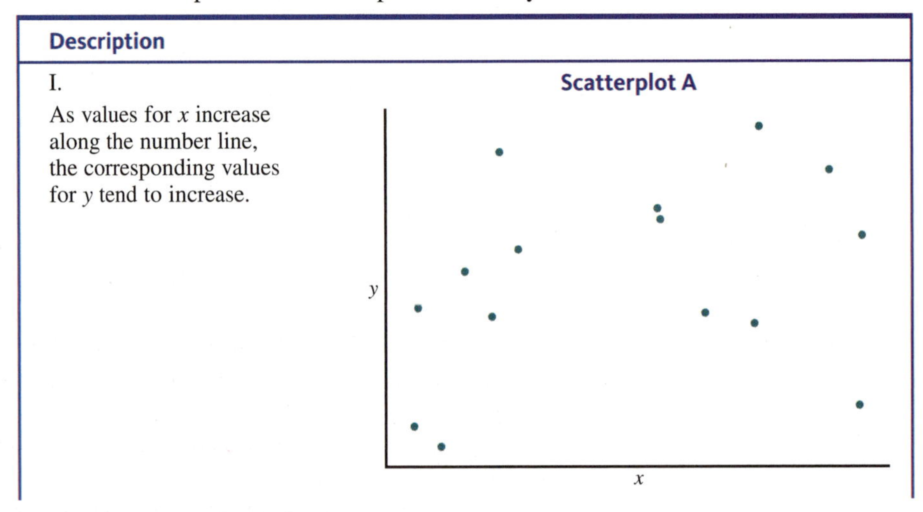

Description	
I. As values for x increase along the number line, the corresponding values for y tend to increase.	Scatterplot A

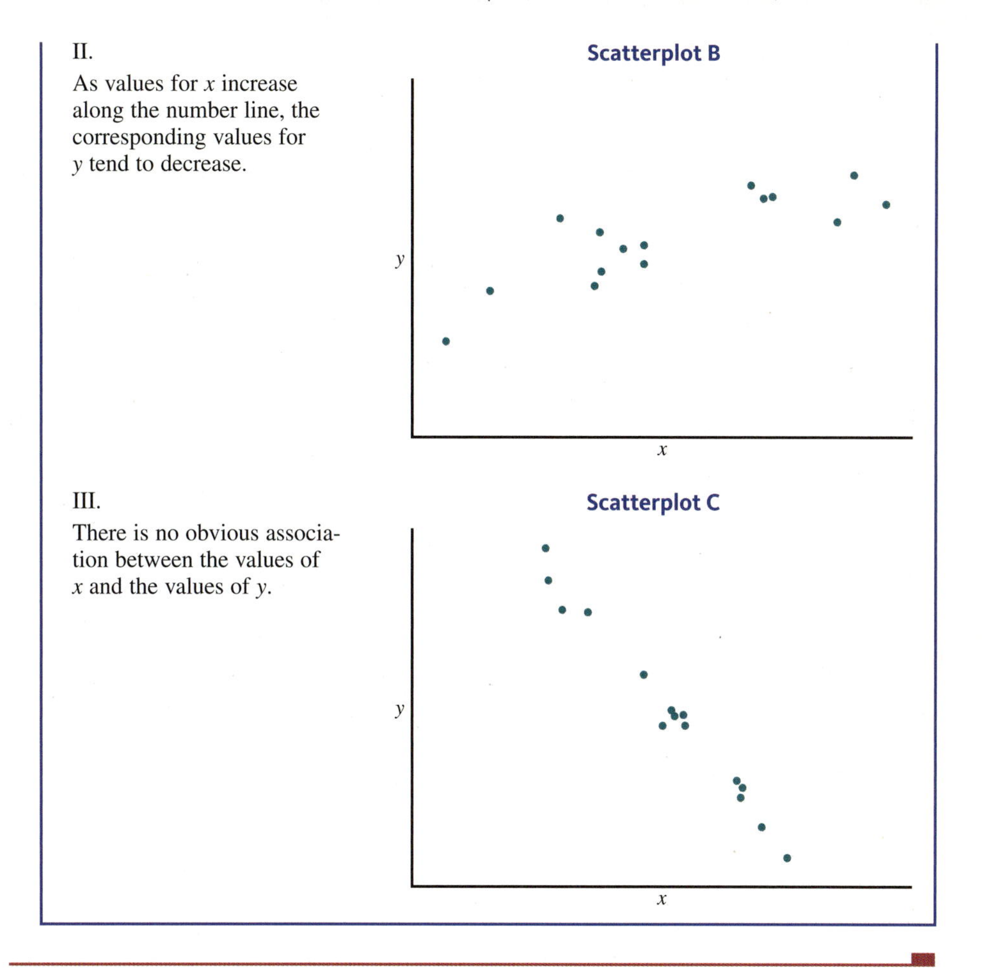

Answers

a.

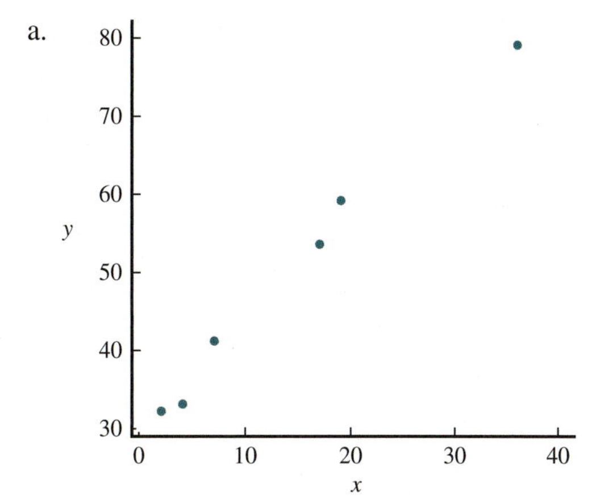

As the x-variable Years since college graduation increases, the y-variable Salary tends to increase.

b. Scatterplot B represents two variables that are not related. In Scatterplot A, larger values of x are associated with larger values of y. But in Scatterplot B, there is no pattern that suggests that there is an association between x and y.

c. Description I matches Scatterplot B, because as the values of x increase the values of y tend to increase.
Description II matches Scatterplot C, because as the values of x increase the values of y tend to decrease.
Description III matches Scatterplot A, because there is no obvious pattern in the scatterplot.

Linear and Nonlinear Patterns

Recall from Chapter 2 that once you have constructed a scatterplot, you want to look at it to assess whether there appears to be a pattern in the plot or whether the points appear to be scattered at random. This provides insight into a possible relationship between the two variables. If there is a pattern, then it means that there may be a relationship between the two variables and that the variables tend to change together in a predictable way.

For example, in the scatterplot of the data on percentage of people with no natural teeth and unemployment rate for eight southern states, there is a pattern in the plot. Unemployment rate tends to increase as the percentage of people with no natural teeth increases. This means that states with higher percentages of people with no natural teeth tended to have higher unemployment rates.

If the points in a scatterplot appear to be scattered at random with no pattern, then there may be no relationship between the two variables. If there is a pattern where the points in the plot appear to be scattering around some straight line, it is described as a linear pattern. If there is a pattern where the points in the plot appear to be scattering around a curve, it is described as a nonlinear pattern.

Linear patterns are either positive or negative, depending on whether the pattern slopes upward or downward. If the linear pattern slopes upward, like in the following scatterplot, then the y value tends to increase as x increases, and the pattern is described as a positive linear pattern.

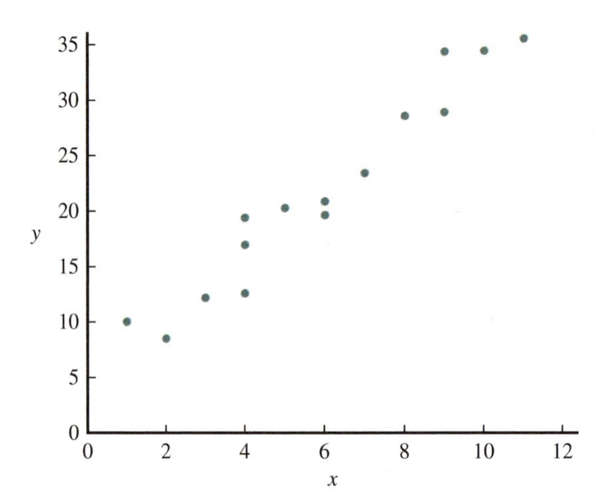

If the pattern slopes downward, like in the following scatterplot, the y value tends to decrease and x increases, and the pattern is described as negative.

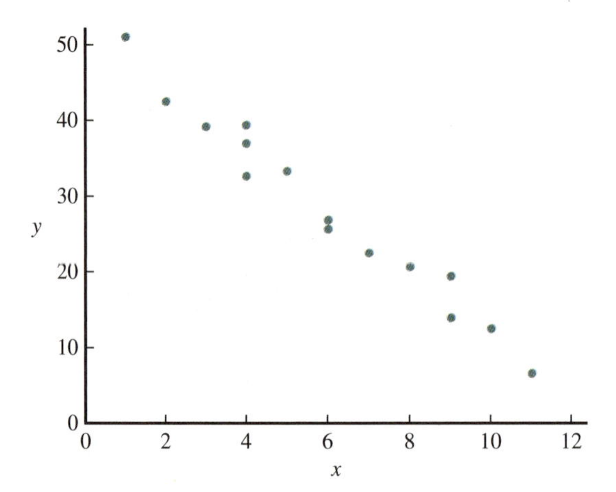

Check It Out!

a. In the following scatterplot, is the association between the variables linear or nonlinear? If the association is linear, it is positive or negative? Check your answer.

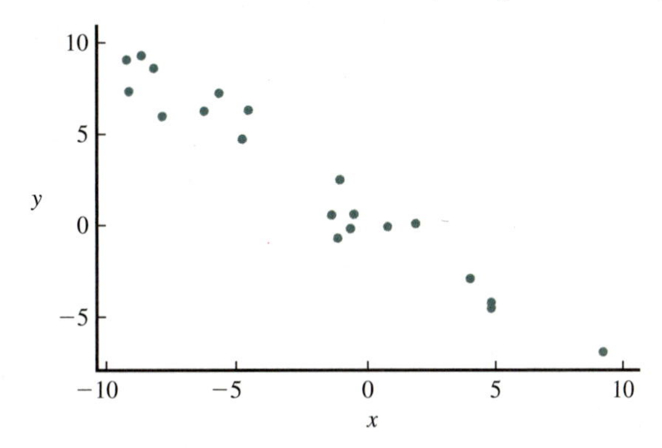

b. In the following scatterplot, is the association between the variables linear or nonlinear? If the association is linear, it is positive or negative? Check your answer.

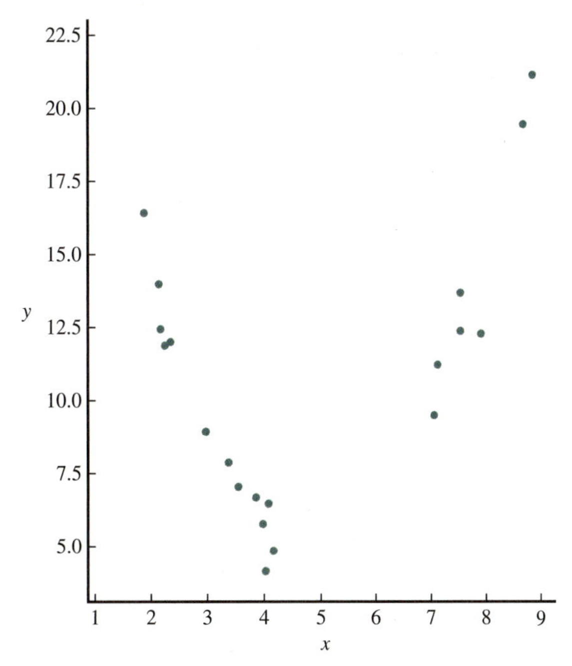

c. Consider the following pairs of variables. Do you think that the association between them would be positive or negative? Explain your reasoning, and then check your answers.

Variables	Positive or Negative?	Explanation
x = high school grade point average y = SAT math score		
x = daily high temperature (deg C) y = daily rainfall amount (in mm)		
x = the number of minutes that have passed since the beginning of a basketball game y = a player's chance of making a free throw		

Answers

a. The association is linear and negative. As the values for x increase, the values for y tend to decrease along a line.

b. The association between x and y is nonlinear. The points in the plot seem to follow a curve. It looks like it might be a parabola, opening upward.

c.

Variables	Answer
x = high school grade point average y = SAT math score	The association is likely to be positive; students with higher high school grade point averages are likely to earn higher scores on the SAT math exam.
x = daily high temperature (deg C) y = daily rainfall amount (in mm)	The association is likely to be negative, since hotter days may tend to have less rain.
x = the number of minutes that have passed since the beginning of a basketball game y = a player's chance of making a free throw	There is likely to be no association or at least only a weak association between how many minutes have passed and a player's chance of making a free throw.

z-Scores

Recall that *z-scores* were introduced in Chapter 3. You will see *z-scores* again in the context of bivariate data when you learn about the correlation coefficient. The correlation coefficient is usually defined in terms of *z-scores*. It is a measure of the strength of the linear relationship between two numerical variables.

To calculate a *z-score* corresponding to a particular value in a data set, you subtract the corresponding mean from that value and then divide by the corresponding standard deviation:

$$z = \frac{\text{data value} - \text{mean}}{\text{standard deviation}}$$

When you are working with bivariate data, it is possible to calculate *z-scores* both for the x values in the data set and for the y values in the data set. For example, consider the following bivariate data set. These data are from a study that investigated whether the amount of a chemical thought to cause cancer that is found in French fries is related to frying time (in seconds).

x = Frying Time (seconds)	y = Acrylamide Concentration (micrograms per kilogram)
150	155
240	120
240	190
270	185
300	140
300	270

Data source: *Food and Chemical Toxicology* [2012]: 3867–3876

For this data set, rounded to three decimal places, the mean of the x values is 250.000 seconds and the standard deviation is 55.857 seconds. Also rounded to three decimal places, the mean of the y values is 176.667 micrograms per kilogram and the standard deviation is 52.883 micrograms per kilogram.

To calculate the *z-scores* for the first (x, y) pair, you start with the x value, which is 150. You would first subtract the mean of the x values from 150 and then divide that number by the standard deviation of the x values:

$$z = \frac{\text{value} - \text{mean}}{\text{standard deviation}} = \frac{150 - 250.000}{55.857} = \frac{-100.000}{55.857} = -1.79$$

(rounded to two decimal places).

To calculate the *z-score* for the y value of the first (x, y) pair, you start with the y value, which is 155. You would first subtract the mean of the y values from 150 and then divide that number by the standard deviation of the y values:

$$z = \frac{\text{value} - \text{mean}}{\text{standard deviation}} = \frac{155 - 176.667}{52.883} = \frac{-21.667}{52.883} = -0.41$$

(rounded to two decimal places).

Check It Out!

a. Recall that in the Frying Time example the mean of the x values is 250.000 seconds and the standard deviation is 55.857 seconds and that the mean of the y values is 176.667 micrograms per kilogram and the standard deviation is 52.883 micrograms per kilogram.

The second x-value in the dataset is $x = 240$ seconds. Will the *z-score* for $x = 240$ seconds be positive or negative? Explain your answer, then check your reasoning.

b. The third y value in the dataset is 190 mcg/kg. Will the *z-score* for $y = 190$ be positive or negative? Explain your reasoning, and check the answer.

c. Complete the table by calculating the *z-scores* for the remaining x-values and the remaining y-values. Check your answers along the way.

x = Frying Time	*z-Score* for x	y = Acrylamide Concentration	*z-Score* for y
150	−1.79	155	−0.41
240		120	
240		190	
270		185	
300		140	
300		270	

d. What does it mean if a *z-score* is equal to zero? Explain your reasoning, and check your answer.

Answers

a. The *z-score* for $x = 240$ seconds will be negative, because the value less than the mean x value, 250.000.

b. The *z-score* for $y = 190$ will be positive, because 190 is greater than the mean y value, 176.667.

c.

x = Frying Time	*z-Score* for x	Answer	y = Acrylamide Concentration	*z-Score* for y	Answer
150	−1.79		155	−0.41	
240		−0.18	120		−1.07
240		−0.18	190		0.25
270		0.36	185		0.16
300		0.90	140		−0.69
300		0.90	270		1.76

d. A *z-score* equal to zero indicates that the original value must have been equal to the mean. When this is the case, the numerator of the *z-score* will be zero, causing the *z-score* to be equal to zero as well.

SECTION 4.1 EXERCISES

For each of the following pairs of variables in Exercises 4.1–4.3, identify which is likely to represent the independent (*x*) variable and which represents the dependent (*y*) variable.

4.1 Variable 1: Days without smoking for a participant in a cessation program
Variable 2: Number of sessions attended by a smoker in a cessation program

4.2 Variable 1: Time spent by a student studying for a final exam
Variable 2: Score on the same final exam

4.3 Variable 1: Annual salary for a baseball player
Variable 2: Number of home runs hit per season by a baseball player

4.4 Asking prices (in dollars) for one model of car for sale in a particular city were recorded, along with the age of the car (in years). The resulting data are given in the accompanying table.

Car	Age	Price
1	3	17,000
2	4	9,200
3	4	13,000
4	5	11,000
5	6	7,350
6	6	5,175
7	6	6,850
8	8	5,125
9	12	3,990

Construct a scatterplot representing the dependent (y) variable, price, plotted against the independent (x) variable, age. Comment on how price (y) changes as age (x) changes.

4.5 Which of the following two scatterplots (A or B) represents two variables that are not related? Explain your reasoning.

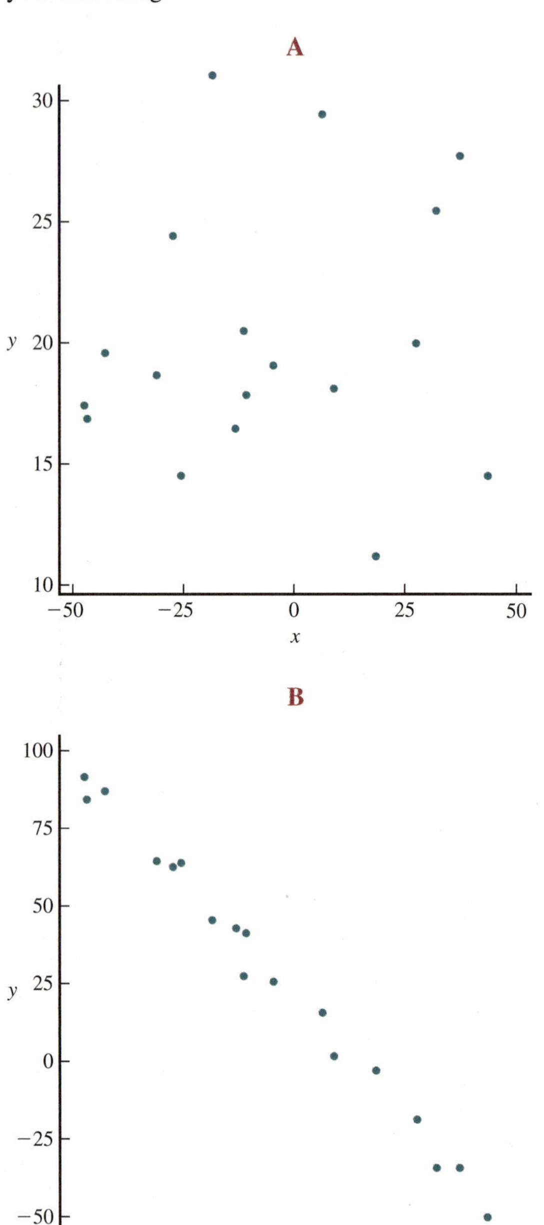

For each description given in Exercises 4.6–4.8, indicate which of the three histograms below (A, B, or C) is consistent with the given description.

4.6 As values for x increase along the number line, the corresponding values for y tend to decrease.

4.7 There is no obvious association between the values of x and the values of y.

4.8 As values for x increase along the number line, the corresponding values for y tend to increase.

Scatterplot A

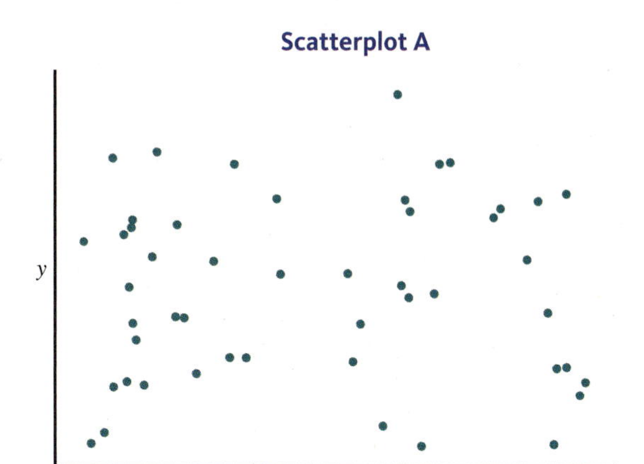

Scatterplot B

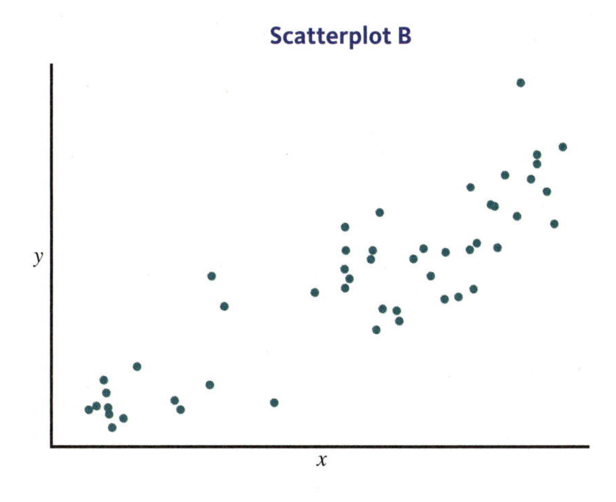

Scatterplot C

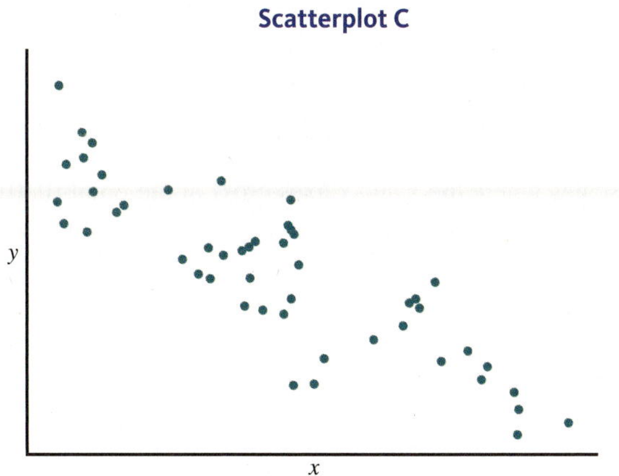

4.9 In the following scatterplot, is the association between the variables linear or nonlinear? If the association is linear, it is positive or negative?

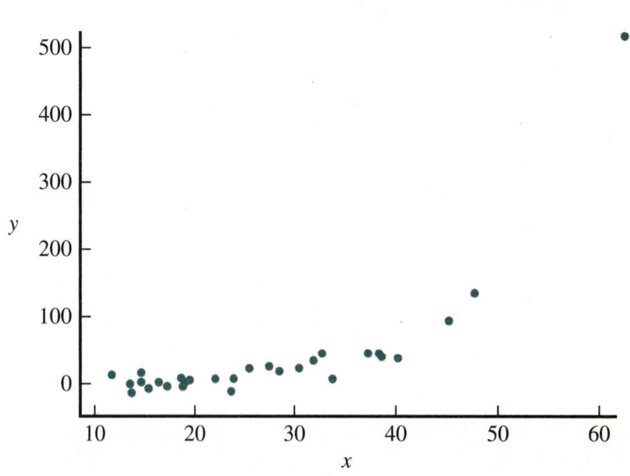

4.10 In the following scatterplot, is the association between the variables linear or nonlinear? If the association is linear, is it positive or negative?

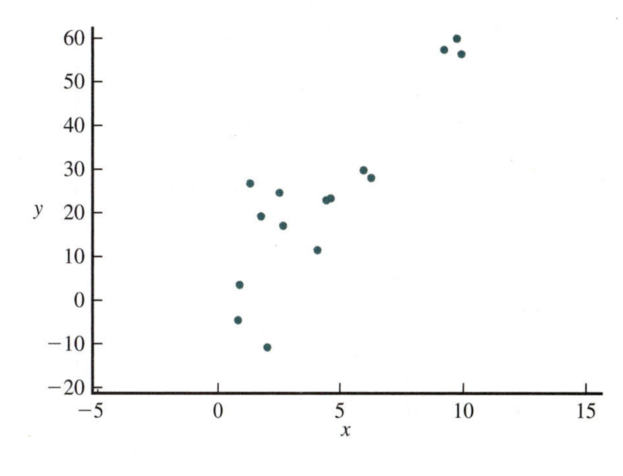

Consider the following pairs of variables given in Exercises 4.11–4.13. Do you believe that the association between them would be positive or negative? Explain your reasoning.

4.11 x = Daily hours of sunlight (minutes)
y = Daily growth of plants (mm)

4.12 x = Number of wolves per square mile
y = Number of elk per square mile

4.13 x = Height of a student in a high school
y = Grade point average for a student in a high school

A teacher decided to calculate *z-scores* for the scores (out of 100 points) on a test that students earned on an exam. The mean score on the exam was 84 points, and the standard deviation was 6.0 points. The teacher also calculated the *z-scores* for the points earned by students on homework assignments, which had a total of 250 points. The points earned on homework had a mean of 217 with a standard deviation of 9.8 points.

Complete the following table by calculating the *z-scores* for the *x* values and the *y* values.

Exercise	x = Homework Points (out of 250)	*z-Score* for x	y = Exam Score (out of 100)	*z-Score* for y
4.14	181		75	
4.15	144		73	
4.16	212		94	

SECTION 4.2 Working with Lines

The Equation of a Line

Consider the line in the following graph.

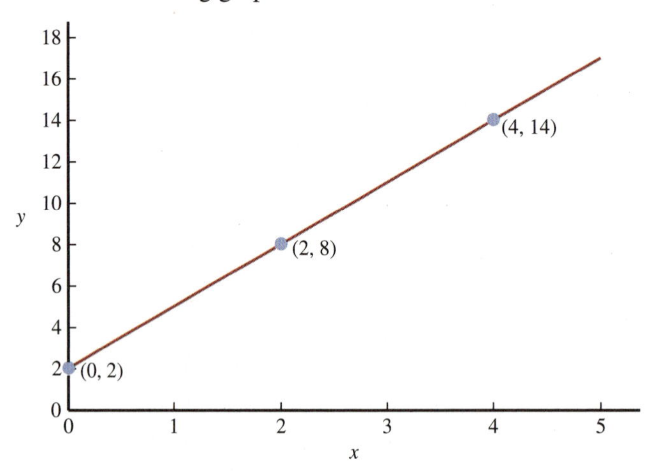

This line can be described mathematically in several ways. One way is to specify any two points on the line. For example, for this line, you could say the line passing through (0, 2) and (2, 8). Or you might say the line passing through (2, 8) and (4, 14). This is enough to describe a particular line because once two points on the line are given, there is only one straight line that goes through these two points.

Lines can also be described using an equation that determines all the points that fall on the line. Although there are several common forms for equations of lines, they all specify the **slope** of the line and the ***y*-intercept** of the line. The slope tells you how quickly the line rises or falls as you move from left to right along the line. The *y-intercept* tells you the value of y when x is 0. In the graph of the line, this is the place that the line crosses the y axis if the graph is drawn using scales where the x and y axes cross at the point (0, 0).

The three most common forms for the equation of a line are shown in the accompanying table.

Equation of a Line	Slope and Intercept	Example
$y = mx + b$	In this form, m represents the slope of the line and b represents the *y-intercept* of the line.	$y = 3x + 2$ $m = 3, b = 2$
$y = ax + b$	In this form, a represents the slope of the line and b represents the *y-intercept* of the line.	$y = 3x + 2$ $a = 3, b = 2$
$y = a + bx$	In this form, b represents the slope of the line and a represents the *y-intercept* of the line.	$y = 2 + 3x$ $a = 2, b = 3$

This can be a bit confusing at first, but notice that all three forms for the line start with $y =$ on the left side of equation. On the right side of the equation, all three forms have a term that is a number multiplied by x and another term that is just a number by itself. The number multiplied by x is the slope of the line and the number that is by itself is the *y-intercept* of the line. The only differences are the order in which the terms are listed and the letters that are used to represent the slope and the *y-intercept*. Notice that all three of the examples given in the table specify the same line—the line with a slope of 3 and a *y-intercept* of 2.

The form $y = mx + b$ is the one that is used most often in algebra, so it is probably the one that you have seen in your previous math classes. However, most graphing calculators use the notation in the second two forms. If you are working with a graphing calculator, you may have the option to specify which of these two forms ($y = ax + b$ or $y = a + bx$) you want to use.

For reasons that are not obvious in the context of the introductory statistics course, in statistics the form that is used is $y = a + bx$. In the rest of this section, we will use this form for the equation of a line, and if you are using a graphing calculator that gives you a choice, you will want to choose this form.

In the notation used in statistics, a line is represented by the equation

$$y = a + bx$$

The letter *a* represents the *y-intercept* of the line.

The letter *b* represents the slope of the line.

In your statistics course, you will need to identify the values of the slope and the *y-intercept* given the equation of a line. For example, in the line

$$y = 24 + 7.2x$$

the slope of the line is $b = 7.2$ and the *y-intercept* is $a = 24$. Remember that the slope is the number that is multiplied by x in the equation of the line and the *y-intercept* is the number that stands alone in the equation.

You will also need to be able to write the equation of a line for a given slope and *y-intercept*. To do this, remember that the equation of a line is $y = a + bx$, where a represents the *y-intercept* and b represents the slope. This means that the equation of the line with slope 7.26 and *y-intercept* 12.48 is

$$y = 12.48 + 7.26x$$

The equation of the line with slope -4.99 and *y-intercept* 3.21 is

$$y = 3.21 + (-4.99)x$$

which is usually written as

$$y = 3.21 - 4.99x$$

Check It Out!

a. For each of the following lines, identify the slope and identify the *y-intercept*. Check your answers as you go.

Equation of Line	Slope	*y-Intercept*
$y = 2.09 - 4.04x$	$b =$	$a =$
$y = 5.1 + 4.5x$	$b =$	$a =$
$y = 5 - 3.1x$	$b =$	$a =$

b. For each of the following values for the slope and *y-intercept*, write the equation of the corresponding line. Check your answers as you go.

Slope	*y-Intercept*	Equation of Line
$b = 0.2$	$a = 9.7$	
$b = 8.8$	$a = 1.4$	
$b = -4.1$	$a = -3$	

Answers

a.

Equation of Line	Answers
$y = 2.09 - 4.04x$	Slope $= b = -4.04$ *y-intercept* $= a = 2.09$
$y = 5.1 + 4.5x$	Slope $= b = 4.5$ *y-intercept* $= a = 5.1$
$y = 5 - 3.1x$	Slope $= b = -3.1$ *y-intercept* $= a = 5$

b.

Slope	*y-Intercept*	Answers
$b = 0.2$	$a = 9.7$	$y = 9.7 + 0.2x$
$b = 8.8$	$a = 1.4$	$y = 1.4 + 8.8x$
$b = -4.1$	$a = -3$	$y = -3 - 4.1x$

Finding Points on a Line

Once you know the equation of a line, it is easy to find the values for points that fall on the line. You can find the (x, y) pair for the point on the line that corresponds to a particular value of x by substituting the value of x into the equation and then evaluating the right side of the equation to determine the corresponding value of y. For example, consider the line that has the following equation:

$$y = 4.53 + 6.72x$$

The y value of the point that falls on this line that has an x value of 3.3 is

$$y = 4.53 + 6.72x = 4.53 + 6.72(3.3) = 4.53 + 22.176 = 26.706$$

This means that the point (3.3, 26.706) is a point on the line.

The y value of the point that falls on this line that has an x value of 5.1 is

$$y = 4.53 + 6.72x = 4.53 + 6.72(5.1) = 4.53 + 34.272 = 38.802$$

This means that the point (5.1, 38.802) is also a point on the line.

Finding points that fall on a line with a negative slope is done in the same way. For example, the y value of the point on the line $y = 12.2 - 4.7x$ that corresponds to an x value of 4 is

$$y = 12.2 - 4.7x = 12.2 - 4.7(4) = 12.2 - 18.8 = -6.6$$

The point $(4, -6.6)$ is a point on this line.

Check It Out!

a. For each line and its accompanying x value, calculate the corresponding y value. Check your answers along the way.

x Value	Equation of Line	*y* Value
$x = 4$	$y = 0.9 + 8.1x$	
$x = 1.6$	$y = 5 + 7.3x$	
$x = -1.09$	$y = 8 - 9x$	

b. For each line, identify a point that is on the line by choosing a value for x and calculating the corresponding y value. Check your answers.

Equation of Line	x Value	y Value
$y = 6.7 + 3.7x$	$x =$	$y =$
$y = -9.8 - 7.5x$	$x =$	$y =$
$y = 6.52 + 2.77x$	$x =$	$y =$

Answers

a.

x Value	Equation of Line	Answer
$x = 4$	$y = 0.9 + 8.1x$	$y = 0.9 + 8.1(4) = 33.3$
$x = 1.6$	$y = 5 + 7.3x$	$y = 5 + 7.3(1.6) = 16.68$
$x = -1.09$	$y = 8 - 9x$	$y = 8 - 9(-1.09) = 17.81$

b.

Equation of Line	Answers Will Vary, Depending on x
$y = 6.7 + 3.7x$	For example, $x = 5.6$, $y = 6.7 + 3.7(5.6) = 27.42$
$y = -9.8 - 7.5x$	For example, $x = -2.56$, $y = -9.8 - 7.5(-2.56) = 9.4$
$y = 6.52 + 2.77x$	For example, $x = 7$, $y = 6.52 + 2.77(7) = 25.91$

What is the Slope of a Line?

The slope of a line tells you whether the line rises or falls and how quickly the line rises or falls as you move from left to right along the line. If the slope is positive, the line goes up as you move from left to right. If the slope is negative the line goes down as you move from left to right. The farther the value of the slope is from 0, the steeper the line (meaning the line rises or falls more quickly).

The graph below shows two lines, labeled Line 1 and Line 2. Both lines have a positive slope. You know this because both lines go up as you move from left to right. One of these lines has slope 3 and the other line has slope 5. Notice that Line 1 is steeper and rises more quickly than Line 2. This means that the slope for Line 1 is farther from 0 than the slope of Line 2, so Line 1 must be the line with slope 5 and Line 2 must be the line with slope 3.

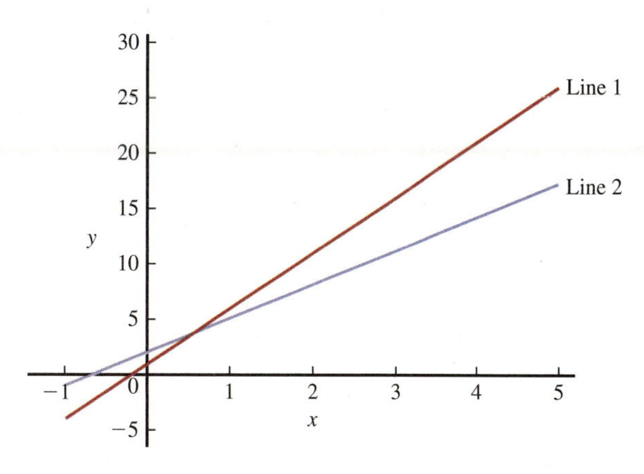

Now consider two lines that have negative slopes (the line goes down as you move from left to right). In the graph below, one line has a slope of −7 and the other has a slope of −2. Line 4 is steeper and goes down more quickly, so it must be the line with slope −7 because −7 is farther from 0 than −2 is.

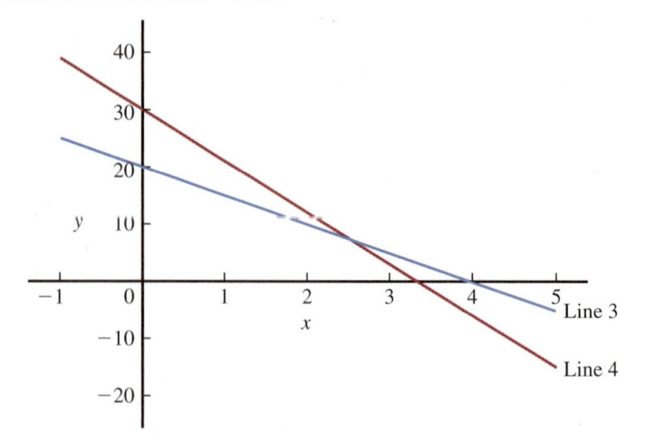

Check It Out!

a. For each of the following lines, indicate whether the slope is positive or negative. Check your answers as you go.

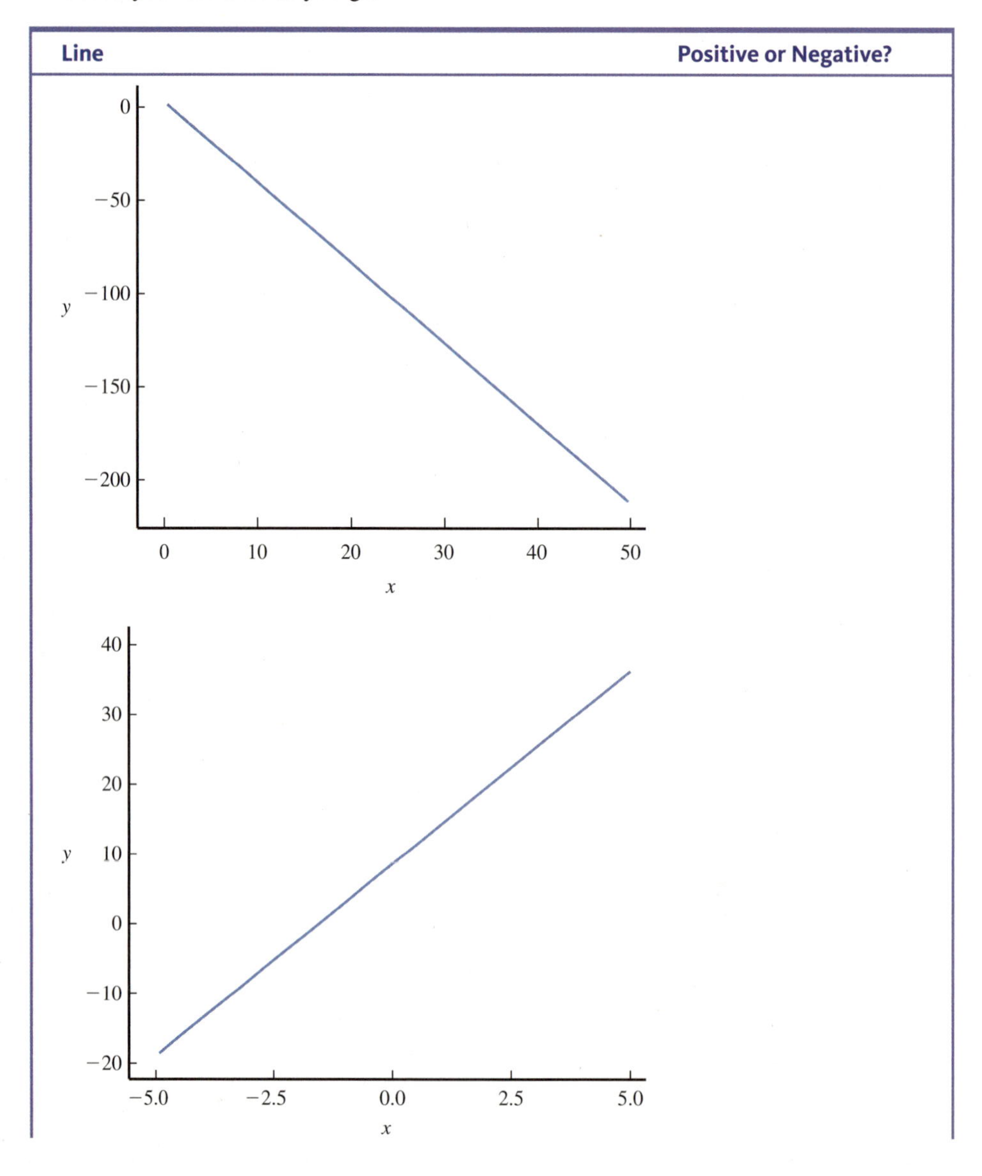

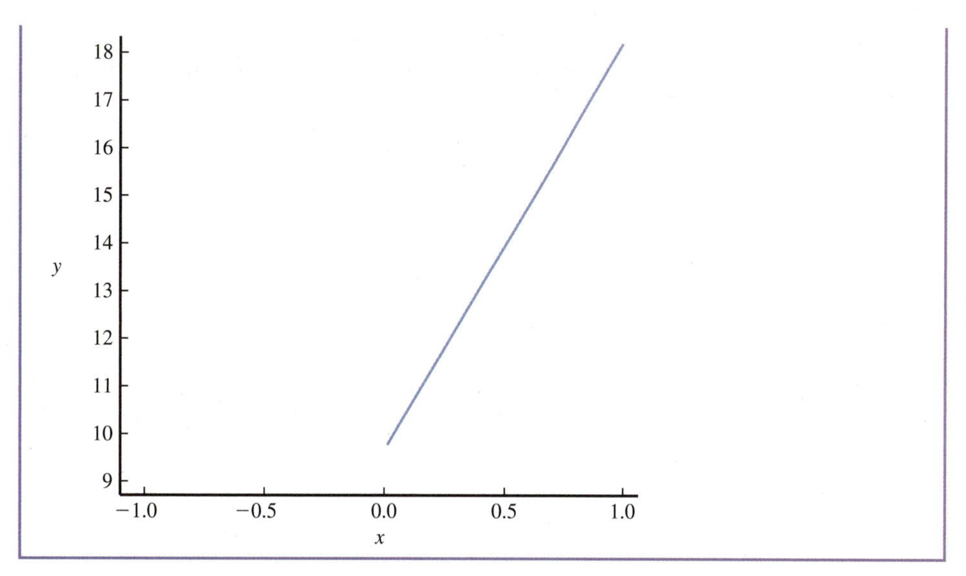

b. Which of the two lines in the scatterplot (A or B) has a slope of 0.73 and which has a slope of 0.13? Check your answers.

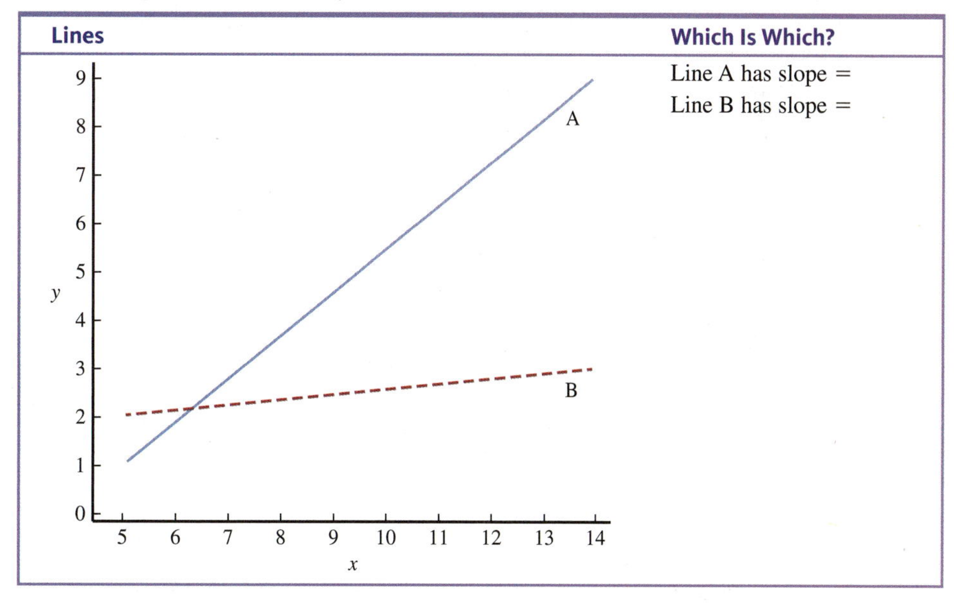

c. Which of the following two lines has a slope of −1.88 and which has a slope of −0.4? Check your answer.

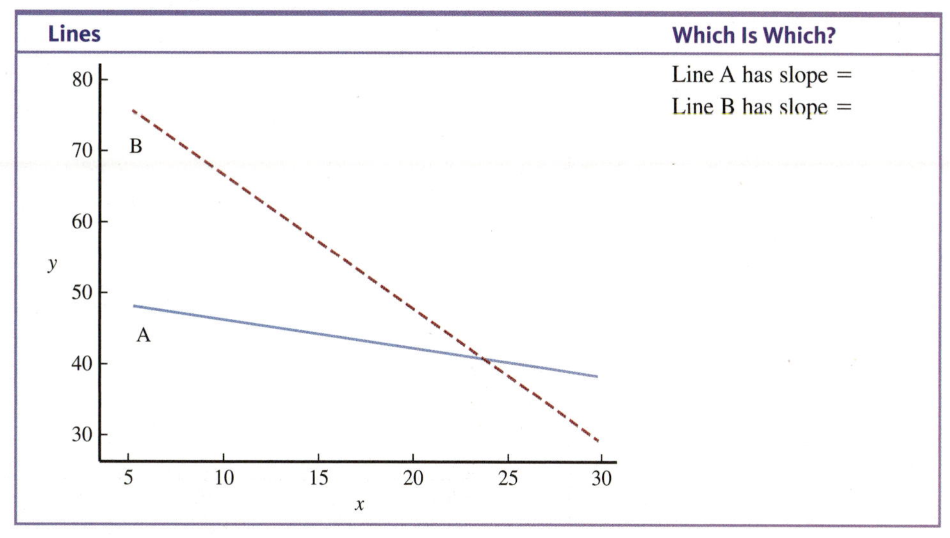

Answers

a.

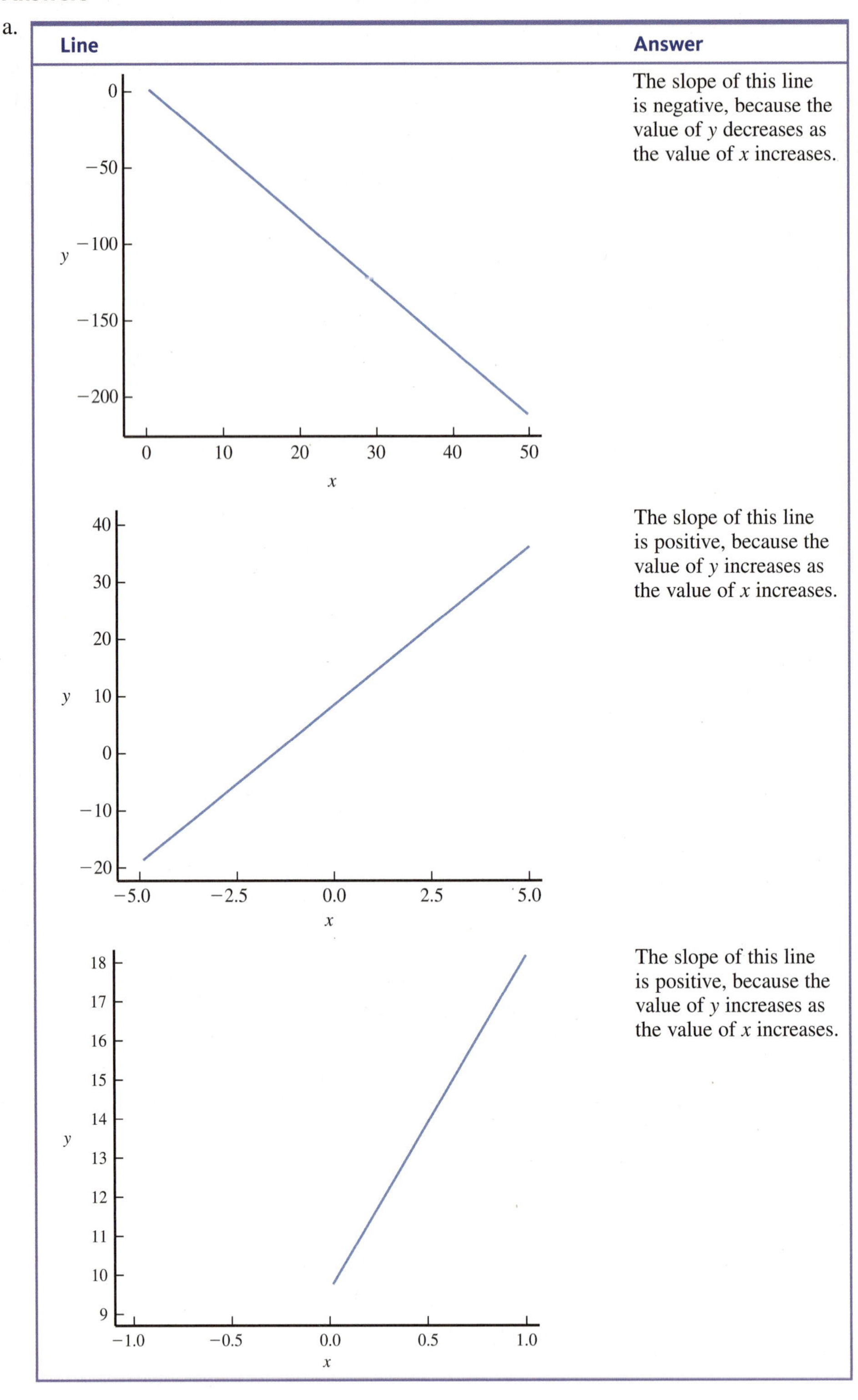

b. Line A has slope = 0.73, because the slope of Line A is positive and Line A is steeper than Line B. Line B has slope = 0.13, because the slope is also positive but the line is not as steep as Line A.

c. Line A has slope = −0.4, because the slope of Line A is negative and Line A is not as steep as Line B. Line B has slope = −1.88, because the slope is also negative but the line is steeper than Line A.

The slope of a line describes how fast the value of y changes as the value of x changes. The slope of a line is defined as

$$\text{slope} = \frac{\text{change in } y}{\text{change in } x}$$

You can calculate the slope using any two points on the line. We will refer to these points as Point 1 and Point 2. First determine the change in x by calculating

$$\text{change in } x = (\text{value of } x \text{ for Point 2}) - (\text{value of } x \text{ for Point 1})$$

and the change in y by calculating

$$\text{change in } y = (\text{value of } y \text{ for Point 2}) - (\text{value of } y \text{ for Point 1})$$

You can then calculate the slope using the formula given above. This is illustrated in the following figure:

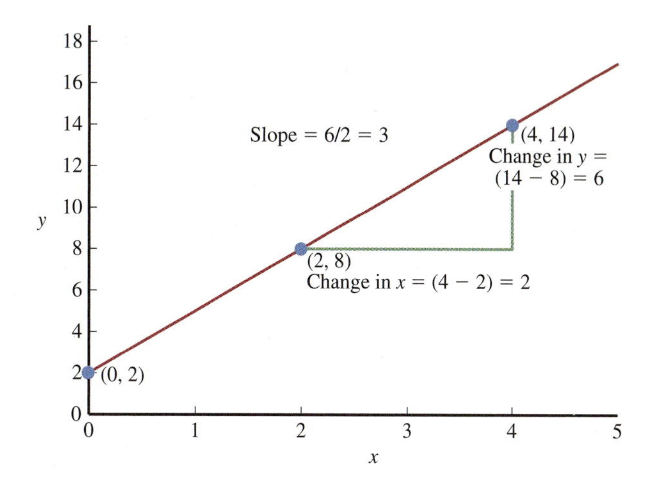

Notice that for this line, when $x = 0$, the value of y is 2. This tells you that the *y-intercept* is 2. With a *y-intercept* of 2 and a slope of 3, the equation of this line is $y = 2 + 3x$.

When the slope of a line is +3, this means that for points on the line, the value of y goes up by 3 when the value of x goes up by one. You can see this in the table below that shows some of the points on the line $y = 2 + 3x$.

x Value	*y* Value ($y = 2 + 3x$)	Point on the Line
1	$y = 2 + 3x = 2 + 3(1) = 5$	(1, 5)
2	$y = 2 + 3x = 2 + 3(2) = 8$	(2, 8)
3	$y = 2 + 3x = 2 + 3(3) = 11$	(3, 11)

What do you think happens to the value of y when x goes up by one when the slope of the line is negative? The following table shows some points on the line $y = 5 - 2x$.

x Value	*y* Value ($y = 5 - 2x$)	Point on the Line
1	$y = 5 - 2x = 5 - 2(1) = 3$	(1, 3)
2	$y = 5 - 2x = 5 - 2(2) = 1$	(2, 1)
3	$y = 5 - 2x = 5 - 2(3) = -1$	(3, −1)

The slope of the line $y = 5 - 2x$ is -2. Notice that the value of y decreases by 2 when the value of x goes up by 1.

Check It Out!

a. Use the two points given to calculate the slope of the line connecting them. Check your answers as you go.

Point 1	Point 2	Slope
(1.382, 4.561)	(5.443, 7.231)	
(−0.347, 0.229)	(−0.689, −0.851)	
(8.5, −1.7)	(−9.7, −4.2)	
(−4, 7)	(−6, −4)	

b. Use the equation of the line to calculate the y value for each of the given x values, and then write the corresponding point on the line.

Equation of Line	x Value	y Value	Point on the Line
$y = -4.9 + 6.5x$	8		
$y = -28 + 6x$	9.4		
$y = 0.9 - 6.2x$	1.8		

Answers

a.

Point 1	Point 2	Answer
(1.382, 4.561)	(5.443, 7.231)	$\text{Slope} = \dfrac{7.231 - 4.561}{5.443 - 1.382} = 0.66$
(−0.347, 0.229)	(−0.689, −0.851)	$\text{Slope} = \dfrac{(-0.851) - (0.229)}{(-0.689) - (-0.347)} = 3.16$
(8.5, −1.7)	(−9.7, −4.2)	$\text{Slope} = \dfrac{(-4.2) - (-1.7)}{(-9.7) - (8.5)} = 0.14$
(−4, 7)	(−6, −4)	$\text{Slope} = \dfrac{(-4) - 7}{(-6) - (-4)} = 5.5$

b.

Equation of Line	x Value	Answers	
$y = -4.9 + 6.5x$	8	$y = -4.9 + 6.5(8) = 47.1$	(8, 47.1)
$y = -28 + 6x$	9.4	$y = -28 + 6(9.4) = 28.4$	(9.4, 28.4)
$y = 0.9 - 6.2x$	1.8	$y = 0.9 - 6.2(1.8) = -10.26$	(1.8, −10.26)

Special Cases—Horizontal and Vertical Lines

Horizontal lines are lines that have a slope of 0. For example, consider the line that passes through the points (4, 5) and (7, 5) shown in the figure below.

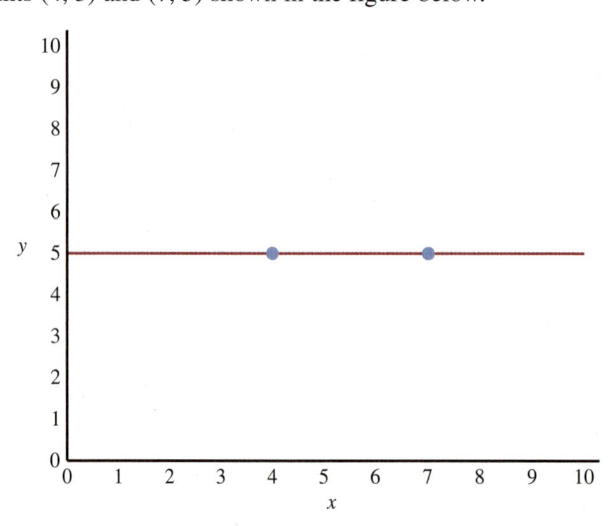

Notice that for every point on this line, the value of y is equal to 5. If you were to calculate the slope of this line, you would get slope $= 0$, because no matter which two points on the line that you choose, the change in y is always 0. It also makes sense that a horizontal line would have a slope of zero because it doesn't slope upward (positive slope) or downward (negative slope).

For horizontal lines, the slope is always 0. This means that the equation of a horizontal line is

$$y = a + bx$$
$$y = a + 0(x)$$
$$y = a$$

where a is the *y-intercept* of the line. For example, the equation of the line in the previous figure is $y = 5$.

In your statistics course, you won't work with vertical lines, but just in case you were wondering, let's look at the equation and slope of a vertical line. For example, consider the vertical line pictured in the following figure. What do all the points on this line have in common?

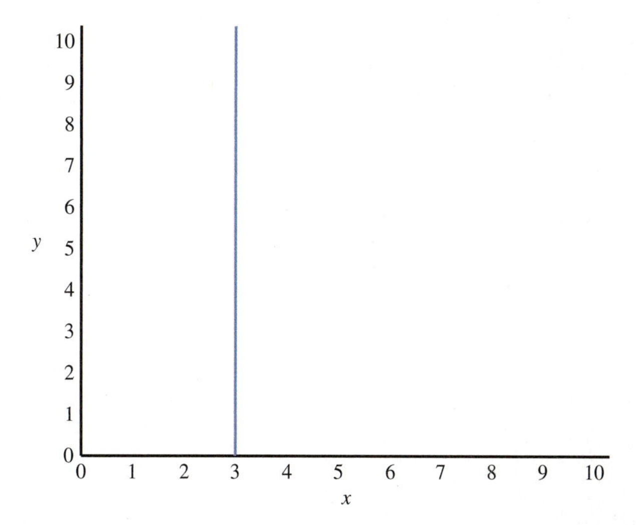

Notice that all the points on this line have an x value of 3. The equation of this line is written as $x = 3$. Vertical lines do not have a *y-intercept* and the slope is undefined for vertical lines. The slope is undefined because the slope is calculated by dividing the change in y by the change in x. The change in x is zero for any two points on this line, and division by 0 is undefined.

Graphing a Line

The easiest way to graph a line is to find two points on the line. Because there is only one line that passes through any two points, once you know two points that are on the line, you can plot those points and then draw the line. For example, suppose that you want to graph the line $y = 18 - 3x$. You can just pick two different values for x and substitute them into the equation of the line to find the corresponding y value. Suppose you choose $x = 2$ and $x = 4$.

***x* Value**	**Corresponding *y* Value**	**Point on the Line**
2	$y - 18 - 3x = 18 - 3(2) = 18 - 6 = 12$	(2, 12)
4	$y - 18 - 3x = 18 - 3(4) = 18 - 12 = 6$	(4, 6)

Plotting the two points (2, 12) and (4, 6) and then drawing the line that passes through these two points results in the following:

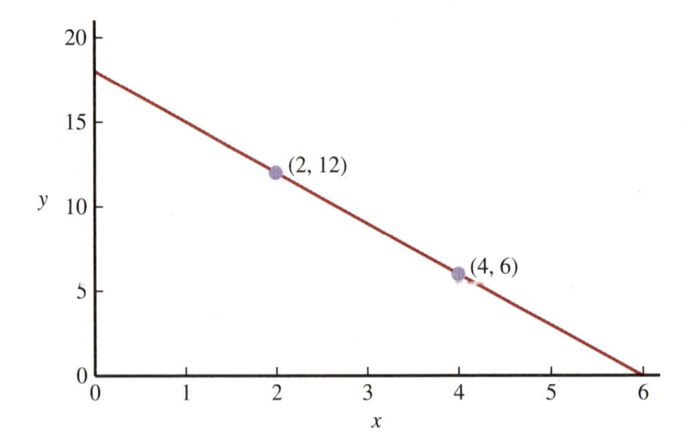

Notice that the line slopes downward as you move from left to right (because the slope of the line is negative).

You could have also graphed this line using the definition of slope and intercept to find two points on the line $y = 18 - 3x$. Because the *y-intercept* is 18, you know that this is the value of y when x is 0. That means that the point (0, 18) is one of the points on the line. Because the slope of the line is -3, you know that the value of y decreases by 3 when the value of x increases by 1. This means that the point (1, 15) is also a point on the line, because if you start with the point (0, 18), which you know is on the line, and increase x by 1 and decrease y by 3, you get (1, 15). Plotting these two points and then drawing the line that passes through them results in the same that was shown in the previous figure.

Check It Out!

a. For the line $y = 59 + 71x$, use the x values in the table to find two points on the line. Then, graph the line. Check your answers.

x Value	Point on the Line
1	
4	

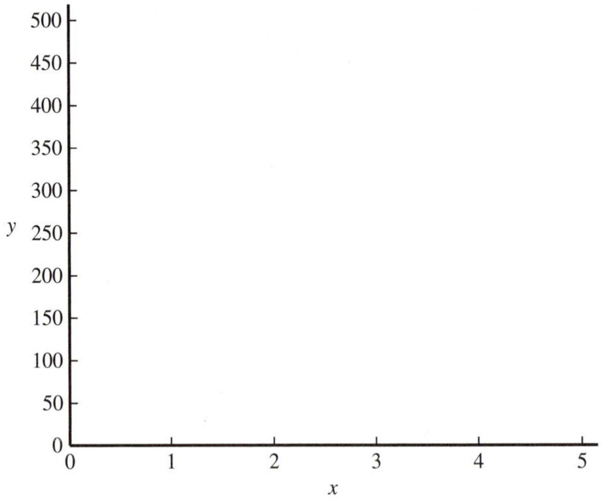

b. For the line $y = 2.4 - 3.1x$, use the x values in the table to find two points on the line. Then, graph the line using the axes on the following page. Check your answers.

x Value	Point on the Line
0.07	
0.24	

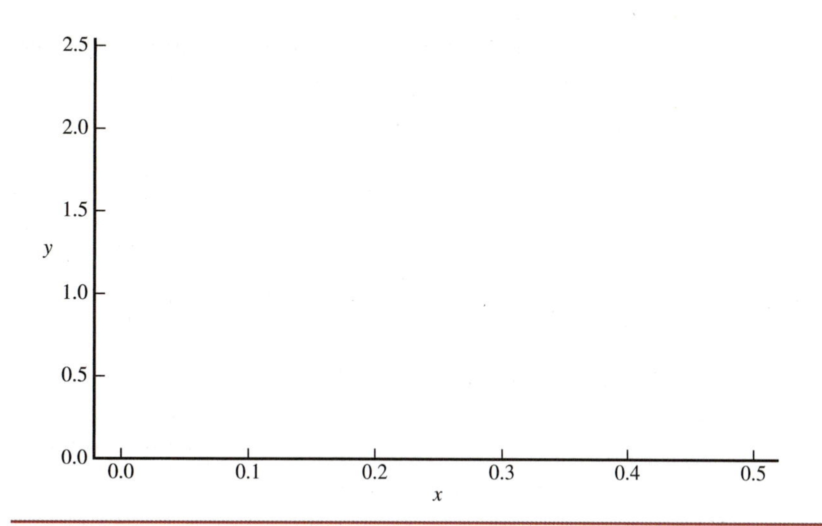

Answers

a.

x Value	Answer
1	(1, 130)
4	(4, 343)

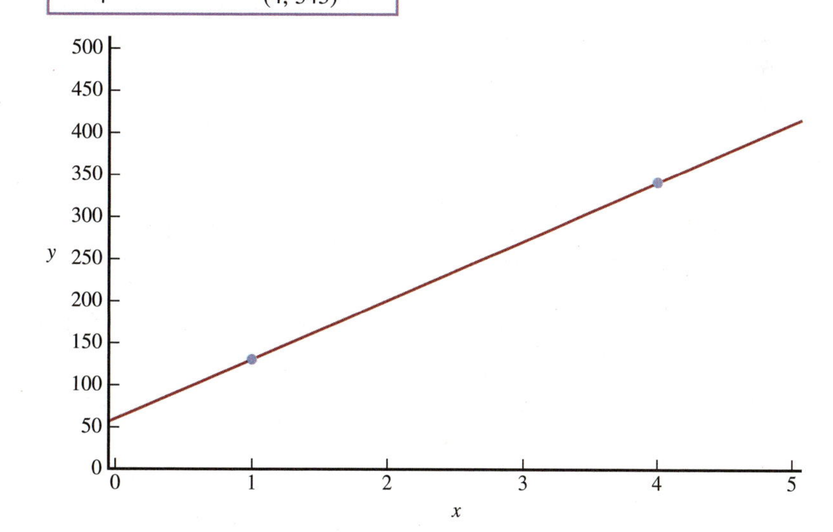

b.

x Value	Answer
0.07	(0.07, 2.18)
0.24	(0.24, 1.66)

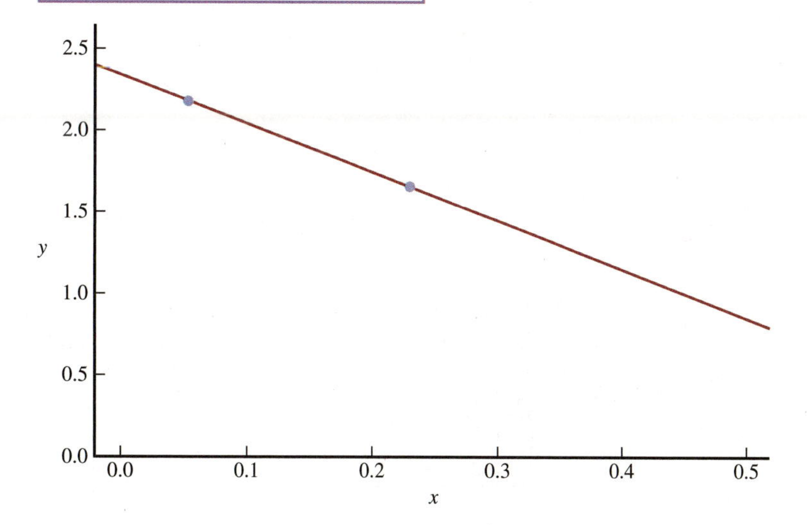

SECTION 4.2 EXERCISES

For each of the lines in Exercises 4.17–4.19, identify the slope (b) and identify the *y-intercept* (a).

4.17 $y = 98.9 - 10x$

4.18 $y = 0.737 + 0.17x$

4.19 $y = 71 + 38x$

For each of the given values for the slope and intercept in Exercises 4.20–4.22, write the equation of the corresponding line.

4.20 $b = 9, a = 4.8$

4.21 $b = 4.1, a = 3$

4.22 $b = 0.8, a = 0.18$

For each line and accompanying x value in Exercises 4.23–4.25, calculate the corresponding y value.

4.23 $x = 20, y = 48.7 - 1.7x$

4.24 $x = 3.1, y = -4.6 + 5.6x$

4.25 $x = 0.94, y = 5 + 3x$

For each line given in Exercises 4.26–4.28, identify a point that is on the line by choosing a value for x and calculating the corresponding y value.

Exercise	Equation of Line	*x* Value	*y* Value
4.26	$y = 9.11 - 0.6x$	$x =$	$y =$
4.27	$y = 6 + 2x$	$x =$	$y =$
4.28	$y = -6.9 + 0.3x$	$x =$	$y =$

For each of the lines given in Exercises 4.29–4.31, indicate whether the slope is positive or negative.

4.29 $y = -0.692 + 0.37x$

4.30 $y = 6 + 6x$

4.31 $y = 3.5 - 7.1x$

4.32 Which of the two lines in the scatterplot (A or B) has a slope of -0.1 and which has a slope of -0.4?

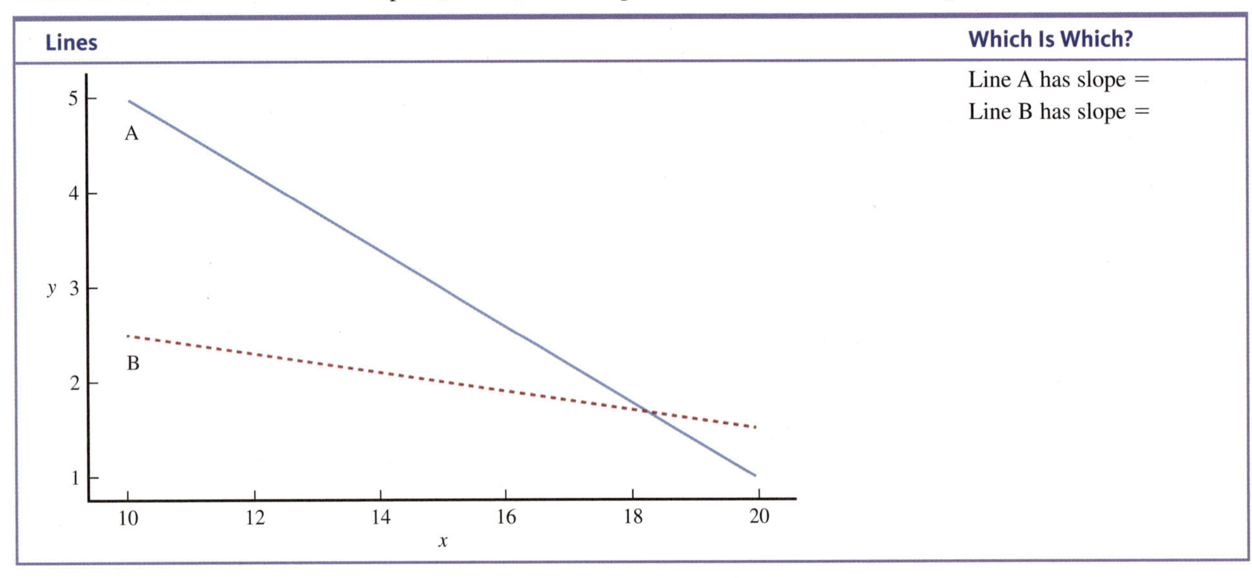

4.33 Which of the following two lines has a slope of 90 and which has a slope of 45?

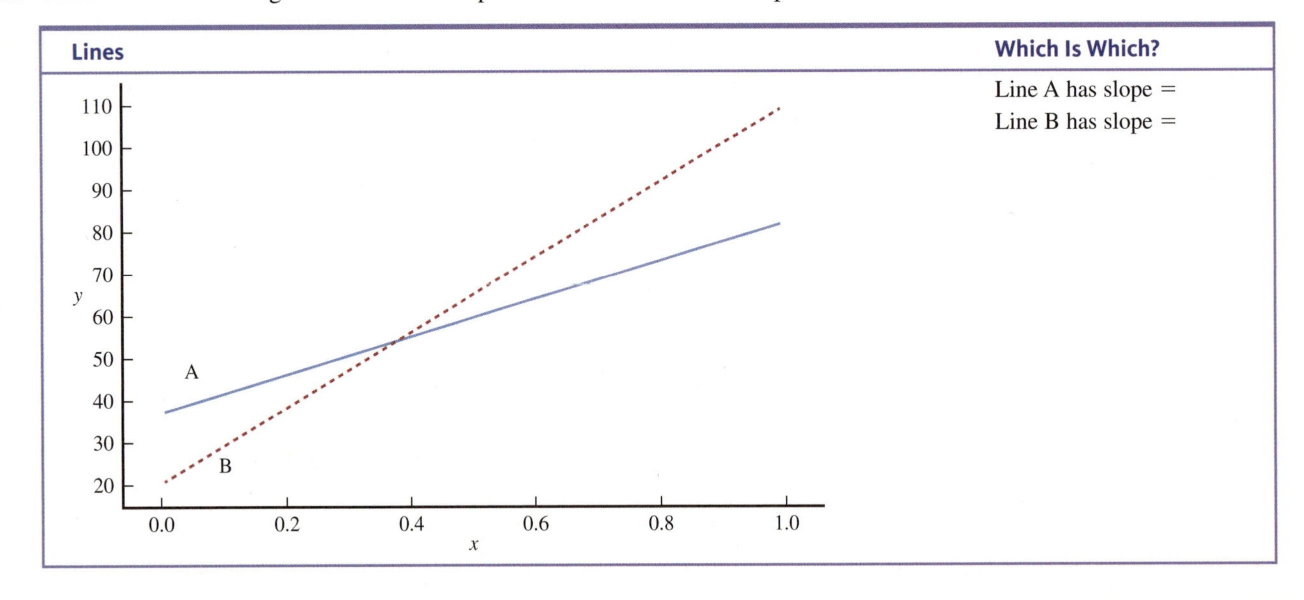

In Exercises 4.34–4.36, use the two points given to calculate the slope of the line connecting them.

4.34 (53.1, 39.0); (65.2, 23.2)

4.35 (6.9, 8.9); (−8.8, −8.9)

4.36 (76, 1); (16, 38)

In Exercises 4.37–4.39, use the equation of the line to calculate the *y* value for the given *x* value, and then give the corresponding point on the line.

4.37 Equation of line: $y = 4 + 3x$; x value: 9

4.38 Equation of line: $y = -0.8 + 0.4x$; x value: 48.1

4.39 Equation of line: $y = 43.8 + 18.6x$; x value: 4.1

4.40 For the line $y = 6 - 8x$, use the x-values in the table to find two points on the line. Then, graph the line on the axes below.

x Value	Point on the Line
0.1	
0.2	

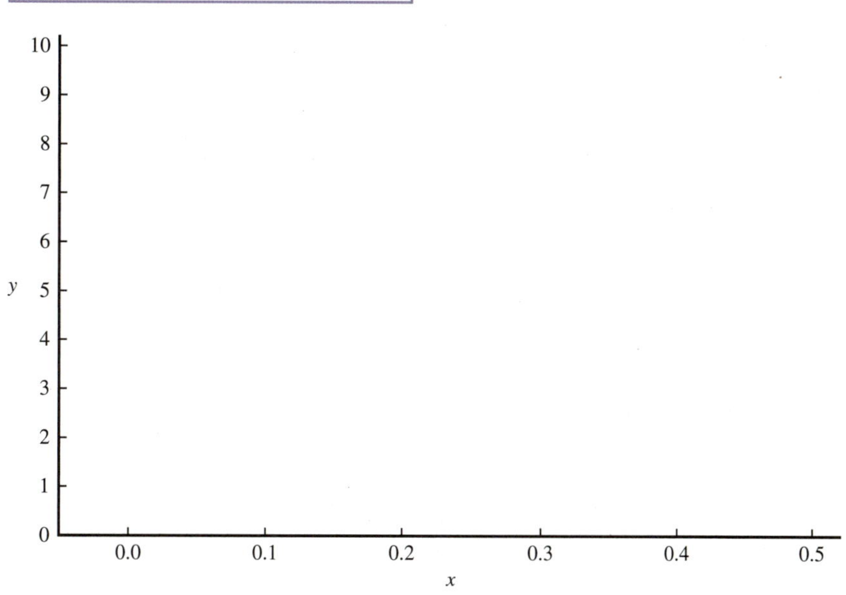

4.41 For the line $y = 19 + 9.3x$, use the x values in the table to find two points on the line. Then, graph the line on the axes below.

x Value	Point on the Line
2	
4	

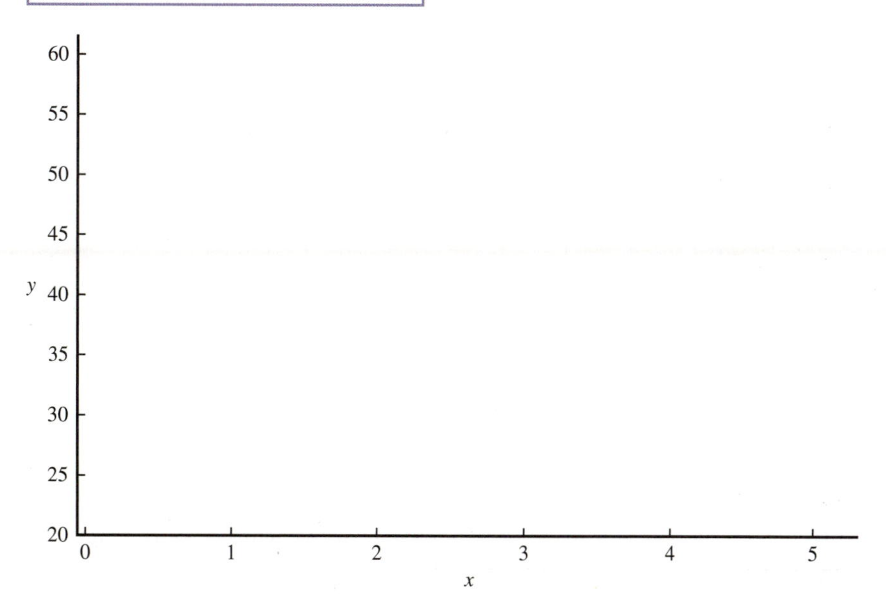

SECTION 4.3 Linear Models and Using a Line to Make Predictions

In statistics, a **model** is a mathematical description (usually in the form of an equation) that represents something in the real world. In your introductory statistics course, models are often used to describe the relationship between two variables. The idea is that if the relationship between two variables can be described mathematically, you can then use that description to predict the value of one of the variables (the dependent variable) based on the value of the other variable (the independent variable).

For example, consider the following data on

$$x = \text{Average weight (in grams)}$$

and

$$y = \text{Average wingspan (in centimeters)}$$

for nine different types of birds.

Bird	Average Weight (in grams)	Average Wingspan (in centimeters)
Barn Owl	460	93
Pigeon	370	70
Jay	190	58
Woodpecker	21	26
Turtle Dove	180	55
Magpie	250	60
Robin	22	22
Quail	135	35
Hummingbird	6	10

(Based on information from naturemappingfoundation.org/natmap/facts/birds.html and Wikipedia.org.)

Looking at the scatterplot of these data (below), it looks like there is a relationship between average weight and average wingspan for these nine types of birds. There is a positive linear pattern in the scatterplot.

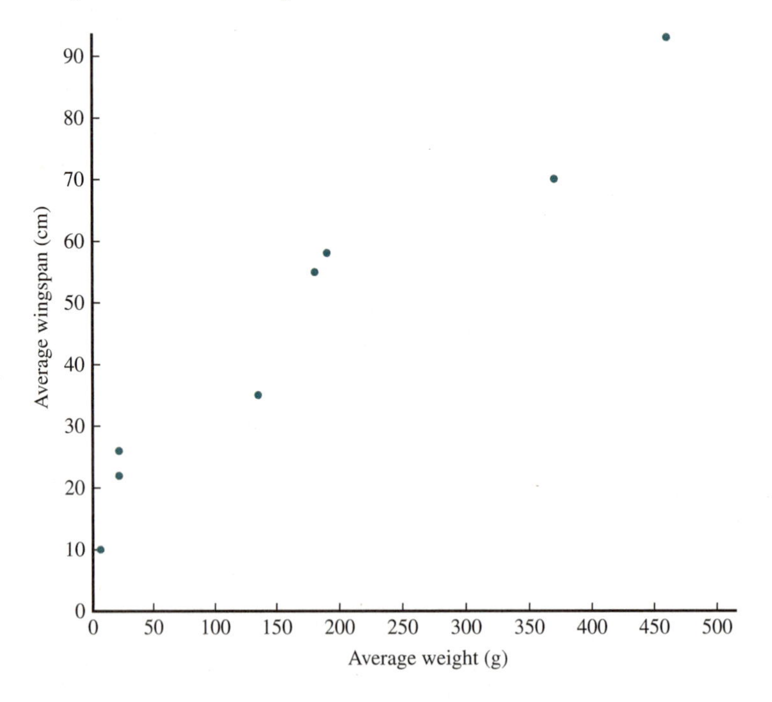

Because the pattern in the scatterplot is linear, you can use a straight line to describe the way in which average wingspan tends to vary with average weight. The line $y = 18.4 + 0.15x$ has been added to the following scatterplot, and you can see that this line does a reasonable job of describing the pattern in the scatterplot.

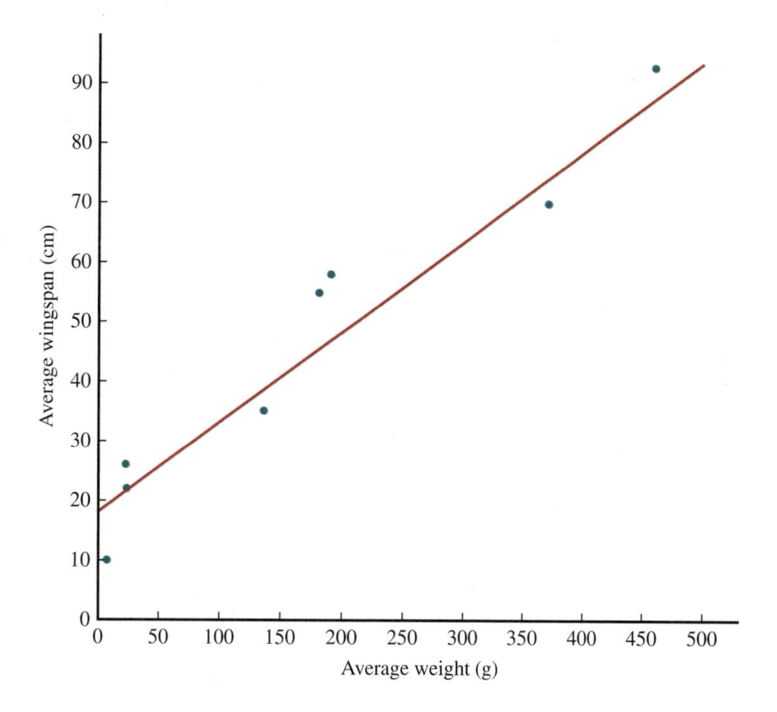

The equation $y = 18.4 + 0.15x$ is a model for the relationship between x = Average weight and y = Average wingspan. Because the model for this relationship is a line, it is called a **linear model**.

For bivariate data, a **model** is an equation that is represents the relationship between two variables.

If the equation used to model a relationship is the equation of a line, the model is called a **linear model**.

If a linear model does a reasonable job of describing the relationship between a dependent variable and an independent variable, you can use the model to predict values of the dependent variable. For example, you can use the model $y = 18.4 + 0.15x$ to predict average wingspan (y) if you know average weight (x). Suppose that the nine types of birds whose information was used to construct the scatterplot of wingspan versus weight are typical of other birds with average weights less than 500 grams. Magpies are a type of bird that have an average weight of 250 grams. The model $y = 18.4 + 0.15x$ can be used to predict the average wingspan of magpies by using the equation of the line to find the average wingspan that corresponds to an average weight of 250 grams:

$$\begin{aligned} y &= 18.4 + 0.15x \\ &= 18.4 + 0.15(250) \\ &= 18.4 + 37.5 \\ &= 55.9 \text{ cm} \end{aligned}$$

When you use a linear model to make predictions, there is one other thing that you should consider. You should be cautious in making predictions for values of the independent variable, x, that are outside the range of x values in the data set. For example, the nine types of birds used to construct the scatterplot of average wingspan versus average weight all had average weights that were between 6 grams and 460 grams. There is no evidence that the linear pattern you see in the scatterplot would continue outside this range of weights. It would not be appropriate to use the linear model $y = 18.4 + 0.15x$ to predict the average wingspan of buzzards, because buzzards have an average weight of 1300 grams, which is far above the range of average weights in the data set.

Check It Out!

a. Would a linear model be appropriate to model the relationships represented in the following scatterplots? Check your answers.

Scatterplot	Is a Linear Model Appropriate?
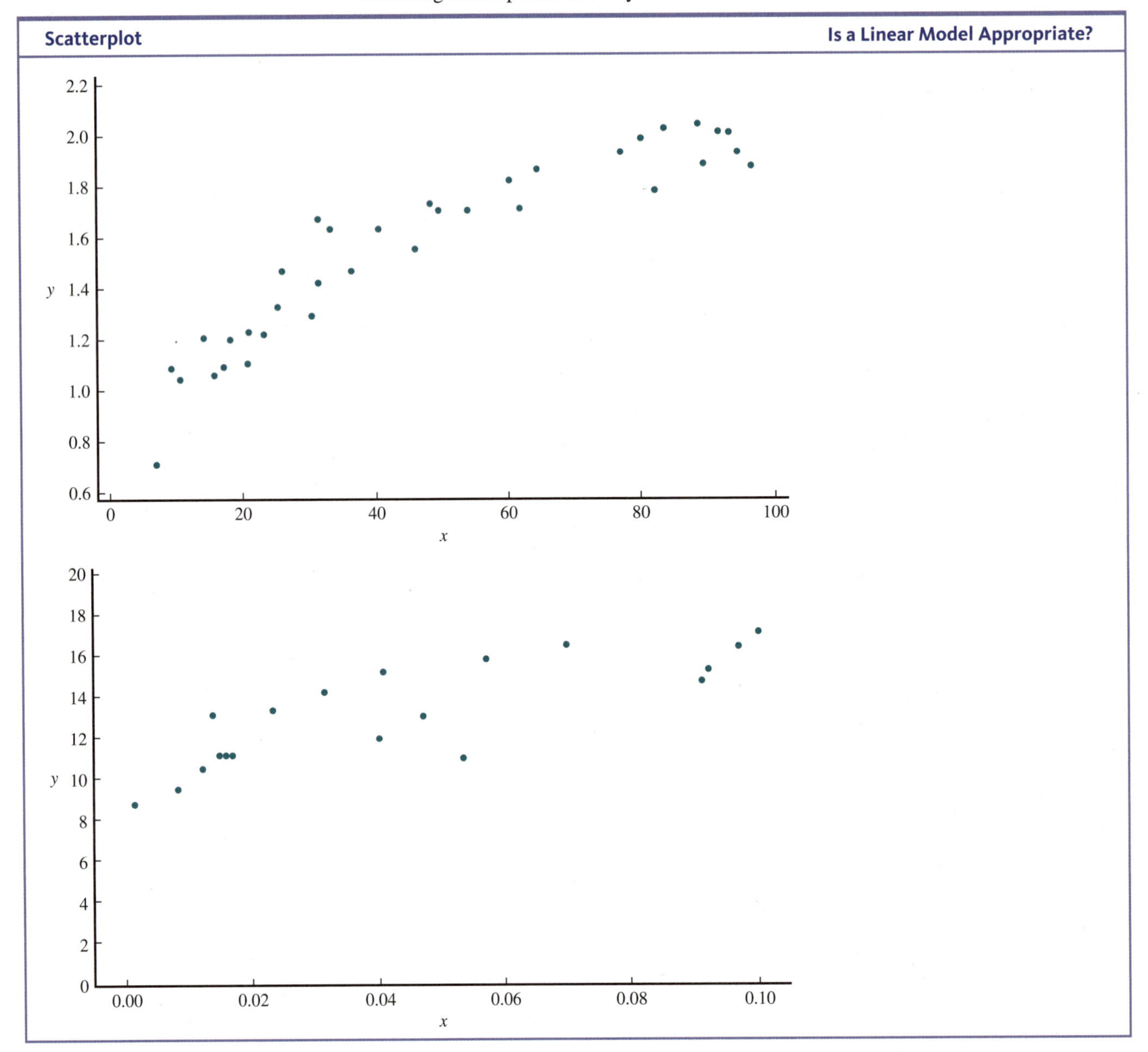	

b. Which of the following equations represent linear models? Explain your reasoning, and then check your answers.

Equation	Does It Represent a Linear Model?
$y = 7.7 \cdot 5^x$	
$y = 7.1 - 0.8x$	

c. For the given linear models, make a prediction using the given value of x. Check your answers as you go.

Linear model	x Value	Prediction
$y = 7.1 - 0.8x$	$x = 0.56$	
$y = 0.4 + 0.7x$	$x = 9$	

d. Examine the following scatterplots. Would it be appropriate to make a prediction for each of the specified x values? Explain your reasoning, and check your answers.

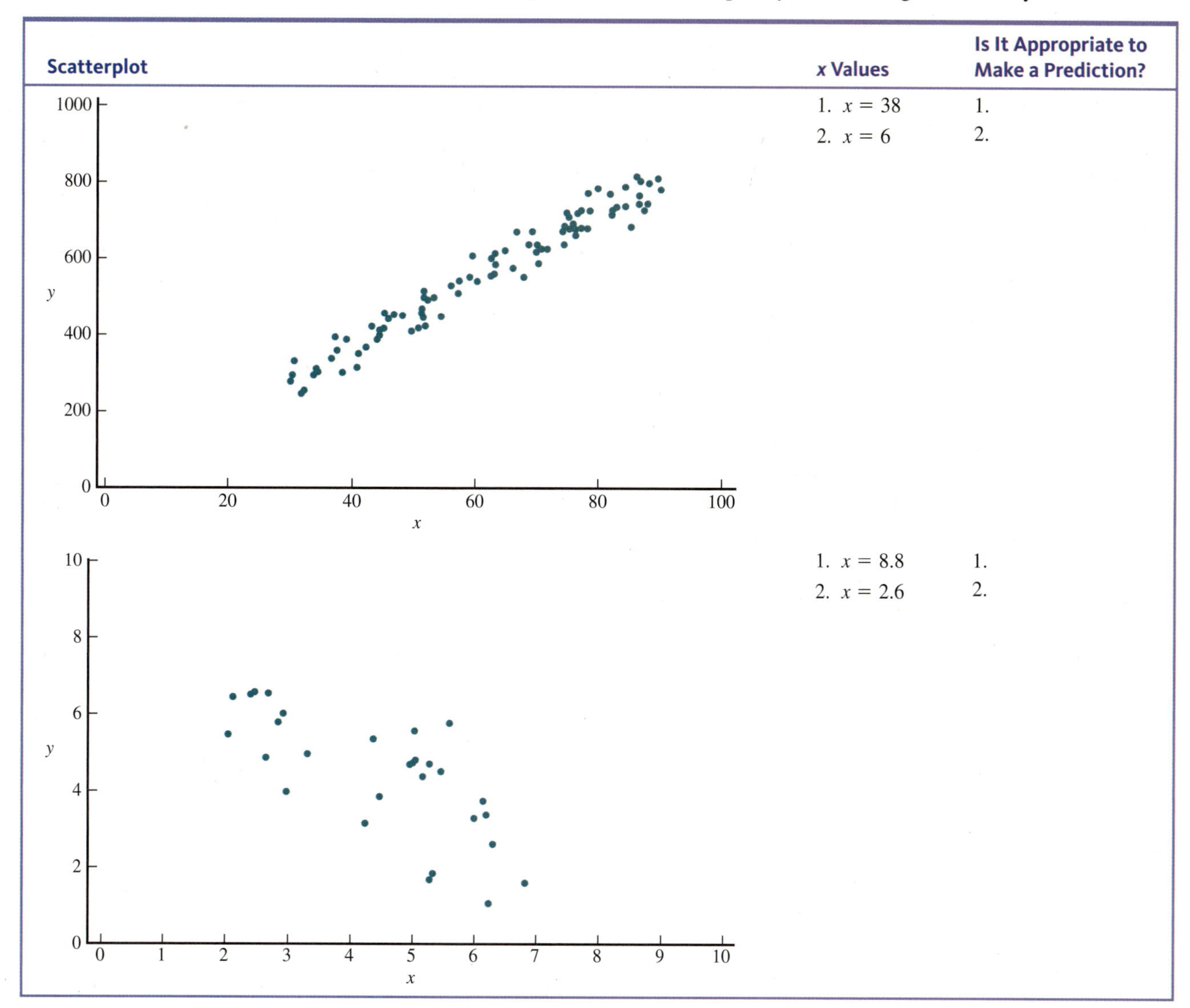

Scatterplot	x Values	Is It Appropriate to Make a Prediction?
(scatterplot 1)	1. $x = 38$ 2. $x = 6$	1. 2.
(scatterplot 2)	1. $x = 8.8$ 2. $x = 2.6$	1. 2.

Answers

a.

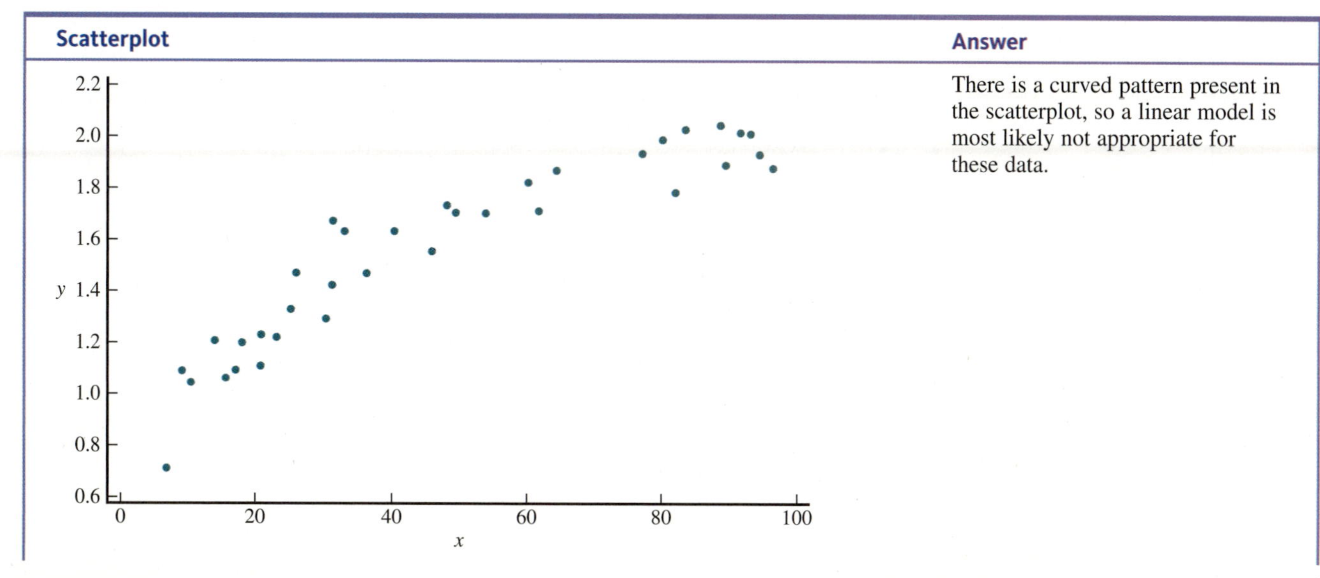

Scatterplot	Answer
(scatterplot)	There is a curved pattern present in the scatterplot, so a linear model is most likely not appropriate for these data.

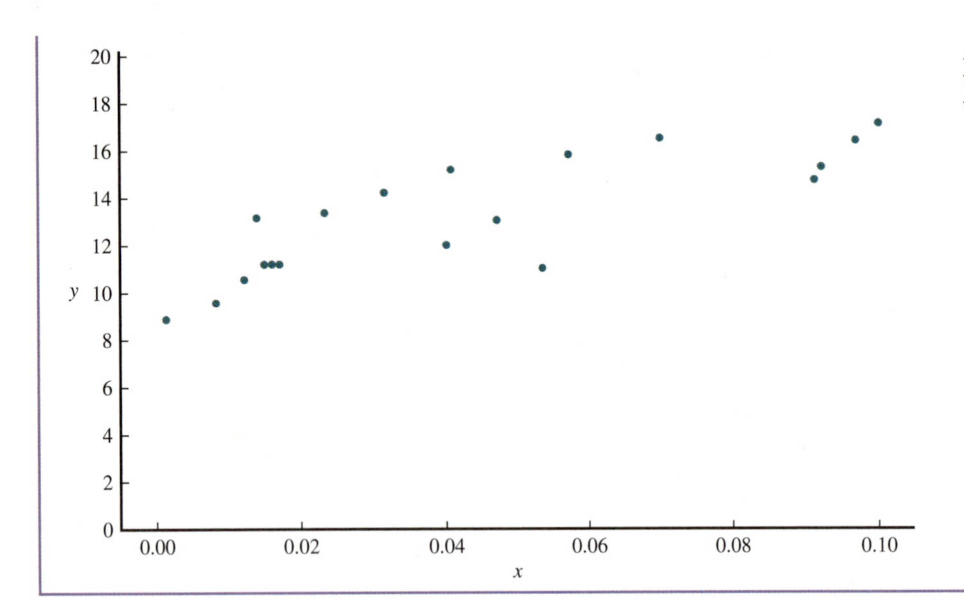

A linear model is appropriate for these data, because the data follow the pattern of a straight line.

b.

Equation	Answer
$y = 7.7 \cdot 5^x$	No, this does not represent a linear model, because the x value is in the exponent. It is not the equation of a line.
$y = 7.1 - 0.8x$	Yes, this model is linear. It has an intercept, and the coefficient on the x value is the slope.

c.

Linear Model	x Value	Answer
$y = 7.1 - 0.8x$	$x = 0.56$	$y = 7.1 - 0.8(0.56) = 6.65$
$y = 0.4 + 0.7x$	$x = 9$	$y = 0.4 + 0.7(9) = 6.7$

d.

Scatterplot	x Values	Answer
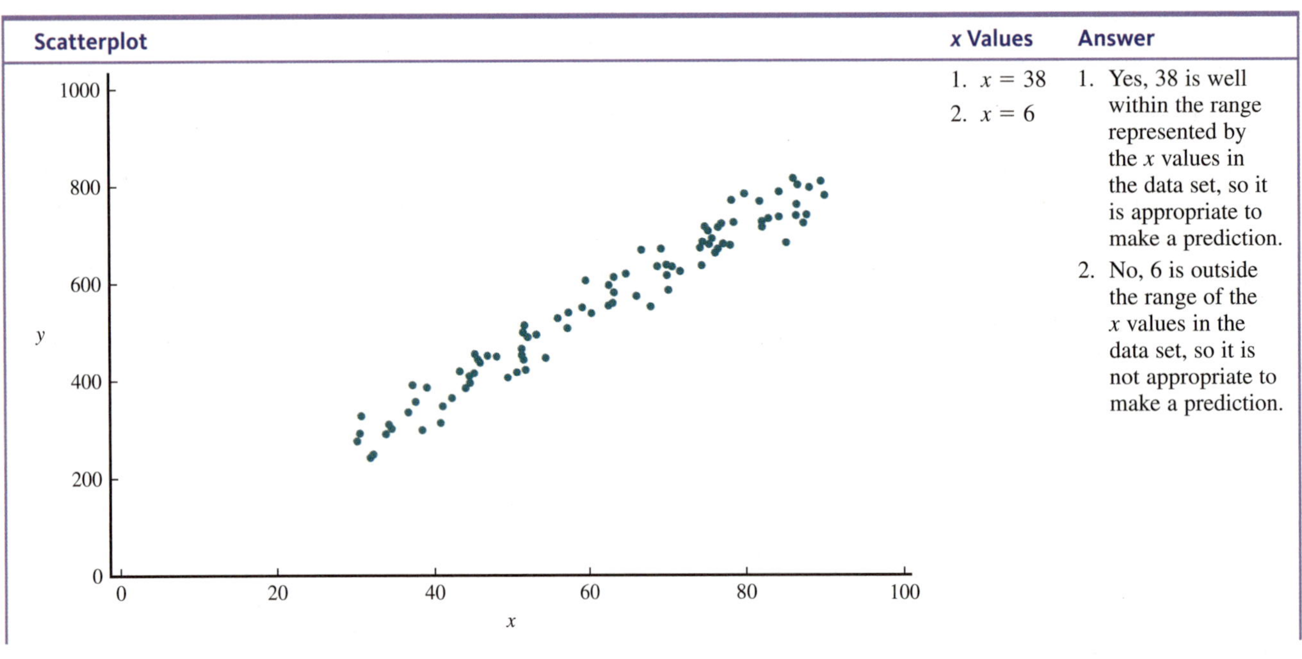	1. $x = 38$ 2. $x = 6$	1. Yes, 38 is well within the range represented by the x values in the data set, so it is appropriate to make a prediction. 2. No, 6 is outside the range of the x values in the data set, so it is not appropriate to make a prediction.

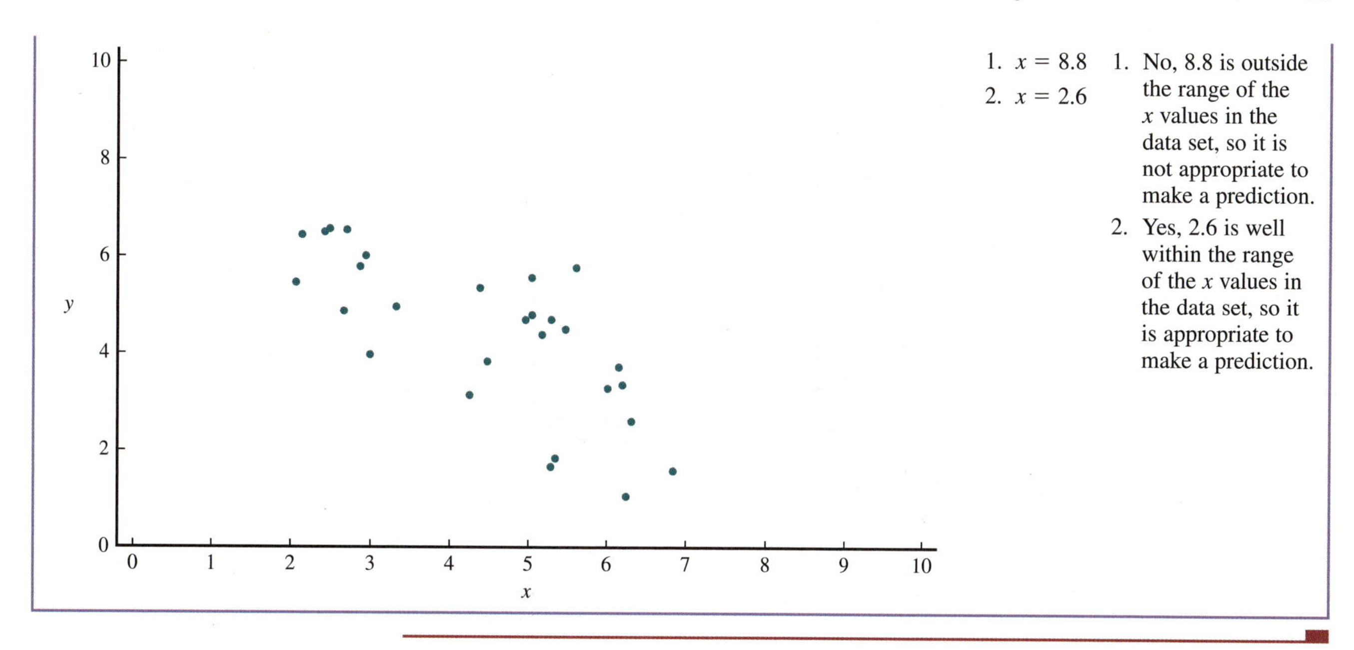

1. $x = 8.8$
2. $x = 2.6$

1. No, 8.8 is outside the range of the x values in the data set, so it is not appropriate to make a prediction.
2. Yes, 2.6 is well within the range of the x values in the data set, so it is appropriate to make a prediction.

Interpreting the Slope and *y-intercept* in a Linear Model

You have already seen how the slope and *y-intercept* of a line can be interpreted in the context of lines as mathematical objects. You must think a bit more carefully before interpreting the slope and *y-intercept* of a linear model that is used to represent a relationship between variables in a bivariate data set. This is because it is generally not the case that there is a perfect linear relationship between the variables that define a bivariate data set.

Look back at the scatterplot for the bird data. Notice that the points in that scatterplot form a linear pattern, but they do not all fall exactly on a straight line. This means that while the line $y = 18.4 + 0.15x$ may provide a reasonable description of the relationship between wingspan and weight, it is not true that wingspan increases by exactly 0.15 centimeters when weight increases by 1 gram. A correct interpretation of the slope of 0.15 would be to say "the predicted wingspan increases by 0.15 centimeters when weight increases by 1 gram."

There is another thing you should consider when a line is being used to model a relationship in bivariate data. Depending on the variables that define the data set, it might not be reasonable to interpret the y intercept in the linear model. This would be the case if an x value of 0 doesn't make sense. For example, it does not make sense to interpret the intercept of 18.4 centimeters as the wingspan of a type of bird that has an average weight of 0 grams! In addition, it is not appropriate to interpret the y intercept if $x = 0$ is far outside the range of the x values in the data set.

Interpreting Slope and *y-Intercept* in a Linear Model

In a linear model, the **slope** is interpreted as the change in the predicted value of the dependent variable associated with an increase of 1 unit in the value of the independent variable. It can also be interpreted as the average change in the value of the dependent variable associated with an increase of 1 unit in the value of the independent variable.

If it is reasonable to interpret the ***y-intercept***, it can be interpreted as the predicted value of the dependent variable when the value of the independent variable is 0.

Let's consider one more example before moving on. Suppose that the linear model $y = 100{,}000 + 100x$ can be used to model the relationship between y = price (in dollars) and x = size (in square feet) of houses in a particular community. The slope of the linear model is $100 and the y intercept is $100,000. Suppose that this model is based on data from a sample of 50 houses that ranged in size from 900 square feet to 3,000 square feet.

Interpretation of the slope: The predicted price of a house increases by $100 when the size increases by 1 square foot.

Interpretation of the *y-intercept:* It is not reasonable to interpret the *y-intercept* in this context because the x values in the data set ranged from 900 square feet to 3000 square feet and 0 square feet is outside that range. In addition, it doesn't make sense to think about predicting the price of a house that has a size of 0 square feet.

Check It Out!

a. Interpret the slope in each of the linear models. Check your answers.

Variables	Linear Model	Interpret the Slope
x = Number of books in a shipment sent from an online bookseller y = Weight (pounds) of books in each shipment	$y = 2.1 + 0.87x$	
x = Age (years) y = Number of text messages sent per day	$y = 197 - 4.3x$	

b. First, for each of the linear models, state whether it is appropriate to interpret the *y-intercept*. If it is appropriate, then provide an interpretation for the intercept. Check your answers.

Variables	Linear Model	If Appropriate, Interpret the *y-Intercept.* If Not Appropriate, Explain Why Not.
x = Body length (inches) for adult falcons captured and released in the wild y = Wingspan (inches) for the same falcons	$y = 12 + 1.3x$	
x = Distance (miles) from city pizza shop to requested delivery location y = Delivery time (minutes)	$y = 0.8 + 9.2x$	

Answers

a.

Variables	Linear Model	Answer
x = Number of books in a shipment sent from an online bookseller y = Weight (pounds) of books in each shipment	$y = 2.1 + 0.87x$	For each increase of one book, the predicted weight of the shipment increases by 0.87 pounds.
x = Age (years) y = Number of text messages sent per day	$y = 197 - 4.3x$	For each increase of one year in age, the predicted number of text messages sent per week decreases by 4.3.

b.

Variables	Linear Model	Answer
x = Body length (inches) for adult falcons captured and released in the wild y = Wingspan (inches) for the same falcons	$y = 12 + 1.3x$	It is not appropriate to interpret the *y-intercept*, because the minimum body length of adult falcons is not near zero.
x = Distance (miles) from city pizza shop to requested delivery location y = Delivery time (minutes)	$y = 0.8 + 9.2x$	It may be appropriate to interpret the *y-intercept*, because there may be locations very close to the pizza shop for which the delivery time would be very small.

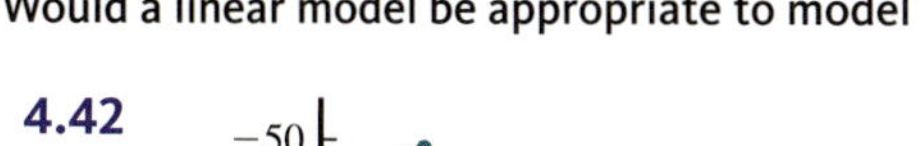

SECTION 4.3 EXERCISES

Would a linear model be appropriate to model the relationships represented in the scatterplots given in Exercises 4.42–4.43?

4.42

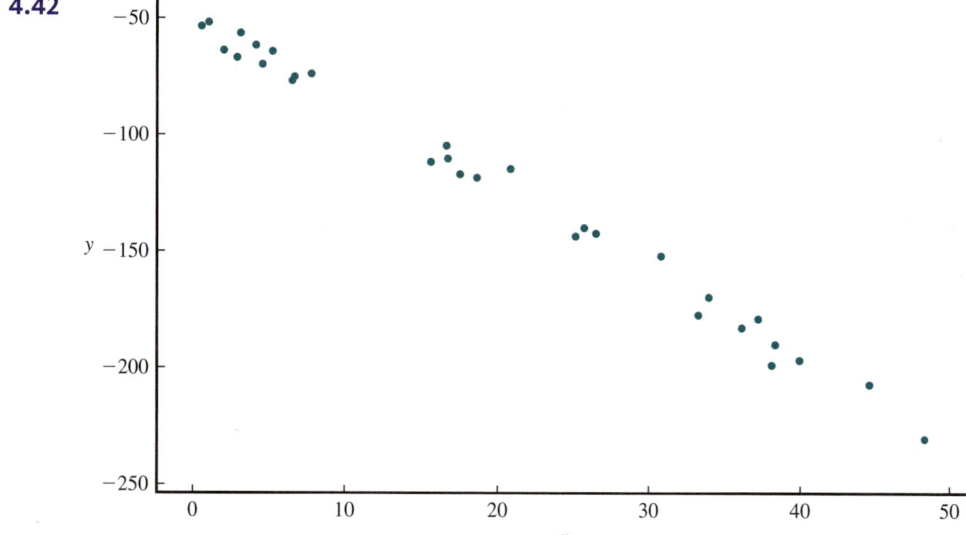

4.43

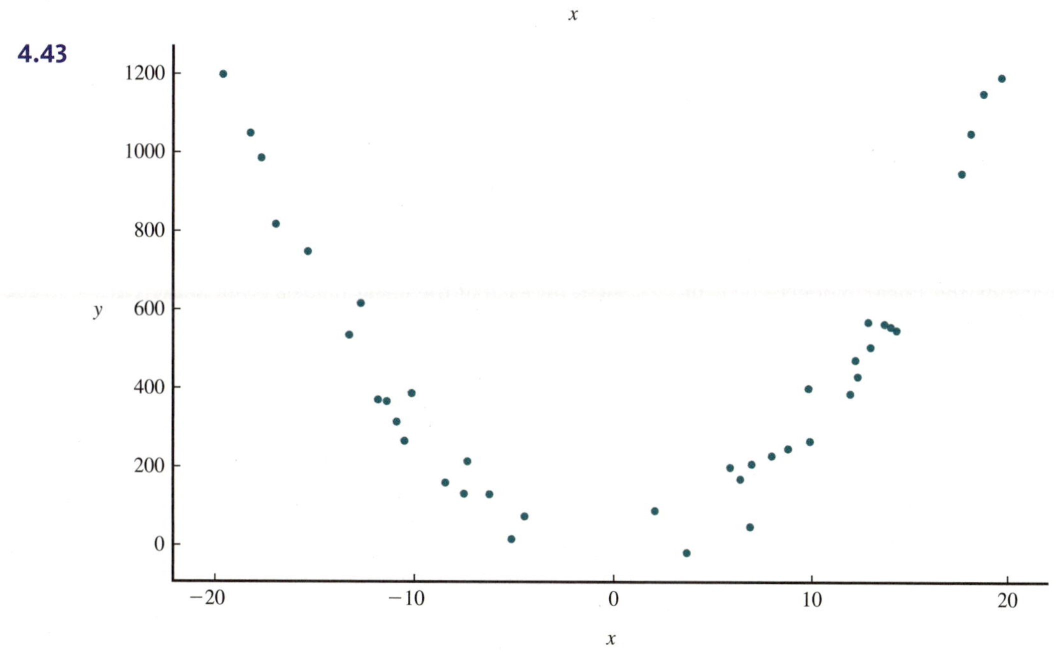

For each equation in Exercises 4.44 and 4.45, does the given equation represent a linear model? Explain your reasoning.

4.44 $y = -0.6 - 0.5x + 0.57x^2$

4.45 $y = 32 + 41x$

For each of the given linear models in Exercises 4.46 and 4.47, make a prediction for the given *x* value.

4.46 $y = 6.5 + 5.7x, \quad x = 2$

4.47 $y = -8.88 - 0.97x, \quad x = 19$

For each scatterplot given in Exercises 4.48–4.49, determine if it would be appropriate to make a prediction for each of the specified *x* values. Explain your reasoning.

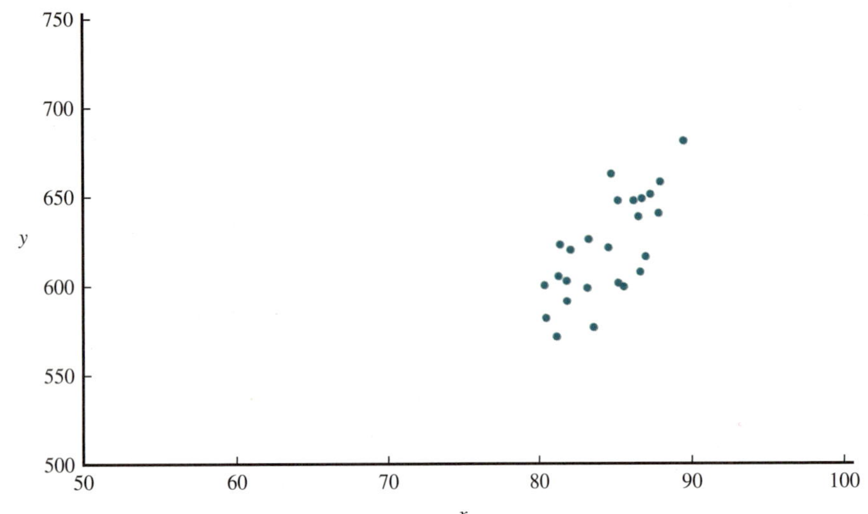

1. $x = 85$
2. $x = 55$

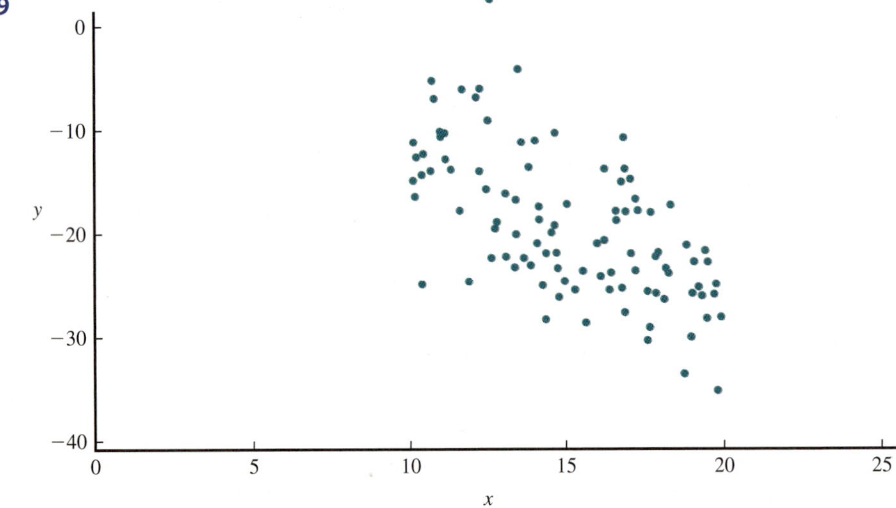

1. $x = 6.3$
2. $x = 18.6$

Interpret the slope in each of the linear models given in Exercises 4.50 and 4.51.

4.50 x = Weight (kg) at age 13 $\quad y = 0.6 + 0.01x$

y = Bone mineral density (g/cm^3) at age 27

4.51 x = Elapsed time (days) since snow melt $\quad y = 107 - 0.64x$

y = Green biomass concentration (g/cm^3)

First, for each of the linear models given in Exercises 4.52 and 4.53, state whether it is appropriate to interpret the *y-intercept*. If it is appropriate, then provide an interpretation for the intercept. If it is not appropriate, explain why not.

4.52 x = Width (mm) of the eggs for species of birds $\quad y = -1.8 + 1.44x$

y = Height (mm) for species of birds

4.53 x = Time (minutes) to exhaustion on a treadmill $\quad y = 90 - 2.4x$

y = 20 km ski time (minutes)

SECTION 4.4 Deviations from a Line and the Sum of Squared Deviations

When you use a linear model to predict the value of a dependent variable, you are basing the predicted y value on a point that falls on the line specified by the linear model. For example, suppose the linear model $y = 3 + 0.38x$ can be used to model the relationship between x = Height (in feet) and y = Top speed (in miles per hour, mph) of roller coasters. This model is based on the data for seven roller coasters in California shown in the following table. A scatterplot of the data is also shown and the line representing the linear model $y = 3 + 0.38x$ is shown on the scatterplot.

Roller Coaster Name	Height (in feet)	Top Speed (in mph)
Riddler's Revenge	156	65
Silver Bullet	146	55
Superman: Ultimate Flight	150	62
Tatsu	170	62
Viper	188	70
X2	175	76
Xcelerator	205	82

(Data from the Roller Coaster Database, rcdb.com)

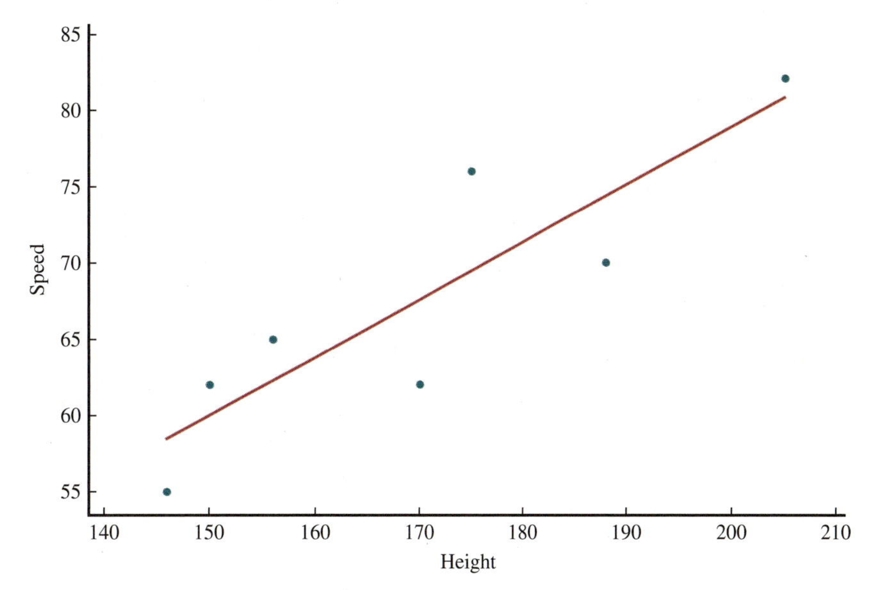

Suppose that this sample of seven roller coasters is representative of roller coasters in California that have heights between 140 feet and 210 feet. Then it is reasonable to consider using the line $y = 3 + 0.38x$ to predict the top speed for a roller coaster in California based on its height. For example, another roller coaster in California is Vertical Velocity. This roller coaster has a height of 150 feet. Using the line $y = 3 + 0.38x$ to predict the top speed for this roller coaster, you get

$$\begin{aligned} y &= 3 + 0.38x \\ &= 3 + (0.38)(150) \\ &= 3 + 57.00 \\ &= 60.00 \text{ mph} \end{aligned}$$

The predicted value of y (top speed) is the y value of the point on the line that is above $x = 150$. The actual value of top speed for Vertical Velocity is 65 mph, so the predicted value based on the line is "off" by 5 mph. This difference is called a prediction error.

Usually, if you are using a line to predict the value of a dependent variable, you don't know the actual y value (if you did, you wouldn't need to predict it!), so you won't know exactly how far off the prediction is. But when a linear model is used for prediction, it is important to have a sense of how accurate predictions will be. In your statistics course, you will do this by looking at the data used to develop the model. For example, in the roller coaster example, you do have data for seven roller coasters, and so you do know the actual y values (top speeds) for these seven roller coasters. You can look at what the line would have predicted for top speed for these seven roller coasters and then compare the actual top speed with the predicted top speeds.

The first roller coaster in the data set is Riddler's Revenge, which has a height of 156 feet. Using the line $y = 3 + 0.38x$ to predict the top speed of this roller coaster results in

$$\begin{aligned} y &= 3 + 0.38x \\ &= 3 + (0.38)(156) \\ &= 3 + 59.28 \\ &= 62.28 \text{ mph} \end{aligned}$$

The difference between the actual top speed for Riddler's Revenge and the predicted value is

$$\text{actual } y \text{ value} - \text{predicted } y \text{ value} = 65 - 62.28 = 2.72 \text{ mph.}$$

This difference is called a **deviation from the line**, and it represents the vertical distance of the point representing the actual height and top speed for this roller coaster to the line.

The difference

$$\text{actual } y \text{ value} - \text{predicted } y \text{ value}$$

is called a **deviation from the line** used to determine the predicted value. It provides information on the vertical distance from an observed data value to the line.

A deviation from a line is positive when the observed data point is above the line and the deviation from a line is negative when the observed data point is below the line.

A deviation from a line can be calculated for each observed value in a data set. The deviation corresponding to the first observation in the roller coaster data set was calculated to be 2.72 mph. To calculate the deviation from the line $y = 3 + 0.38x$ for the second observed value (Silver Bullet, with a height of 146 feet and a top speed of 55 mph), first find the predicted top speed for a height of 146 feet:

$$\begin{aligned} y &= 3 + 0.38x \\ &= 3 + (0.38)(146) \\ &= 3 + 55.48 \\ &= 58.48 \text{ mph} \end{aligned}$$

Then calculate the deviation from the line:

$$\text{deviation from the line} = \text{observed } y \text{ value} - \text{predicted } y \text{ value} = 55 - 58.48 = -3.48$$

Notice that this deviation is negative because the data point (146, 55) is below the line. The deviations for the first two data points have been entered in the table below and are illustrated in the accompanying plot.

Roller Coaster Name	Height (in feet)	Top Speed (in mph)	Predicted Top Speed	Deviation from the Line $y = 3 + 0.38x$
Riddler's Revenge	156	65	62.28	2.72
Silver Bullet	146	55	58.48	−3.48
Superman: Ultimate Flight	150	62		
Tatsu	170	62		
Viper	188	70		
X2	175	76		
Xcelerator	205	82		

(Data from the Roller Coaster Database, rcdb.com)

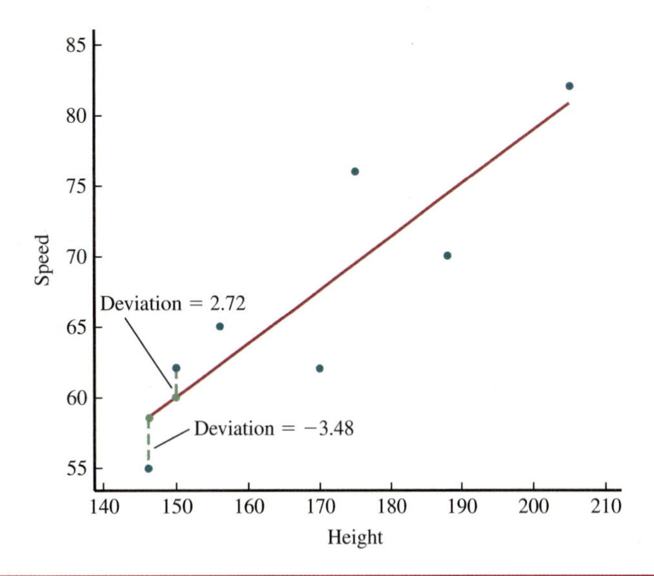

Check It Out!

Use the linear model $y = 3 + 0.38x$ to calculate the remaining Predicted Top Speed values in the table below, and then calculate the corresponding deviations. Check your answers as you go.

Roller Coaster Name	Height	Top Speed	Predicted Top Speed	Deviation from the Line $y = 3 + 0.38x$
Riddler's Revenge	156	65	62.28	2.72
Silver Bullet	146	55	58.48	−3.48
Superman: Ultimate Flight	150	62		
Tatsu	170	62		
Viper	188	70		
X2	175	76		
Xcelerator	205	82		

Answers

Roller Coaster Name	Height	Top Speed	Predicted Top Speed	Deviation from the Line $y = 3 + 0.38x$
Riddler's Revenge	156	65	62.28	2.72
Silver Bullet	146	55	58.48	−3.48
Superman: Ultimate Flight	150	62	60.00	2.00
Tatsu	170	62	67.60	−5.60
Viper	188	70	74.44	−4.44
X2	175	76	69.50	6.50
Xcelerator	205	82	80.90	1.10

In statistics, deviations from a line are also sometimes called **residuals.** In your statistics course, you will calculate and use deviations from a line to construct a graph called a residual plot. You will also use the **sum of the squared deviations** (also called the **sum of squared residuals**) to help assess the fit of a linear model and to learn about the accuracy of predictions based on a linear model.

To calculate the **sum of the squared deviations**:

1. Calculate the predicted values corresponding to each observation in the data set.
2. Calculate the deviation from the line for each observation in the data set.
3. Square each of the deviations.
4. Add up the squared deviations

Let's look at an example. The table below shows the deviations from the line $y = 3 + 0.38x$ and the squared deviations (rounded to three decimal places) for the roller coaster data set. The numbers in the last column in the table were calculated by squaring the deviations in the previous column.

Roller Coaster Name	Height (in feet)	Top Speed (in mph)	Predicted Top Speed	Deviation from the Line $y = 3 + 0.38x$	Squared Deviation
Riddler's Revenge	156	65	62.28	2.72	7.398
Silver Bullet	146	55	58.48	−3.48	12.110
Superman: Ultimate Flight	150	62	60.00	2.00	4.000
Tatsu	170	62	67.60	−5.60	31.360
Viper	188	70	74.44	−4.44	19.714
X2	175	76	69.50	6.50	42.250
Xcelerator	205	82	80.90	1.10	1.210

(Data from the Roller Coaster Database, rcdb.com)

Now you can calculate the sum of the squared deviations from the line $y = 3 + 0.38x$ for this data set:

$$\text{sum of squared deviations} = 7.398 + 12.110 + 4.000 + 31.360 + 19.714 + 42.250 + 1.210 = 118.042$$

Check It Out!

a. The study "Physiological Characteristics and Performance of Top U.S. Biathletes" (*Medicine and Science in Sports and Exercise* [1995]: 1302–1310) explored the relationship between cardiovascular fitness (measured by time to exhaustion running on a treadmill) and performance in a 20 km ski race. The x-variable is Treadmill time (minutes), and the y-variable is 20 km ski time (minutes). A linear model for these data is $y = 88.8 - 2.33x$.

Some of the Predicted Ski Times, Deviations from the Line $y = 88.8 - 2.33x$, and Squared Deviations have been calculated. Complete the others. Check your answers.

Observation	Treadmill	Ski Time	Predicted Ski Time	Deviation from the Line $y = 88.8 - 2.33x$	Squared Deviation
1	7.7	71.0	70.8	0.2	0.030
2	8.4	71.4	69.2	2.2	4.866
3	8.7	65.0			
4	9.0	68.7			
5	9.6	64.4			
6	9.6	69.4			
7	10.0	63.0			
8	10.2	64.6			
9	10.4	66.9			
10	11.0	62.6			
11	11.7	61.7			

b. Calculate the sum of the squared deviations for the data given in part (a). Check your answer.

Answers

a.

Observation	Treadmill	Ski Time	Predicted Ski Time	Deviation from the Line $y = 88.8 - 2.33x$	Squared Deviation
1	7.7	71.0	70.8	0.2	0.030
2	8.4	71.4	69.2	2.2	4.866
3	8.7	65.0	68.5	−3.5	12.209
4	9.0	68.7	67.8	0.9	0.821
5	9.6	64.4	66.4	−2.0	3.976
6	9.6	69.4	66.4	3.0	9.036
7	10.0	63.0	65.5	−2.5	6.054
8	10.2	64.6	65.0	−0.4	0.155
9	10.4	66.9	64.5	2.4	5.630
10	11.0	62.6	63.1	−0.5	0.278
11	11.7	61.7	61.5	0.2	0.043

b. sum of squared deviations $= 0.030 + \cdots + 0.043 = 43.1$

SECTION 4.4 EXERCISES

4.54 Data consistent with results published in the article "Weight-Bearing Activity During Youth Is a More Important Factor for Peak Bone Mass than Calcium Intake" (*Journal of Bone and Mineral Research* [1994], 1089–1096) appear in the accompanying table. The x variable is Weight at age 13 and the y variable is Bone mineral density (BMD) at age 27. A linear model for the data is $y = 0.6 + 0.01x$.

Calculate the Predicted BMD values, the deviations from the line $y = 0.6 + 0.01x$, and the squared deviations.

Observation	Weight	BMD	Predicted BMD	Deviation from the Line $y = 0.6 + 0.01x$	Squared Deviation
1	54.4	1.15			
2	59.3	1.26			
3	74.6	1.42			
4	62.0	1.06			
5	73.7	1.44			
6	70.8	1.02			
7	66.8	1.26			
8	66.7	1.35			
9	64.7	1.02			
10	71.8	0.91			
11	69.7	1.28			
12	64.7	1.17			
13	62.1	1.12			
14	68.5	1.24			
15	58.3	1.00			

4.55 Calculate the sum of the squared deviations for the linear model and data given in Exercise 4.54.

4.56 Using the information given in Exercises 4.54 and 4.55, examine the squared deviations and their sum. What does it tell you about the observed y-values and the corresponding predicted y-values if the sum of the squared deviations is small? Explain your reasoning.

SECTION 4.5 The Least-Squares Line

One of the topics that you will discuss in your statistics course is how to choose which line you will use to model a linear relationship between a dependent variable and an independent variable. Some lines are clearly better choices than others. For example, look at the two lines shown in the following scatterplot of y = Top speed and x = Height for the roller coaster data set.

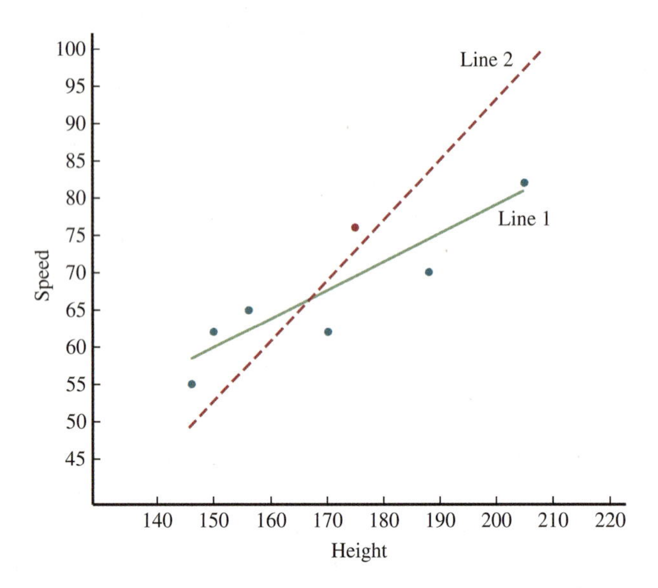

Which line do you think does a better job of describing the linear pattern in the scatterplot? Keep in mind that if a line is used to make predictions, the prediction will be a point on the line. Which line do the points tend to be closer to overall? Notice that the point shown in red in the scatterplot is closer to line 2 than to Line 1, but that all the other points are closer to line 1. This means that overall, for these roller coasters, Line 1 would tend to produce more accurate predictions of top speed than Line 2.

Recall from Section 4.4 that deviations from a line provide information about prediction errors, which are vertical distances of points from the line. The sum of squared deviations from the line provides an overall measure of the prediction errors for the line. Because the deviations from Line 1 tend to be smaller than the deviations from Line 2, Line 1 would have a smaller sum of squared deviations.

The sum of squared deviations from a line can be used to compare lines, as was just done with Line 1 and Line 2 for the roller coaster data. Because a smaller sum of squared deviations means predictions will tend to be more accurate, the "best" line to describe a relationship between two numerical variables would be the line with the smallest possible sum of squared deviations. The line with the smallest possible sum of squared deviations is called the **least-squares line**. The "least-squares" is a reference to the fact that this line has a smaller sum of squared deviations than any other line. In your statistics course, you will see how to determine the slope and the *y-intercept* of the least-squares line for a given data set and how this line can be used to predict values of the dependent variable.

In the scatterplot of the roller coaster data, Line 1 is the line $y = 3 + 0.38x$. This is the least-squares line for the roller coaster data set. This means that the sum of squared deviations for this line (calculated to be 118.042 in the previous section) is smaller than the sum of the squared deviations for any other line.

Let's look at one more example. The scatterplot below was constructed using data on

$$x = \text{Average weight (in grams)}$$

and

$$y = \text{Average wingspan (in centimeters)}$$

for 9 different types of birds. Two lines are shown on this plot. One of the lines is the least-squares line. Which line is it?

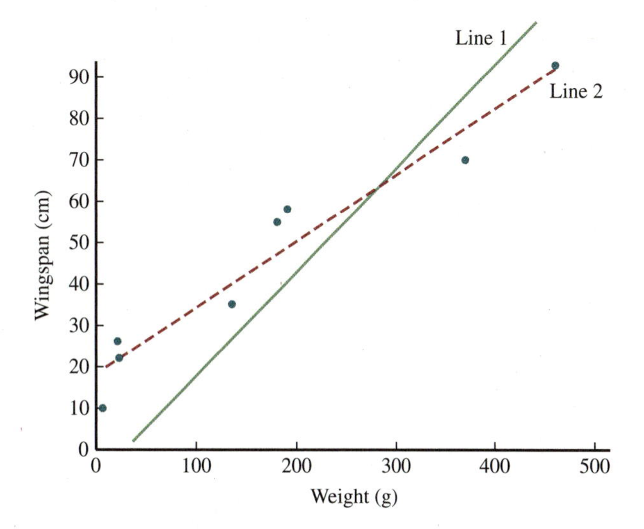

The points tend to fall closer to Line 2 than they do to Line 1. The sum of squared deviations from Line 2 is smaller than the sum of squared deviations for Line 1. This means that if one of these lines is the least-squares line, it must be Line 2.

Check It Out!

Recall the study "Physiological Characteristics and Performance of Top U.S. Biathletes" (*Medicine and Science in Sports and Exercise* [1995]: 1302–1310) that explored the relationship between cardiovascular fitness (measured by time to exhaustion running on a treadmill) and performance in a 20 km ski race. The x-variable is Treadmill time (minutes), and the y-variable is 20 km ski time (minutes).

In the following scatterplot, 20 km ski time is plotted against Treadmill time, but two different lines are represented on the display. Examine the relationships between the lines and the data points, and choose which line you believe to be the least-squares line, Line A or Line B? Explain your reasoning, then check your answer.

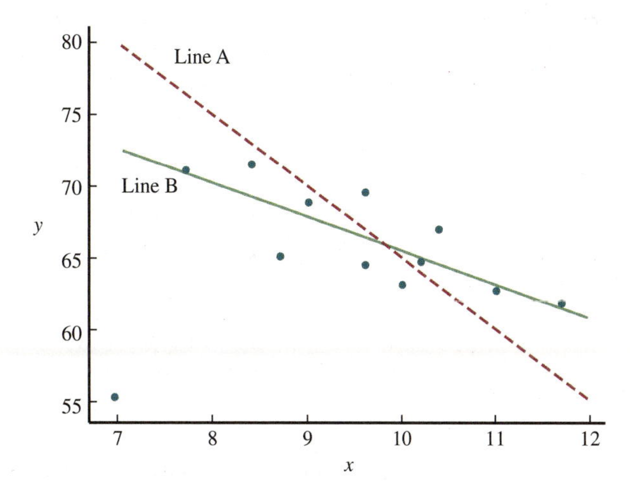

Answers

The data points appear to be much more evenly distributed around Line B, with apparently smaller deviations overall. This evidence indicates that Line B is the least-squares regression line.

SECTION 4.5 EXERCISES

4.57 Recall the data from the article "Weight-Bearing Activity During Youth Is a More Important Factor for Peak Bone Mass than Calcium Intake" (*Journal of Bone and Mineral Research* [1994], 1089–1096). The x variable is Weight at age 13 and the y variable is Bone mineral density at age 27, and the data were used to construct the accompanying scatterplot.

Two lines are also represented in the scatterplot. Examine the relationships between the lines and the data points, and choose which line you believe to be the least-squares line, Line A or Line B? Explain your reasoning.

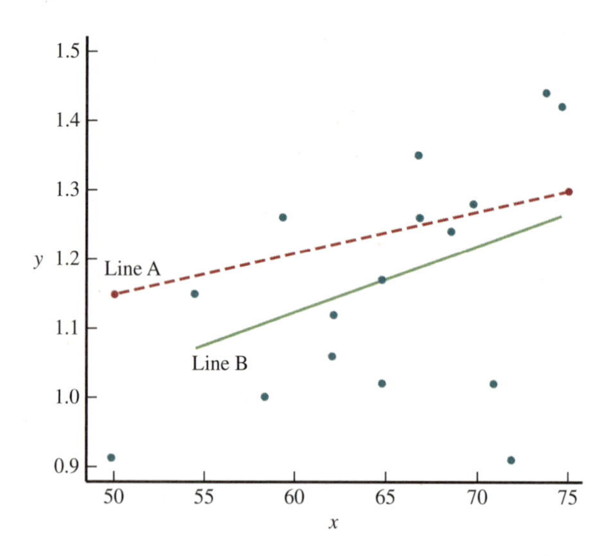

SECTION 4.6 Evaluating Expressions

There are many different expressions that you will encounter as you learn about methods for summarizing bivariate numerical data.

Note: Depending on your access to technology, such as a graphing calculator or a statistics software package, you may not need to evaluate most of these expressions by hand. But just in case your statistics class is not relying on technology to do these calculations, this section looks at how to evaluate these expressions.

The following data set (based on information from the Bureau of Labor Statistics Consumer Expenditure Survey) will be used to illustrate how the expressions considered in this section are evaluated.

Suppose five pet owners were asked their ages and the average amount of money they spend per month to feed and care for their pets. With x = Age (in years) and y = Amount spent on pets per month (in dollars), you might be interested in whether there is a relationship between age and how much people spend on their pets.

Age	Amount Spent on Pets per Month
20	25
27	36
38	39
50	52
55	53

Here is a scatterplot of these data that also shows the least-squares line.

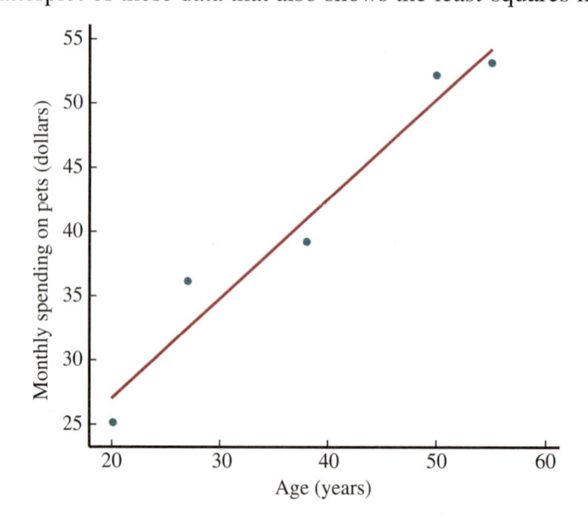

Correlation Coefficient *r*

$$r = \frac{\sum z_x z_y}{n - 1}$$

To calculate the correlation coefficient, you first need to calculate the *z-scores* for the x values in the data set and the *z-scores* for the y values in the data set. Recall that you calculate a *z-score* by subtracting the mean and dividing by the standard deviation, so you will also have to calculate the mean and standard deviation of the x values and the mean and standard deviation of the y values.

For the data set given above, the mean and standard deviation of the x values are

$$\bar{x} = \frac{\sum x}{n} = \frac{190}{5} = 38 \qquad s = \sqrt{\frac{\sum (x - \bar{x})^2}{n - 1}} = \sqrt{\frac{878}{4}} = \sqrt{219.5} = 14.816$$

The mean and the standard deviation of the y values are

$$\bar{y} = \frac{\sum y}{n} = \frac{205}{5} = 41 \qquad s = \sqrt{\frac{\sum (y - \bar{y})^2}{n - 1}} = \sqrt{\frac{550}{4}} = \sqrt{137.5} = 11.726$$

(For a review of how to evaluate the expressions for the mean and standard deviation, see Section 3.5.)

You can now calculate the *z-scores*. For the first observation in the data set, (20, 25), the *z-score* for x is

$$z = \frac{x - \text{mean}}{\text{standard deviation}} = \frac{20 - 38}{14.816} = \frac{-18}{14.816} = -1.215$$

The *z-score* for y is

$$z = \frac{y - \text{mean}}{\text{standard deviation}} = \frac{25 - 41}{11.726} = \frac{-16}{11.726} = -1.364$$

The *z-scores* for the other four observations in the data set have also been calculated and appear in the table below.

Age	Amount Spent on Pets per Month	*z-Score* for *x*	*z-Score* for *y*
20	25	−1.215	−1.364
27	36	−0.742	−0.426
38	39	0.000	−0.171
50	52	0.810	0.938
55	53	1.147	1.023

Now look at the formula for the correlation coefficient:

$$r = \frac{\sum z_x z_y}{n - 1}$$

The numerator indicates that you need to multiply the *z-score* of the x value by the *z-score* of the y value for each observation and then add up those values. For the first observation,

$$z_x = -1.215 \qquad z_y = -1.364 \qquad z_x z_y = (-1.215)(-1.364) = 1.657$$

This value and the values for the other four observations have been added to the table:

Age	Amount Spent on Pets per Month	*z*-Score for *x*	*z*-Score for *y*	$z_x z_y$
20	25	−1.215	−1.364	1.657
27	36	−0.742	−0.426	0.316
38	39	0.000	−0.171	0.000
50	52	0.810	0.938	0.760
55	53	1.147	1.023	1.173

Finally, you now have what you need to evaluate the expression for the correlation coefficient!

$$r = \frac{\sum z_x z_y}{n-1} = \frac{1.657 + 0.316 + 0.000 + 0.760 + 1.173}{5-1} = \frac{3.096}{4} = 0.977$$

The value of the correlation coefficient is positive and close to 1. This is what you would expect because the scatterplot suggests that there is a strong linear relationship between x = Age and y = Amount spent on pets per month.

Check It Out!

a. Recall that in the Frying Time example of Section 4.1 you calculated the *z-scores* for all values of x and for all values of y and these *z-scores* appear in the table below.

Complete the table by calculating the rest of the $z_x z_y$ values. Check your answers along the way.

x = Frying Time	*z*-Score for *x*	y = Acrylamide Concentration	*z*-Score for *y*	$z_x z_y$
150	−1.790	155	−0.410	0.734
240	−0.18	120	−1.07	
240	−0.18	190	0.25	
270	0.36	185	0.16	
300	0.90	140	−0.69	
300	0.90	270	1.76	

b. Use the information in Part (a) to calculate the sum $\sum z_x z_y$, and then calculate the value of the correlation coefficient r. Check your answers.

Answers

a.

x = Frying Time	*z*-Score for *x*	y = Acrylamide Concentration	*z*-Score for *y*	$z_x z_y$
150	−1.790	155	−0.410	0.734
240	−0.18	120	−1.07	0.193
240	−0.18	190	0.25	−0.045
270	0.36	185	0.16	0.058
300	0.90	140	−0.69	−0.621
300	0.90	270	1.76	1.584

b. $\sum z_x z_y = 1.902$

$$r = \frac{\sum z_x z_y}{n-1} = \frac{0.734 + \cdots + 1.76}{6-1} = \frac{1.902}{5} = 0.380$$

Slope of the Least-Squares Line

$$b = \frac{\sum(x - \bar{x})(y - \bar{y})}{\sum(x - \bar{x})^2}$$

To calculate the slope of the least-squares line, denoted by b, first evaluate the expression in the numerator of the formula for b. This expression indicates that you need to sum up the values of $(x - \bar{x})(y - \bar{y})$ for each of the observations in the data set. One way to approach this calculation is to set up a table with five columns labeled x, y, $(x - \bar{x})$, $(y - \bar{y})$, and $(x - \bar{x})(y - \bar{y})$. List the (x, y) values from the data set in the first two columns. Then fill in the values in the $(x - \bar{x})$ column by subtracting the mean of the x values, $\bar{x}$, from each x value. The values in the $(y - \bar{y})$ column are obtained by subtracting $\bar{y}$ from each y value. Then you can fill in the $(x - \bar{x})(y - \bar{y})$ column by multiplying the values in the third and fourth columns.

For the data set on x = Age and y = Amount spent on pets per month described at the beginning of this section, the first observation was (20, 25). In the discussion of the correlation coefficient, the values of $\bar{x}$ and $\bar{y}$ were calculated to be $\bar{x} = 38$ and $\bar{y} = 41$. This means that for the first observation in the data set,

$$(x - \bar{x}) = 20 - 38 = -18 \qquad (y - \bar{y}) = 25 - 41 = -16$$

$$(x - \bar{x})(y - \bar{y}) = (-18)(-16) = 288$$

These values, along with those for the other observations in the data set, are shown in the following table.

x = Age	y = Amount Spent on Pets	$(x - \bar{x})$	$(y - \bar{y})$	$(x - \bar{x})(y - \bar{y})$
20	25	−18	−16	288
27	36	−11	−5	55
38	39	0	−2	0
50	52	12	11	132
55	53	17	12	204

The value of $\sum(x - \bar{x})(y - \bar{y})$ is obtained by adding up all the $(x - \bar{x})(y - \bar{y})$ values. For this data set,

$$\sum(x - \bar{x})(y - \bar{y}) = 288 + 55 + 0 + 132 + 204 = 679$$

The expression in the denominator of the formula for b is $\sum(x - \bar{x})^2$. To calculate this value, you need to find $(x - \bar{x})^2$ for each observation in the data set. You already found the values of $(x - \bar{x})$ (they are in the third column of the table above). So now you just need to square each of these values and add them up. The table below adds a sixth column for the $(x - \bar{x})^2$ values.

x = Age	y = Amount Spent on Pets	$(x - \bar{x})$	$(y - \bar{y})$	$(x - \bar{x})(y - \bar{y})$	$(x - \bar{x})^2$
20	25	−18	−16	288	324
27	36	−11	−5	55	121
38	39	0	−2	0	0
50	52	12	11	132	144
55	53	17	12	204	289

Then

$$\sum(x - \bar{x})^2 = 324 + 121 + 0 + 144 + 289 = 878$$

Now that you have calculated the value for the numerator and the value for the denominator in the formula for b, you can calculate the value of the slope of the least-squares line

$$b = \frac{\sum(x-\bar{x})(y-\bar{y})}{\sum(x-\bar{x})^2} = \frac{679}{878} = 0.773$$

The value of the slope of the least-squares line is 0.773.

Check It Out!

Complete the following table for the Frying Time data, and then use the values in the table to calculate the slope of the least-squares line, b. Check your answers.

x = Frying Time	y = Acrylamide Concentration	$(x-\bar{x})$	$(y-\bar{y})$	$(x-\bar{x})(y-\bar{y})$	$(x-\bar{x})^2$
150	155	−100	−21.7	2,170	10,000
240	120				
240	190				
270	185				
300	140				
300	270				

$\sum(x-\bar{x})(y-\bar{y}) =$

$\sum(x-\bar{x})^2 =$

$$b = \frac{\sum(x-\bar{x})(y-\bar{y})}{\sum(x-\bar{x})^2} =$$

Answers

x = Frying Time	y = Acrylamide Concentration	$(x-\bar{x})$	$(y-\bar{y})$	$(x-\bar{x})(y-\bar{y})$	$(x-\bar{x})^2$
150	155	−100	−21.7	2,170	10,000
240	120	−10	−56.7	567	100
240	190	−10	13.3	−133	100
270	185	20	8.3	166	400
300	140	50	−36.7	−1,835	2,500
300	270	50	93.3	4,665	2,500

$\sum(x-\bar{x})(y-\bar{y}) = 5600$

$\sum(x-\bar{x})^2 = 15,600$

$$b = \frac{\sum(x-\bar{x})(y-\bar{y})}{\sum(x-\bar{x})^2} = \frac{5600}{15,600} = 0.359$$

y-intercept of the Least-Squares Line

$$a = \bar{y} - b\bar{x}$$

Calculating the *y-intercept* of the least-squares line is easy once you have calculated the mean of the x values, $\bar{x}$, the mean of the y values, $\bar{y}$, and the slope of the least-squares line, b. You would first multiply $\bar{x}$ by the value of the slope, b, and then subtract this value from $\bar{y}$.

For the data set on x = Age and y = Amount spent on pets per month described at the beginning of this section, the values of $\bar{x}$ and $\bar{y}$ were calculated to be $\bar{x} = 38$ and $\bar{y} = 41$.

The slope of the least-squares line was $b = 0.773$. Substituting these values into the equation for a, the *y-intercept* of the least-squares line, results in

$$a = 41 - (0.773)(38) = 41 - 29.374 = 11.626$$

Predicted *y* value, $\hat{y}$

$$\hat{y} = a + bx$$

To distinguish predicted y values from observed values of y, the notation $\hat{y}$ is used to represent a predicted y value. Predicted y values are calculated using the equation of the least-squares line, so you need to know the values of the slope and the *y-intercept* before you can calculate predicted y values.

Once you know the values of the slope and the *y-intercept*, you can use them to write the equation of the least-squares line. For the data set described at the beginning of this section, where x is age and y is the amount spent on pets per month, the slope was calculated to be $b = 0.773$ and the *y-intercept* was calculated to be $a = 11.626$. For this data set, the equation of the least-squares line is

$$\hat{y} = a + bx = 11.626 + 0.773x$$

This is the equation that you would use to predict the amount of money spent on pets per month based on the age of the pet owner. For example, to predict the amount of money spent on pets per month for a pet owner who is 40 years old, you would substitute 40 into the equation and then evaluate the expression on the right-hand side of the equation:

$$\hat{y} = a + bx = 11.626 + 0.773x = 11.626 + 0.773(40) = 11.626 + 30.920 = \$42.546$$

Based on the least-squares line, you would predict that a 40-year-old pet owner spends \$42.55 per month on his or her pet.

Check It Out!

For the Frying Time data, calculate the intercept a, write the equation of the least-squares line, and use the least-squares line to predict the y = Acrylamide concentration when the Frying time is $x = 250$ seconds. Check your answers.

$a =$

$\hat{y} = a + bx =$

Answers

$a = \bar{y} - b\bar{x} = 176.7 - 0.359(250) = 86.95$

$\hat{y} = a + bx = 86.95 + 0.359x$

$\hat{y} = a + b(250) = 86.95 + 0.359(250) = 176.70$

Residuals and Sum of Squared Residuals

$$\text{residual} = y - \hat{y}$$

$$\text{SSResid} = \sum (y - \hat{y})^2$$

When you use the least-squares line to describe the relationship between two numerical variables, calculating residuals provides information on how accurate you can expect predictions based on the line to be. There is a residual corresponding to each observation in the data set. To calculate a residual for an observation (x, y), you first use the x value for the observation to calculate the corresponding predicted value, and then you calculate the residual by subtracting the predicted value from the observed y value for that observation.

For example, for the data set on x = Age and y = Amount spent on pets per month, described at the beginning of this section, the first observation was (20, 25). The equation of the least-squares line was $\hat{y} = 11.626 + 0.773x$. To find the residual for this observation, first calculate the predicted y value when $x = 20$:

$$\hat{y} = 11.626 + 0.773x = 11.626 + 0.773(20) = 11.626 + 15.460 = 27.086$$

The residual is calculated by subtracting this predicted value from the observed y value for this observation, which was 25:

$$\text{residual} = 25 - 27.086 = -2.086$$

Another quantity that you might need to calculate is the sum of the squared residuals, which is sometimes abbreviated as SSResid. The sum of the squared residuals is used to calculate two important measures that are used to evaluate a linear model, r^2 and s_e. You calculate SSResid using the formula

$$\text{SSResid} = \sum (y - \hat{y})^2$$

This formula tells you that you should add up the squared residuals. To do this, you would first calculate the residual for each observation in the data set. Then you square each of the residuals and add up those numbers to get the sum of the squared residuals.

You can use a table like the one that follows to organize these calculations. The first two columns are just the x and the y values in the data set. The third column contains the predicted y value corresponding to each observation, and the last two columns contain the residuals and then the squared residuals. Adding the squared residuals in the last column will produce the value of SSResid, the sum of squared residuals. The table that follows shows the calculations for the data set on x = Age and y = Amount spent on pets per month described at the beginning of this section.

x = Age	**y = Amount Spent on Pets**	**Predicted Value, $\hat{y}$**	**Residual $(y - \hat{y})$**	**Squared Residual $(y - \hat{y})^2$**
20	25	27.086	−2.086	4.351
27	36	32.497	3.503	12.271
38	39	41.000	−2.000	4.000
50	52	50.276	1.724	2.972
55	53	54.141	−1.141	1.302

Adding the squared residuals in the last column gives

$$\text{SSResid} = 4.351 + 12.271 + 4.000 + 2.972 + 1.302 = 24.896$$

Check It Out!

For the Frying Time data, complete the table of predicted values, residuals, and squared residuals. Then use the values in the table to calculate SSResid. Check your answers.

x = Frying Time	**y = Acrylamide Concentration**	**Predicted Value, $\hat{y}$**	**Residual $(y - \hat{y})$**	**Squared Residual $(y - \hat{y})^2$**
150	155	140.8	14.2	202.51
240	120			
240	190			
270	185			
300	140			
300	270			

SSResid =

Answers

x = Frying Time	y = Acrylamide Concentration	Predicted Value, $\hat{y}$	Residual $(y - \hat{y})$	Squared Residual $(y - \hat{y})^2$
150	155	140.8	14.2	202.51
240	120	173.1	−53.1	2817.20
240	190	173.1	16.9	286.39
270	185	183.8	1.2	1.33
300	140	194.6	−54.6	2982.80
300	270	194.6	75.4	5682.80

$\text{SSResid} = 202.51 + \cdots + 5682.80 = 11{,}973.08$

Total Sum of Squares

$$\text{SSTotal} = \sum(y - \bar{y})^2$$

The total sum of squares is sometimes abbreviated as SSTotal. Like the sum of squared residuals, it is also used in the calculation of r^2.

The formula for SSTotal should look familiar. It is just the sum of the squared deviations from the mean for the y values in the data set, and it is also used to calculate the standard deviation of the y values.

To calculate SSTotal, you would first calculate the deviations from the mean of the y values, $\bar{y}$, by subtracting $\bar{y}$ from each of the y values. Then you square each of the deviations and add up those numbers to get the sum of the squared deviations.

You can use a table like the one that follows to organize these calculations. This table uses the data on x = Age and y = Amount spent on pets per month, described at the beginning of this section. The first two columns are just the x and the y values in the data set. The third column contains the values of $(y - \bar{y})$, and the last column contains the squared deviations, $(y - \bar{y})^2$. Adding the squared deviations in the last column will produce the value of SSTotal. The table that follows shows the calculations for this data set.

x = Age	y = Amount Spent on Pets	$(y - \bar{y})$	$(y - \bar{y})^2$
20	25	−16	256
27	36	−5	25
38	39	−2	4
50	52	11	121
55	53	12	144

Adding the squared deviations in the last column gives

$$\text{SSTotal} = 256 + 25 + 4 + 121 + 144 = 550$$

Check It Out!

For the Frying Time data, complete the table of deviations and squared deviations. Then use the values in the table to calculate SSTotal. Check your answers.

x = Frying Time	y = Acrylamide Concentration	$(y - \bar{y})$	$(y - \bar{y})^2$
150	155	−21.7	470.89
240	120		
240	190		
270	185		
300	140		
300	270		

SSTotal =

Answers

x = Frying Time	y = Acrylamide Concentration	$(y - \bar{y})$	$(y - \bar{y})^2$
150	155	−21.7	470.89
240	120	−56.7	3,214.9
240	190	13.3	176.89
270	185	8.3	68.89
300	140	−36.7	1,346.9
300	270	93.3	8,704.9

$\text{SSTotal} = 470.89 + \cdots + 8704.9 = 13{,}983.34$

The Coefficient of Determination, r^2

$$r^2 = 1 - \frac{\text{SSResid}}{\text{SSTotal}}$$

Once you have calculated the values of SSResid and SSTotal (discussed earlier in the section), it is easy to calculate the value of r^2. First, divide the value for SSResid by the value for SSTotal, and then subtract this number from 1.

For the data on x = Age and y = Amount spent on pets per month, described at the beginning of this section, SSResid = 24.896 and SSTotal = 550. Substituting these values into the formula for r^2 results in

$$r^2 = 1 - \frac{\text{SSResid}}{\text{SSTotal}} = 1 - \frac{24.986}{550} = 1 - 0.045 = 0.955$$

If you have already calculated the correlation coefficient, r, then a quick way to calculate r^2 is to just square the value for r. For example, for the data on x = Age and y = Amount spent on pets per month, described at the beginning of this section, the value of the correlation coefficient was $r = 0.977$, so $r^2 = (0.977)^2 = 0.955$.

Check It Out!

Recall that in the Frying Time example, SSResid = 11,973.08 and SSTotal = 13,983.34. Use these values to calculate the value of r^2 for the Frying Time data, and check your answer.

$r^2 =$

Answer

$$r^2 = 1 - \frac{\text{SSResid}}{\text{SSTotal}} = 1 - \frac{11{,}973.08}{13{,}983.34} = 1 - 0.856 = 0.144$$

Standard Deviation About the Least-Squares Line, s_e

$$s_e = \sqrt{\frac{\text{SSResid}}{n - 2}}$$

Once you have calculated the values of SSResid (discussed earlier in the section), it is easy to calculate the value of s_e. First divide the value for SSResid by the value of $n - 2$, where n is the number of observations in the data set. Then take the square root of that number.

For the data on x = Age and y = Amount spent on pets per month described at the beginning of this section, SSResid = 24.896 and $n = 5$. Substituting these values into the formula for s_e results in

$$s_e = \sqrt{\frac{\text{SSResid}}{n-2}} = \sqrt{\frac{24.896}{5-2}} = \sqrt{\frac{24.896}{3}} = \sqrt{8.299} = 2.881$$

Check It Out!

Calculate the value of s_e for the Frying Time data, and check your answer.

Recall that in the Frying Time example, SSResid = 11,973.08.

$s_e =$

Answer

$$s_e = \sqrt{\frac{\text{SSResid}}{n-2}} = \sqrt{\frac{11{,}973.08}{6-2}} = \sqrt{\frac{11{,}973.08}{4}} = \sqrt{2993.27} = 54.7$$

SECTION 4.6 EXERCISES

For Exercises 4.58–4.61, use the data on x = Weight at age 13 and y = Bone mineral density at age 27 given in the table below.

Observation	Weight	BMD
1	54.4	1.15
2	59.3	1.26
3	74.6	1.42
4	62.0	1.06
5	73.7	1.44
6	70.8	1.02
7	66.8	1.26
8	66.7	1.35
9	64.7	1.02
10	71.8	0.91
11	69.7	1.28
12	64.7	1.17
13	62.1	1.12
14	68.5	1.24
15	58.3	1.00

4.58 Calculate the equation of the least-squares line, along with the predicted values and the residuals.

4.59 Calculate the sums of squares SSResid and SSTotal.

4.60 Calculate the value of r^2.

4.61 Calculate the value of s_e.

5 Probability—The Math You Need to Know

SECTION 5.1 Review—Proportions, Decimal Numbers and Percentages, Ordering Decimal Numbers

Several of the topics covered earlier will also be important as you begin your study of probability. This section provides a quick review of these topics.

Proportions, Decimal Numbers, and Percentages

In your work with probability, there will be many situations that will require that you convert between proportions, decimal numbers, and percentages. Recall that in statistics, a **proportion** is a number between 0 and 1 (including 0 and 1, which are also possible values for a proportion) that represents a part of a whole group. Proportions are usually calculated by converting a fraction, such as $\frac{34}{80}$, to a decimal number by dividing the number in the numerator of the fraction by the number in the denominator. For example, the fraction $\frac{34}{80}$ would be converted to a decimal number by dividing 34 by 80, resulting in $\frac{34}{80} = 0.425$.

Probabilities are usually expressed as either a fraction or a decimal number. If you will be using a probability expressed as a fraction in other calculations, it might make those calculations easier if you first convert the fraction to a decimal number.

As you study probability, you will learn that probabilities can be interpreted as long-run relative frequencies. For example, if the probability of winning a prize in a carnival game is 0.12, this is interpreted as the proportion of the time, in the long run, that you would win a prize if you were to play the game repeatedly.

Because many people have an easier time understanding percentages than proportions, it is common to convert a probability that has been expressed as a decimal number to a percentage before giving an interpretation of the probability.

To convert a probability to a percentage:

1. If the probability is not already expressed as a decimal number, convert the fraction to a decimal number.
2. Multiply the probability by 100 and add a percent symbol (%).

To convert a percentage to a probability:

1. Drop the % symbol.
2. Divide the percentage by 100 to get a decimal number.

For example, if the probability of winning a prize is 0.12, you can convert this probability to a percentage by multiplying it by 100 and adding the % symbol, resulting in 12%. This would be interpreted as the long-run percentage of the time that a prize would be won for this game. You would say that if you were to play the game a large number of times, you would expect to win a prize about 12% of the time.

In your work with probability, there will also be situations where you are given a probability expressed as a percentage and will need to convert that percentage into a decimal number. To convert a percentage to a probability, drop the % symbol and then divide the percentage by 100. For example, suppose that you are told that 73.4% of the students at a particular community college are first-year students. You can use this information to determine the probability that a first-year student will be selected if a student were to be selected at random from this college. Because 73.4% of the students at the college are first-year students, you would expect to get a first-year student about 73.4% of the time in the long-run if you were to repeat the process of selecting a student at random. To express this long-run percentage as a probability, you would drop the % symbol and divide by 100, resulting in

$$\frac{73.4}{100} = 0.734$$

Note that because probabilities must be numbers between 0 and 1 (including 0 and 1), only percentages that are between 0% and 100% are valid representations of a probability.

Check It Out!

a. Convert the proportions to percentages. Check your answers.

i. 0.015

ii. 0.6

b. Convert the percentages to proportions. Check your answers.

i. 67%

ii. 3.0%

Answers

a. i. 1.5%

ii. 60%

b. i. 0.67

ii. 0.030

Ordering Decimal Numbers

In Chapter 1, you saw how to order decimal numbers. As you study probability, there will also be situations where you might need to decide which of two probabilities is greater. Because probabilities are always numbers between 0 and 1 (including 0 and 1), let's review how you would order decimal numbers that are between 0 and 1.

The easiest way to decide which of two decimal numbers between 0 and 1 is greater is to compare them digit by digit. Start by writing the two numbers you want to order so that they have the same number of digits after the decimal place. You can do this by adding 0's to the end of a number as needed—this doesn't change the value of the number. For example, if you want to compare 0.02416 and 0.0237, you would start by adding a 0 to the end of the second number so that both numbers have one digit before the decimal place and five digits after the decimal place.

First number: 0.02416
Second number: 0.02370

Once you have done this, you compare the numbers digit by digit starting with the leftmost digit. As soon as you find a place where the digits differ, the number that has the greater value in this position is the greater number. For the example above,

First number: 0.02416
Second number: 0.02370

This is the first place where the digits differ. Because 4 is greater than 3, the first number is greater than the second number.

Note:
When you order decimal numbers that are between 0 and 1, be sure to do this comparison digit by digit. A common mistake to think that 0.001 is smaller than 0.0005. You could make this mistake if you just look at the first non-zero digits and compare them.

Check It Out!

Identify which decimal number in each of the following pairs is greater, and then check your answers.

a. 0.20 or 0.30

b. 0.54 or 0.53

c. 0.614 or 0.616

Answers

a. 0.30

b. 0.54

c. 0.616

SECTION 5.1 EXERCISES

For Exercises 5.1 and 5.2, convert the given proportions to percentages.

Exercise	Proportion
5.1	0.42
5.2	0.006

For Exercises 5.3 and 5.4, convert the given percentages to proportions.

Exercise	Percentage
5.3	88.9%
5.4	0.88%

Identify which decimal number is greater for each of the pairs in Exercises 5.5–5.7.

Exercise	Which is Greater?
5.5	0.07 or 0.09
5.6	0.52 or 0.54
5.7	0.60 or 0.50

SECTION 5.2 Sets and Set Notation

A **set** is a collection of items, and the items in a set are called elements or members of the set. One way to define a set is to list the elements of the set inside "curly brackets." For example, the set that has the numbers 1, 2, 3, and 5 as its elements can be written as

{1, 2, 3, 5}. It is common to use capital letters to name sets, and to write something like $A = \{1, 2, 3, 5\}$.

In probability, sets are used to represent the possible outcomes of a chance experiment (called the sample space of the chance experiment) and to define events that might occur when a chance experiment is carried out. You will see formal definitions of the terms chance experiment, sample space, and event in your statistics course. But since sample space and events will be defined in terms of sets, this is a good time to review set notation and operations with sets.

Just as numbers can be combined using arithmetic operations (such as addition, subtraction, multiplication and division) to obtain another number, there are operations for sets that allow you to combine sets to obtain another set. Two commonly encountered set operations are called "union" and "intersection."

Operation	Symbol	Meaning	Example	In Words
Union	$\cup$	"or"	$A \cup B$	The union of the sets A and B
Intersection	$\cap$	"and"	$A \cap B$	The intersection of the sets A and B

Union of Two Sets

Consider two sets that are denoted by A and B. The union of these two sets is denoted by $A \cup B$ and is the set that consists of all the elements that are in the set A or in the set B or that are in both A and B. For example, if the sets A and B are defined as

$$A = \{2, 4, 6, 9\}$$
$$B = \{2, 3, 5, 10\}$$

then the union of A and B is the set that contains 2, 4, 6, 9, 3, 5, and 10—all the elements in A and B combined. You can write

$$A \cup B = \{2, 3, 4, 5, 6, 9, 10\}$$

Intersection of Two Sets

Consider two sets that are denoted by A and B. The intersection of these two sets is denoted by $A \cap B$ and is the set that consists of all the elements that in both the set A and the set B. For example, if the sets A and B are defined as

$$A = \{2, 4, 6, 9\}$$
$$B = \{2, 3, 5, 10\}$$

then the intersection of A and B is the set that contains 2. The only member of $A \cap B$ is 2 because 2 is the only element that is in both sets. You can write

$$A \cap B = \{2\}$$

Check It Out!

a. Find the union and the intersection for each pair of sets. Check your answers as you go.

Sets	Union ($\cup$)	Intersection ($\cap$)
$A = \{3, 6, 7, 12\}$ $B = \{1, 4, 12\}$		
$C = \{2, 3, 4, 6, 7, 9, 11, 12\}$ $D = \{1, 5, 7, 11\}$		
$E = \{\text{A, E, H, K, L}\}$ $F = \{\text{A, C, D, E, G, L}\}$		

b. Give two sets A and B that have the following intersection. Check your answer.
$A \cap B = \{1, 3, 8, 9\}$

c. Give two sets C and D that have the following union. Check your answer.
$C \cup D = \{1, 3, 4, 5, 6, 9, 10, 12\}$

Answers

a.

Sets	Answers
$A = \{3, 6, 7, 12\}$ $B = \{1, 4, 12\}$	$A \cup B = \{1, 3, 4, 6, 7, 12\}$ $A \cap B = \{12\}$
$C = \{2, 3, 4, 6, 7, 9, 11, 12\}$ $D = \{1, 5, 7, 11\}$	$C \cup D = \{1, 2, 3, 4, 5, 6, 7, 9, 11, 12\}$ $C \cap D = \{7, 11\}$
$E = \{\text{A, E, H, K, L}\}$ $F = \{\text{A, C, D, E, G, L}\}$	$E \cup F = \{A, C, D, E, G, H, K, L\}$ $E \cap F = \{A, E, L\}$

b. Answers will vary, but one possibility is $A = \{1, 2, 3, 4, 5, 6, 7, 8, 9\}$ and $B = \{1, 3, 8, 9, 10\}$

c. Answers will vary, but one possibility is $C = \{1, 3, 4, 5, 9\}$ and $D = \{1, 4, 6, 10, 12\}$.

The Universal Set and the Complement of a Set

When working with sets, the **universal set** is a set that contains all possible elements of the sets under consideration in a given problem. In the context of probability, the sample space is the universal set—it consists of all possible outcomes of a chance experiment.

Consider a set that is denoted by A. The complement of the set A is denoted by A^C and is the set that consists of all the element of the universal set that are *not* in the set A.

Operation	Symbol	Meaning	Example	Example in Words
Complement	superscript C	"not"	A^C	The complement of the set A

For example, if the universal set and the set A are defined as

$$\text{Universal set} = \{1, 2, 3, 4, 5, 6, 7, 8, 9\}$$
$$A = \{2, 3, 5, 9\}$$

then the complement of A is the set that contains 1, 4, 6, 7, and 8. These are all the elements in the universal set that are not in the set A. You can write

$$A^C = \{1, 4, 6, 7, 8\}$$

Note:
In addition to A^C, you will sometimes see the complement of a set A denoted as A' or $\overline{A}$.

Check It Out!

For each universal set and set, find the complement of the set. Check your answers as you go.

Universal Set	Set	Complement
$\{4, 5, 9\}$	$A = \{4, 5\}$	
$\{2, 3, 5, 7, 9\}$	$A = \{2, 3, 9\}$	
$\{3, 4, 5, 9\}$	$A = \{9\}$	
$\{A, B, C, E, G, I\}$	$A = \{A, B, C, I\}$	

Answers

Universal Set	Set	Answer
$\{4, 5, 9\}$	$A = \{4, 5\}$	$A^C = \{9\}$
$\{2, 3, 5, 7, 9\}$	$A = \{2, 3, 9\}$	$A^C = \{5, 7\}$
$\{3, 4, 5, 9\}$	$A = \{9\}$	$A^C = \{3, 4, 5\}$
$\{A, B, C, E, G, I\}$	$A = \{A, B, C, I\}$	$A^C = \{E, G\}$

SECTION 5.2 EXERCISES

Find

For Exercises 5.8–5.10, find the union and the intersection for the given pair of sets.

5.8 $A = \{6, 7, 8, 9\}$
$B = \{2, 4, 6, 8\}$

5.9 $C = \{2, 7, 8, 14\}$
$D = \{2, 3, 4, 5, 9, 10, 11\}$

5.10 $E = \{\text{D, E, G, H, J}\}$
$F = \{\text{A, C, G, H, I, J}\}$

5.11 Give two sets A and B that have the following intersection.
$A \cap B = \{3, 5, 7, 8, 10\}$

5.12 Give two sets C and D that have the following union.
$C \cup D = \{1, 2, 3, 6\}$

For each universal set and set *A* in Exercises 5.13–5.16, find the complement of the set *A*.

5.13 $\{0, 2, 4, 6, 7, 9\}$, $A = \{4\}$

5.14 $\{31, 70, 72\}$, $A = \{31, 72\}$

5.15 $\{D, E, H, K, O, T, X\}$, $A = \{D, O, X\}$

5.16 $\{13, 55, 109\}$, $A = \{55, 109\}$

SECTION 5.3 Evaluating Expressions

When calculating probabilities, you may encounter situations where you work with fractions that have arithmetic expressions in the numerator or the denominator or both. When this is the case, you should evaluate each of these expressions to convert to a fraction that has a single number in the numerator and a single number in the denominator. Once that is done, you can convert the fraction to a decimal number.

For example, suppose that you want to calculate a probability that involves evaluating the following expression:

$$\frac{68 + 14 + 122}{1000}$$

You would begin by evaluating the arithmetic expression in the numerator ($68 + 14 + 122 = 204$) to get $\frac{204}{1000}$. Converting to a decimal number, you would get $\frac{204}{1000} = 0.204$.

To calculate a probability that involves evaluating the expression

$$\frac{0.68}{0.68 + 0.120}$$

you would first evaluate the arithmetic expression in the denominator ($0.68 + 0.120 = 0.80$) to get $\frac{0.68}{0.80} = 0.85$.

If you are using probability formulas to calculate probabilities, you will need to substitute probability values into these formulas and then evaluate the resulting expressions. Once you have substituted values into these formulas, you will either have a simple arithmetic expression or you will have a fraction that has a form that is like one of the two just considered.

Check It Out!

Evaluate the following expressions. Give your answer as a decimal number rounded to three decimal places. Check your answers.

Expression	Evaluate
$\frac{55 + 10 + 29}{135}$	
$\frac{32}{73} + \frac{4}{73}$	
$\frac{32 + 28 + 11}{96}$	

Answers

Expression	Answer
$\frac{55 + 10 + 29}{135}$	0.696
$\frac{32}{73} + \frac{4}{73}$	0.493
$\frac{32 + 28 + 11}{96}$	0.740

SECTION 5.3 EXERCISES

Evaluate the expressions in Exercises 5.17–5.19. Give your answers as decimal numbers rounded to three decimal places.

5.17 $\frac{44 + 8 + 5}{77}$

5.18 $\frac{89 + 219}{782}$

5.19 $\frac{74}{120} + \frac{16}{60}$

6 Random Variables and Probability Distributions—The Math You Need to Know

SECTION 6.1 Review—Powers of Numbers, Square Roots, Intervals, Proportions, Decimals and Percentages, and *z-Scores*

Many of the topics covered earlier are also relevant to your study of probability distributions. This section provides a quick review of these topics.

Powers and Square Roots

Raising a number to a power is a way to indicate multiplying a number by itself some number of times. For example, 8^4 is read as 8 to the fourth power, and it is shorthand for $8 \times 8 \times 8 \times 8$. The **exponent** is the small number written as a superscript (for example, in 8^4, the exponent is 4), and it specifies how many times the number being multiplied appears. When you multiply a number by itself, this is called squaring the number, and is expressed using an exponent of 2. For example, 7 squared means 7×7 and can be written 7^2. Some other examples are shown in the table below.

Expression	Means
6^5	$6 \times 6 \times 6 \times 6 \times 6$
5^6	$5 \times 5 \times 5 \times 5 \times 5 \times 5$
4^2	4×4

Remember that the square root of a number is represented using the square root symbol, $\sqrt{}$. When you calculate the square root of a number, you are finding a value that can be squared to obtain that number. For example, if you want to know the value of $\sqrt{36}$, you are looking for a number that can be squared (multiplied by itself) to get 36. Because $6 \times 6 = 36$, the square root of 36 is 6.

For some numbers, like 4 and 9, you may be able to determine the square root of the number without even using your calculator, but for other numbers, you will use your calculator. This is the case when you are working with a number that has a square root that is not a whole number or if you want to find the square root of a decimal number.

Check It Out!

a. Evaluate the following expressions, checking your answers as you go.

i. 3^9

ii. 4^5

iii. 9^0

b. Calculate the following square roots, checking your answers along the way. Round your answers to two decimal places.

i. $\sqrt{45}$

ii. $\sqrt{97}$

iii. $\sqrt{32}$

Answers

a. i. 19,683

ii. 1,024

iii. 1

b. i. 6.71

ii. 9.85

iii. 5.66

Intervals

When working with probability distributions, there will be situations where you want to evaluate the probability of observing a value in a particular interval. Recall that an interval is a range of values along the number line.

Intervals are represented by two numbers, which are called the endpoints of the interval. The interval itself is made up of all the numbers between the two endpoints. Some intervals include one or both endpoint values.

The usual way to represent an interval is to list the two endpoints inside parentheses and separated by a comma. The smaller endpoint is listed first. For example, (3.0, 6.5) represents the interval that consists of all the numbers between 3.0 and 6.5. To indicate that you want to include the endpoint value in the interval, you can use a square bracket instead of a parenthesis. Here are a few examples:

Interval	Description of Interval
(2, 9)	All the numbers between 2 and 9
[2, 9)	2 and all the numbers between 2 and 9
[6, 14]	6, 14, and all the numbers between 6 and 14
(−2, 2)	All the numbers between −2 and +2

Proportions, Decimal Numbers, and Percentages

As was the case in the previous chapter where you calculated and interpreted probabilities, you will encounter situations in your study of probability distributions where you may want to convert between fractions, decimal numbers, and percentages.

Because many people have an easier time understanding percentages than proportions, it is common to convert a probability that has been expressed as a decimal number to a percentage before giving an interpretation of the probability.

To convert a probability to a percentage:

1. If the probability is not already expressed as a decimal number, convert the fraction to a decimal number.
2. Multiply the probability by 100 and add a percent symbol (%).

To convert a percentage to a probability:

1. Drop the % symbol.
2. Divide the percentage by 100 to get a decimal number.

Also, remember that because probabilities must be numbers between 0 and 1 (including 0 and 1), only percentages that are between 0% and 100% are valid representations of a probability.

Check It Out!

a. Convert the following probabilities to percentages. Check your answers as you go.

Probability	Percentage
0.821	
0.161	

b. Convert the following percentages to probabilities, and check your answers.

Percentage	Probability
95.5%	
31%	

c. If the probability of observing a value in the interval (21, 53) is 0.219, what percentage of the time in the long run do you expect to observe the value between 21 and 53? Check your answer.

d. If you expect to see a value between 7.4 and 8.3 99.7% of the time in the long run, what is the probability of observing a value in the interval (7.4, 8.3)? Check your answer.

Answers

a.

Probability	Answer
0.821	$0.821 \times 100\% = 82.1\%$
0.161	$0.161 \times 100\% = 16.1\%$

b.

Percentage	Answer
95.5%	$\frac{95.5}{100} = 0.955$
31%	$\frac{31}{100} = 0.310$

c. $0.219 \times 100\% = 21.9\%$

d. $\frac{99.7}{100} = 0.997$

z-Scores

Recall that *z-scores* were introduced in Chapter 3, and you will encounter *z-scores* again as you work with probability distributions. To calculate a *z-score* corresponding to a value, you subtract the mean and then divide by the standard deviation:

$$z = \frac{\text{value} - \text{mean}}{\text{standard deviation}}$$

In the context of probability distributions, the mean and standard deviation used to calculate a *z-score* will be the mean and standard deviation of a probability distribution. A positive *z-score* tells you that the value is greater than the mean and the *z-score* is interpreted as the number of standard deviations the value is above the mean. A negative *z-score* tells you that the value is less than the mean and the *z-score* is interpreted as the number of standard deviations the value is below the mean.

For example, if you are working with a probability distribution that has a mean of 48 and a standard deviation of 9, the *z-score* corresponding to the value 60 is

$$z = \frac{60 - 48}{9} = \frac{12}{9} = 1.33$$

Interpreting this *z-score*, you would say that 60 is 1.33 standard deviations above the mean.

Check It Out!

For each value, mean, and standard deviation, calculate the z-score corresponding to the given value. Check your answers as you go.

x Value	Mean	Standard Deviation	*z-score*
3.9	9.7	7.1	
7.0	3.9	2.6	
6.4	3.5	5.0	

Answers

x Value	Mean	Standard Deviation	Answer
3.9	9.7	7.1	$\frac{3.9 - 9.7}{7.1} = -0.82$
7.0	3.9	2.6	$\frac{7 - 3.9}{2.6} = 1.19$
6.4	3.5	5.0	$\frac{6.4 - 3.5}{5} = 0.58$

SECTION 6.1 EXERCISES

Evaluate the following expressions in Exercises 6.1–6.4.

6.1 9^4

6.2 4^6

6.3 $\sqrt{51}$

6.4 $\sqrt{44}$

Convert the probabilities given in Exercises 6.5 and 6.6 to percentages.

6.5 0.286

6.6 0.579

Convert the percentages given in Exercises 6.7 and 6.8 to probabilities.

6.7 61.2%

6.8 0.8%

6.9 If the probability of observing a value in the interval (7, 9) is 0.104, what percentage of the time in the long run do you expect to observe the value between 7 and 9?

6.10 If you expect to see a value between 3.0 and 15.8 37% of the time in the long run, what is the probability of observing a value in the interval (3.0, 15.8)?

For each x value, mean, and standard deviation given in Exercises 6.11–6.13, calculate the *z-score* corresponding to the given x value.

6.11 x value: 3.3 Mean: 9.9 Standard deviation: 7.0

6.12 x value: 65 Mean: 47 Standard deviation: 7.8

6.13 x value: 6.6 Mean: 4.4 Standard deviation: 0.9

SECTION 6.2 Random Variables

Random variables are not the same as the variables that you previously considered in the context of algebra and algebraic expressions. In algebra, a variable is a letter that is used to represent a number. In statistics, the term variable is also used to represent the value of some characteristic (such as wing length of a bird or weight of an orange) that is measured when data are collected.

Random variables are a bit different. A **random variable** is a quantity whose value depends on the outcome of a chance experiment. This means that a random variable may take on different values, depending on the outcome. For example, you might have a chance experiment that consists of rolling a pair of six-sided dice. The quantity x = Sum of the numbers showing on the upper faces of the dice is a number that depends on the outcome of the chance experiment. The value of x might be 4 or it might be 8. In fact, it could be any integer value from 2 to 12. There is uncertainty about the value of x, because its value depends on the outcome of the experiment.

Many of the random variables that you will encounter in your statistics course are based on chance experiments that involve random selection. For example, suppose you are interested in the commute times of students at a college. The population of interest is all students at the college. If you were to select a student at random from this population, that would be a chance experiment with possible outcomes corresponding to each of the students that might be selected. Then x = Commute time is a random variable, because it associates a number with each possible outcome of the chance experiment.

In your study of probability distributions, you will see how a probability distribution describes the long-run behavior of a random variable. Knowing the probability distribution of a random variable allows you to see what values would be expected and what values would be unusual and surprising.

Check It Out!

a. Consider the chance experiment of a observing a basketball player who will take 10 shots during a game. Some of the shots are 2-point shots, and some are 3-point shots. One random variable is the percentage of the 10 shots that the player makes—that is, scores a basket. Describe two other random variables that could be defined in the context of this chance experiment.

b. In a chance experiment, seven playing cards will be dealt from a standard deck of playing cards (including cards Ace through King, in each of four suits). One random variable is the number of face cards (Jack, Queen, or King) out of the seven cards. Describe two other random variables that could be defined in the context of this chance experiment.

c. A quality control engineer plans to test a piece of duct tape by stretching the tape until it breaks and recording the force required to break the tape. Identify the chance experiment, and describe the random variable in the chance experiment. Check your answers.

d. A researcher plans to test a new antibiotic on a bacteria specimen and record the percentage of bacteria in the specimen that are killed by the antibiotic. Identify the chance experiment, and describe the random variable in the chance experiment. Check your answers.

Answers

a. Answers will vary. Two examples of other random variables are the number of shots made out of 10 shots, and the total number of points scored on the 10 shots taken.

b. Answers will vary. Two examples of other random variables are the number of hearts dealt in seven cards, and the percentage of cards out of the seven that are black cards.

c. The chance experiment is stretching a piece of duct tape until it breaks. The random variable is the force required to break the tape.

d. The chance experiment is applying the antibiotic to the bacteria specimen, and the random variable is the percentage of bacteria in the specimen that are killed.

SECTION 6.2 EXERCISES

6.14 A spinner is divided into 10 equal segments, with one segment labeled 10 points, one segment labeled 5 points, one segment labeled 2 points, and the remaining segments each labeled 0 points. Consider the chance experiment of spinning the spinner once. Describe two random variables that could be defined in the context of this chance experiment.

6.15 A student will be selected at random from the students at a college. This selected student will be asked if they exercised during the previous week and how many minutes that he or she spent exercising. Describe two random variables that could be defined in the context of this chance experiment.

6.16 A botanist plans to measure the volume of nitrogen (as a percentage) remaining in a leaf selected at random from a plant. Identify the chance experiment, and describe the random variable in the chance experiment.

6.17 A person will be selected at random from the people working at a car dealership. The selected person will be asked to rate their job satisfaction on a scale from 1 = Poor to 10 = Outstanding. Identify the chance experiment, and describe the random variable in the chance experiment.

SECTION 6.3 More on Intervals

Recall that intervals are represented by two numbers, which are called the endpoints of the interval. The interval itself is made up of all the numbers between the two endpoints. Some intervals include one or both endpoint values. Previously you represented intervals by listing the lower and upper endpoints of the interval in parentheses. For example, the interval that includes all the numbers between 8 and 14 could be represented as (8, 14). As you continue your study of statistics, you will see that there are other ways to represent intervals.

Representing Intervals on the Number Line

Intervals are sometimes represented using a number line and drawing a line that shows the values that are included in the interval. For example, the interval (8, 14) could be represented as

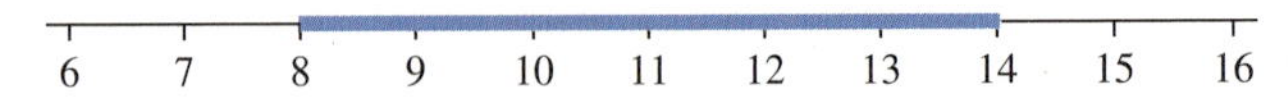

Notice that the colored line extends from 8 to 14 and represents the values in the interval. Now consider the following:

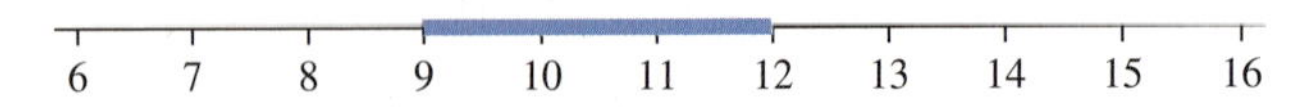

The interval represented on this number line is (9, 12).

Check It Out!

a. Use the accompanying number line to represent the interval (1.3, 5.5).

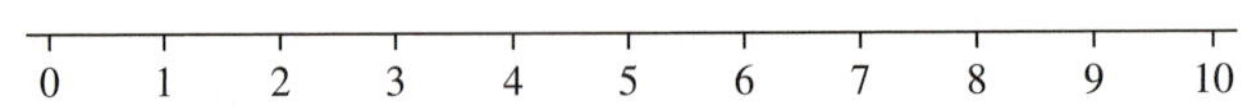

b. Use the accompanying number line to represent the interval (10.9, 72.9).

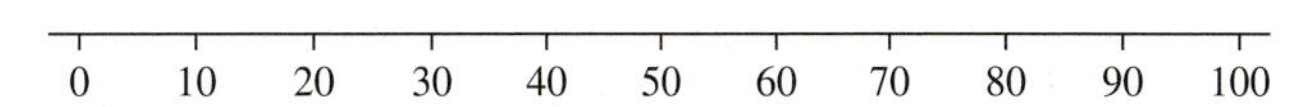

c. What interval is represented on the following number line?

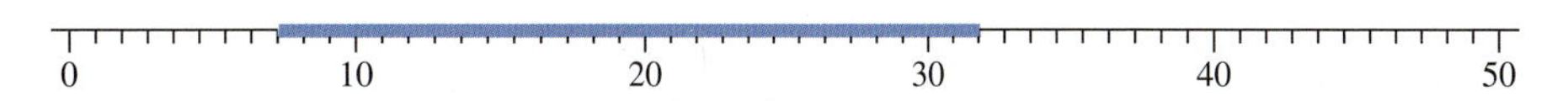

d. What interval is represented on the following number line?

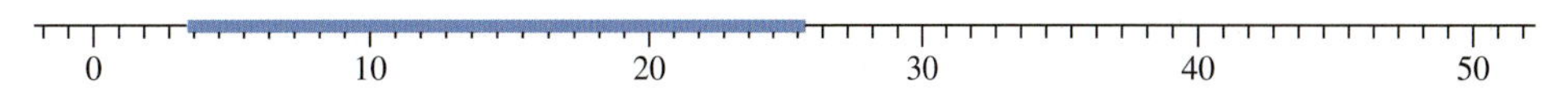

Answers

a.

b.

0 10 20 30 40 50 60 70 80 90 100

c. (7, 32)

d. (4, 26)

Another way to specify an interval is to give the midpoint of the interval and the distance the interval extends on either side of the midpoint. The usual way to represent this is using the symbol ±, which is read as plus and minus. For example, 15 ± 6 represents an interval where the upper endpoint is 15 + 6 and the lower endpoint is 15 − 6. This is the interval (9, 21). Notice that the interval width is twice the number that appears after the ± symbol, because the interval extends that far in both directions.

The interval (94, 106) can be written using ± notation as 100 ± 6. You can arrive at this expression by first finding the midpoint of the interval, which is the average of the two endpoints. You would calculate $\frac{94 + 106}{2} = \frac{200}{2} = 100$. Then you can determine the ± number by subtracting the interval midpoint from the upper endpoint of the interval: 106 − 100 = 6.

You have now seen several different ways to represent an interval. For example, the interval that includes all the numbers between 840 and 925 can be represented in all three of the following ways:

(840, 925)

830 835 840 845 850 855 860 865 870 875 880 885 890 895 900 905 910 915 920 925 930 935 940 945 950

882.5 ± 42.5

Check It Out!

a. Write the interval (7, 46) in "plus or minus" (±) form, and also represent it on the number line provided. Check your answers.

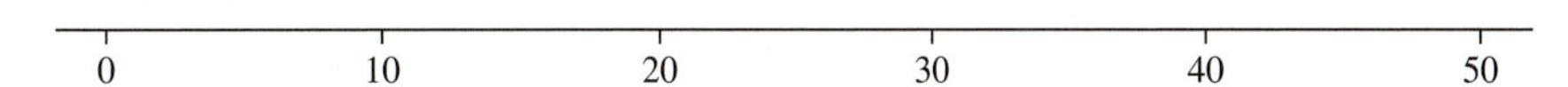

b. Write the interval 4.2 ± 4.6 in the form (lower endpoint, upper endpoint), and also represent it on the number line provided. Check your answers as you go.

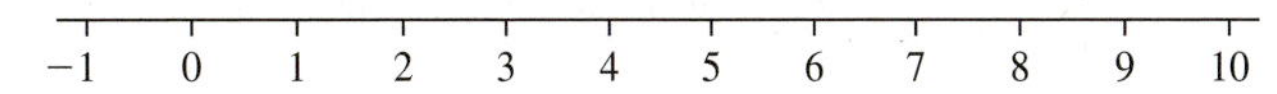

c. An interval is represented on the accompanying number line. Write the interval in the form (lower endpoint, upper endpoint), and also in "plus or minus" (±) form. Check your answers.

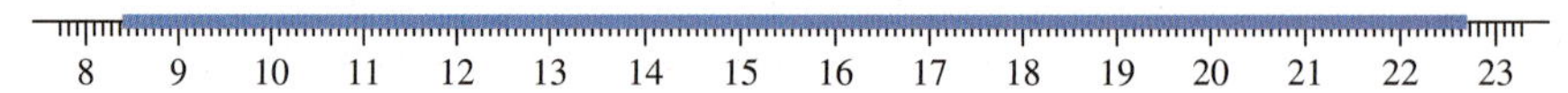

Answers

a. $\frac{7+46}{2} = 26.5$, and $46 - 26.5 = 19.5$, so the interval can be represented as 26.5 ± 19.5.

b. $4.2 - 4.6 = -0.4$, and $4.2 + 4.6 = 8.8$, so the interval can be represented as $(-0.4, 8.8)$.

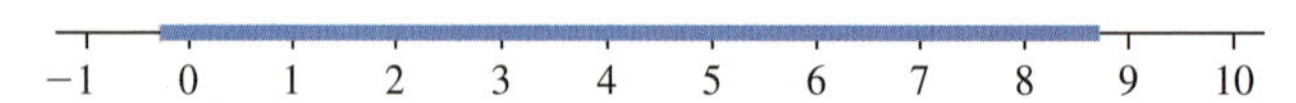

c. (8.3, 22.8)

$\frac{8.3 + 22.8}{2} = 15.55$, and $22.8 - 15.55 = 7.25$, so the interval can be represented as 15.55 ± 7.25.

SECTION 6.3 EXERCISES

In Exercises 6.18 and 6.19, use the accompanying number lines to represent the given intervals.

6.18 Interval: (96, 160)

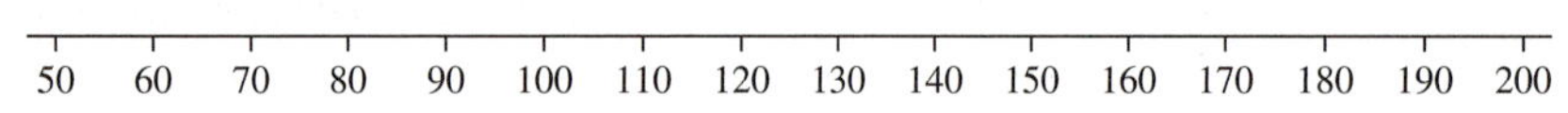

6.19 Interval: (9.4, 43.6)

What intervals are represented on the number lines given in Exercises 6.20 and 6.21?

6.20

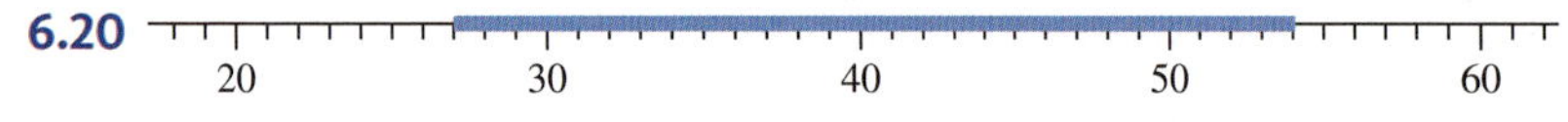

6.21

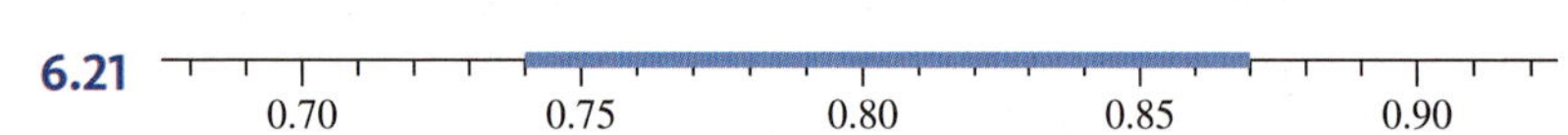

6.22 Write the interval (48.0, 50.7) in "plus or minus" (±) form, and also represent it on the number line provided.

6.23 Write the interval 85.5 ± 3.5 in the form (lower endpoint, upper endpoint), and also represent it on the number line provided.

6.24 An interval is represented on the accompanying number line. Write the interval in the form (lower endpoint, upper endpoint), and also in "plus or minus" (±) form.

SECTION 6.4 Equations and Inequalities

An **equation** is a mathematical expression that states that two quantities are equal. For example, $6 + 4 = 12 - 2$ is an equation because it states that $6 + 4$ is equal to $12 - 2$. Equations are easy to recognize because they contain the equal symbol (=). Equations can also involve algebraic expressions, such as $x + 7 = 38$ or $y = 14 + 3x$.

In your statistics course, you will also encounter situations where you will work with inequalities. An **inequality** is a mathematical expression that states how two quantities compare. Instead of containing the equal symbol (=), an inequality contains one of the following symbols: $<, \leq, >, \geq, \neq$. These symbols are described in the following table.

Symbol	Meaning of the Symbol	Example	Example in Words
$<$	Less than	$3 < 8$	3 is less than 8
$\leq$	Less than or equal to	$x \leq 10$	The value of x is less than or equal to 10
$>$	Greater than	$14 > 3$	14 is greater than 3
$\geq$	Greater than or equal to	$x \geq 2 + 5$	The value of x is greater than or equal to 7
$\neq$	Not equal to	$21 \neq 9$	21 is not equal to 9

If you wanted to translate the statement x is greater than 29 into symbols, you would write $x > 29$. Similarly, the statement x is less than or equal to 100 would be written as $x \leq 100$. Notice that this could also be written as $100 \geq x$. The inequality symbol ($>$ or $<$) always opens up toward the greater value.

You can also use inequalities to represent an interval. For example, the inequality $10 < x < 17$ states that the value of x is between 10 and 17. You could also say that x is greater than 10 but also less than 17.

Check It Out!

a. Choose the correct symbol in place of "?" to indicate the relationship between the numbers given. Check your answers as you go.

Relationship	Symbol
53 ? 86	
2×9 ? $5 \times (9 - 7)$	
$9 \div 5$? $\frac{3}{6} \times 4$	

b. For the given descriptions of inequalities, write the inequalities in symbols. Check your answers along the way.

Description	Inequality
The value of x is greater than 75.	
The value of y is between 3.2 and 8.4, inclusive.	
The value of x is less than fourteen squared.	

c. Write the meanings in words for the inequalities given. Check your answers.

Inequality	Meaning
$\frac{710}{254} < w$	
$3.2 < y < 8.4$	
$x \neq 98 - 55$	

Answers

a.

Relationship	Answer
53 ? 86	$53 < 86$
2×9 ? $5 \times (9 - 7)$	$2 \times 9 > 5 \times (9 - 7)$
$9 \div 5$? $\frac{3}{6} \times 4$	$9 \div 5 < \frac{3}{6} \times 4$

b.

Description	Answer
The value of x is greater than 75.	$x > 75$
The value of y is between 3.2 and 8.4, inclusive.	$3.2 \leq y \leq 8.4$
The value of x is less than fourteen squared.	$x < 14^2$

c.

Inequality	Answer
$\frac{710}{254} < w$	The value of w is greater than $\frac{710}{254} = 2.8$.
$3.2 < y < 8.4$	The value of y is between 3.2 and 8.4.
$x \neq 98 - 55$	The value of x is not equal to the difference between 98 and 55.

Sometimes statements that translate into inequalities use words like "at most" or "at least" or "no more than." When you see these statements, you should think carefully about their meaning if you need to translate statements using these terms into symbols. For example, if a statement says that x is at least 7, does this mean greater than or less than 7 and is 7 included? One way to think about statements like these is to put them into a context that is easy to understand. For example, suppose someone says that they are going to give you at least 7 dollars. Could they give you \$8? Could they give you \$7? How about \$6? Reasoning it out this way helps to see that the statement "at least 7" means 7 or more, or equivalently, "greater than or equal to 7."

What does "no more than" mean? Suppose that you are told that you can eat some cookies, but that you can have no more than 3. Can you have 4? Can you have 2? How about 3? Reasoning this way, you can see that "no more than 3" means "less than or equal to 3." Similar reasoning would lead you to conclude that "at most 3" also means "less than or equal to 3." Similar phrases include "no larger than" and "no smaller than."

Some additional examples are shown in the table below.

Statement	Means	Written in Symbols
x is at most 10	x is less than or equal to 10	$x \leq 10$
x is at least 10	x is greater than or equal to 10	$x \geq 10$
x is no greater than 10 x is no larger than 10 x is no more than 10	x is less than or equal to 10	$x \leq 10$
x is no less than 10 x is no smaller than 10	x is greater than or equal to 10	$x \geq 10$
x is 10 or less	x is less than or equal to 10	$x \leq 10$
x is 10 or more	x is greater than or equal to 10	$x \geq 10$
x is more than 10	x is greater than 10	$x > 10$

Check It Out!

a. For the given descriptions, write the inequalities in symbols, and check your answers along the way.

Description	Inequality
The value of x is at most 1.6.	
The value of u is no larger than 96, and no smaller than 51.	
The value of y is greater than 6.7, but no greater than 21.1.	

b. Write the meanings in words for the inequalities given. Check your answers.

Inequality	Meaning
$6.6 \le x < 6.7$	
$38.5 - 2.8 \le y$	
$(w < 86)$ and $(w > 21)$	

c. Write two different ways to word the meanings for the inequalities given. Check your answers.

Inequality	Meanings
$u \ge 65$	1.
	2.
$2.7 < y < 4.5$	1.
	2.
$50 - 76 \ge x$	1.
	2.

Answers

a.

Description	Answer
The value of x is at most 1.6.	$x \le 1.6$
The value of u is no larger than 96, and no smaller than 51.	$51 \le u \le 96$
The value of y is greater than 6.7, but no greater than 21.1.	$6.7 < y \le 21.1$

b.

Inequality	Answer
$6.6 \le x < 6.7$	The value of x is at least 6.6, but it is also smaller than 6.7.
$38.5 - 2.8 \le y$	The value of y is at least the difference between 38.5 and 2.8, which is 35.7.
$(w < 86)$ and $(w > 21)$	The value of w is between 21 and 86.

c.

Inequality	Possible Answers
$u \ge 65$	1. The value of u is at least 65.
	2. The value of u is not less than 65.
$2.7 < y < 4.5$	1. The value of y is between 2.7 and 4.5.
	2. The value of y is larger than 2.7, but it is also smaller than 4.5.
$50 - 76 \ge x$	1. The value of x is no larger than the difference between 50 and 76.
	2. The value of x is at most -26.

The Meaning of $\approx$

The "not equal to" symbol, $\neq$, is used in mathematical expressions to indicate that two numerical quantities are not equal. For example, you could write $6 \neq 4$ or $x \neq 12$. But sometimes you will see a statement that uses a "curly equal sign," $\approx$. This is the symbol used to represent "approximately equal to." The symbol indicates that two numerical quantities are nearly equal. For example, the number 8.7654 is not equal to 8.77, but they are nearly equal, so you could write $8.7654 \approx 8.77$. In probability, you may see statements like probability ≈ 0. This would mean that although the probability is not exactly 0, it is quite small and very close to 0.

Check It Out!

Write out the meaning in words for each of the given expressions. Check your answers as you go.

Expression	Meaning
$2.96 \approx 3.0$	
$x \approx 43 \div 14$	

Answers

Expression	Answer
$2.96 \approx 3.0$	2.96 is approximately equal to 3.0
$x \approx 43 \div 14$	The value of x is approximately equal to 43 divided by 14, or 3.07 (rounded).

SECTION 6.4 EXERCISES

In Exercises 6.25–6.27, choose the correct symbol in place of "?" to indicate the relationship between the numbers given.

6.25 $5 \times 7 ? 35 ? 78$

6.26 $9 \div 4 ? 15 \div 9$

6.27 $\sqrt{50} ? 8$

For the written descriptions in Exercises 6.28–6.30, write the inequalities in symbols.

6.28 The value of y is between 415 and 505.

6.29 The value of x is less than 6.0.

6.30 The value of w is at least 81.7.

For each of the inequalities given in Exercises 6.31–6.33, write the meaning in words.

Exercise	Inequality	Meaning
6.31	$5.6 < x < 7.6$	
6.32	$9.8 \leq v$	
6.33	$m \leq 5 \times \frac{33}{19}$	

In Exercises 6.34–6.36, write two different ways to word the meanings for the inequalities given.

6.34 $b \leq 300$

6.35 $c > 2.86$

6.36 $82 \leq y < 214$

Write out the meaning in words for each of the expressions in Exercises 6.37 and 6.38.

6.37 $w \approx \sqrt{1.96 \times 5.1}$

6.38 $23 \approx \frac{93}{4}$

SECTION 6.5 Areas Under a Curve

Calculating probabilities for continuous random variables often involves finding the probability that a random variable takes a value in a particular interval. For example, you might want to determine the probability that a random variable x takes a value in the interval (49, 52). This would be expressed in symbols as $P(49 < x < 52)$. This probability is represented by an area under the curve that represents the probability distribution of x, and it is the area under the curve and above the interval (49, 52).

For example, suppose that the curve shown below represents the probability distribution of the random variable x.

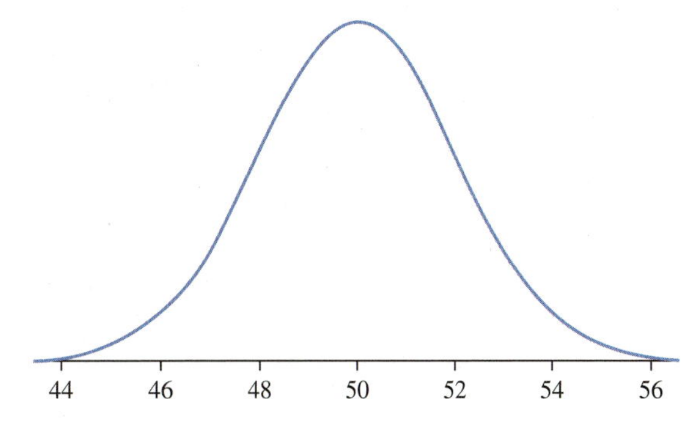

Then $P(49 < x < 52)$ is the area under the curve and above the interval from 49 to 52, as shown here:

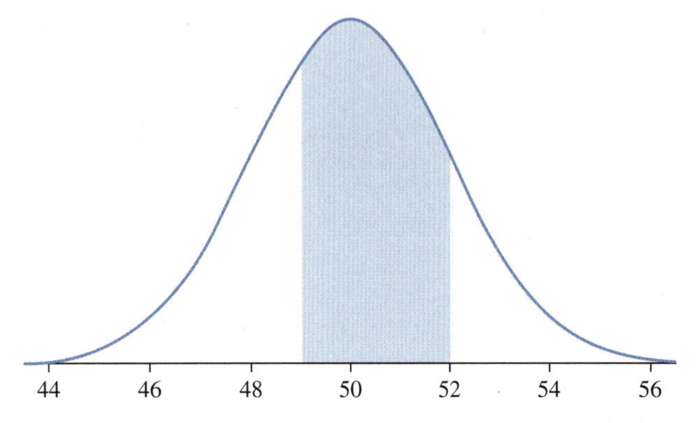

For continuous random variables, the probability that the variable takes on a value in a particular interval is the same whether neither, one, or both of the interval endpoints is included in the interval. For example, this means that the shaded area under the curve above also represents $P(49 \leq x < 52)$, $P(49 < x \leq 52)$, and $P(49 \leq x \leq 52)$.

The probability that x takes a value that is less than 46 can be written in symbols as $P(x < 46)$, and is represented by the area under the curve and to the left of 46:

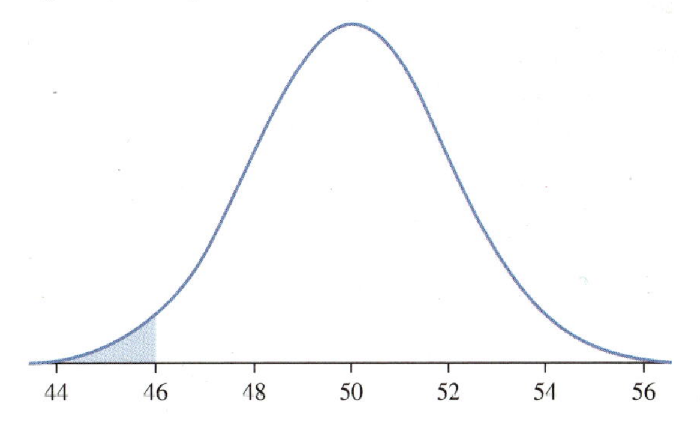

Some other examples are shown in the following table.

Probability	Written in Symbols	Represented as an Area
Probability that x is between 48 and 50	$P(48 < x < 50)$	44 46 48 50 52 54 56

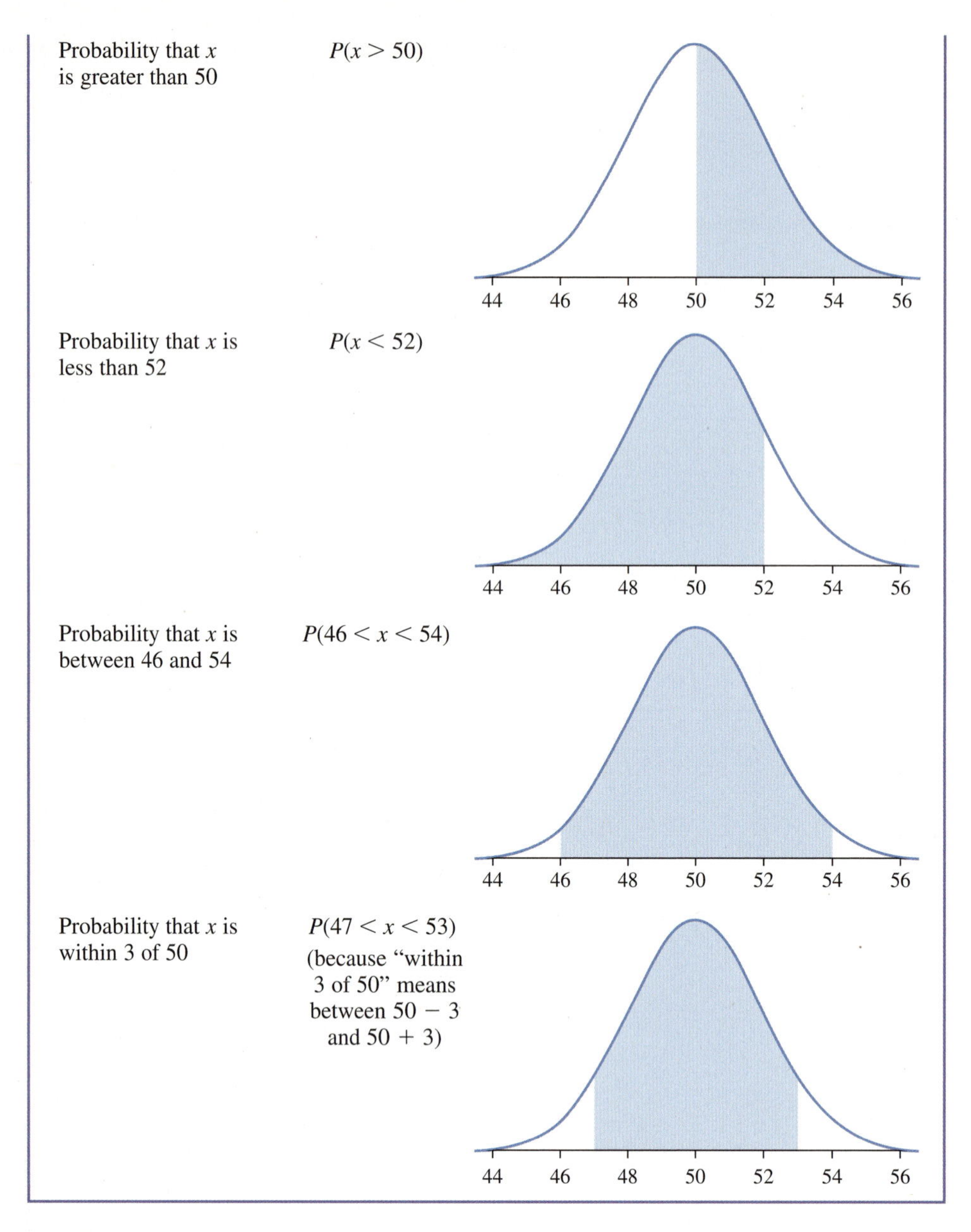

Not all continuous probability distributions have the symmetric bell shape of the probability distribution in the previous example. For example, the random variable y might have the probability distribution shown here:

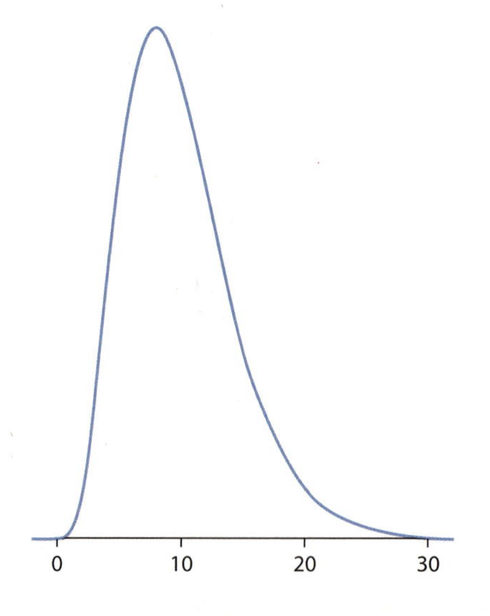

Probabilities such as $P(8 < y < 14)$ or $P(y > 17)$ are represented by areas under this curve and above the interval of interest. For example, $P(8 < y < 14)$ is represented by the following shaded area:

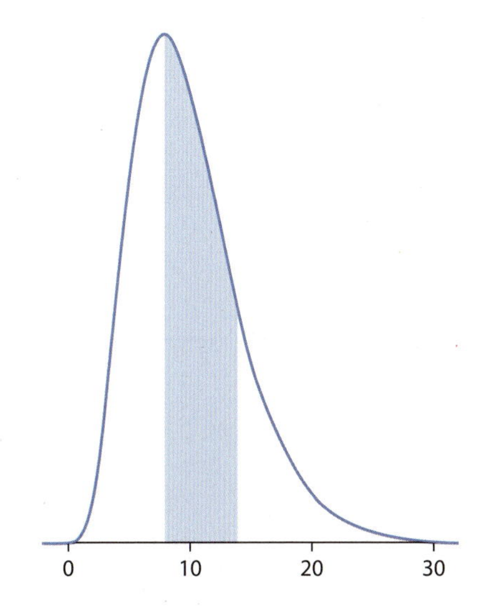

And $P(y > 17)$ is represented by this shaded area:

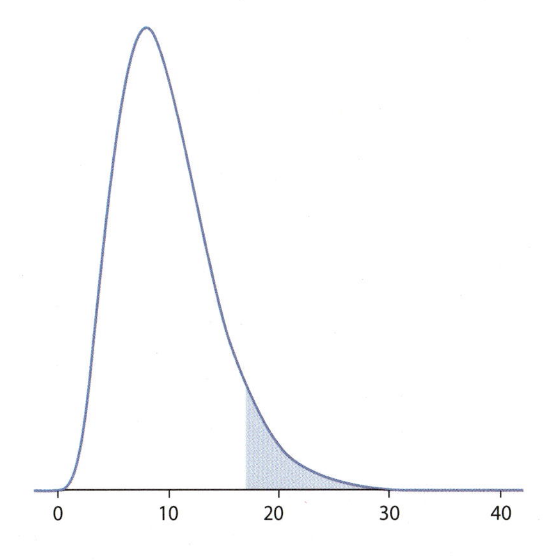

Check It Out!

a. Write the given statements using probability notation. Check your work with the answers provided.

Statement	Probability in Symbols
The probability that x takes a value between 23 and 93.	
The probability that x takes a value that is greater than or equal to 1.05.	
The probability that x takes a value that is at most 97.1.	

b. Mark and shade the areas represented by the given probability statements. Check your answers along the way.

Statement 1: The probability that x takes value that is less than 9.

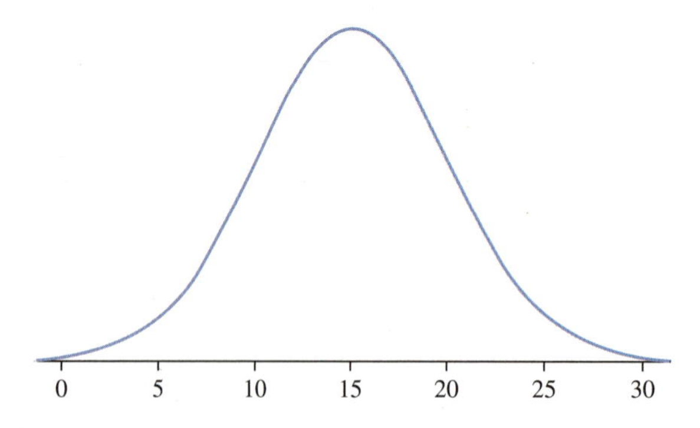

Statement 2: $P(x < 0.80)$

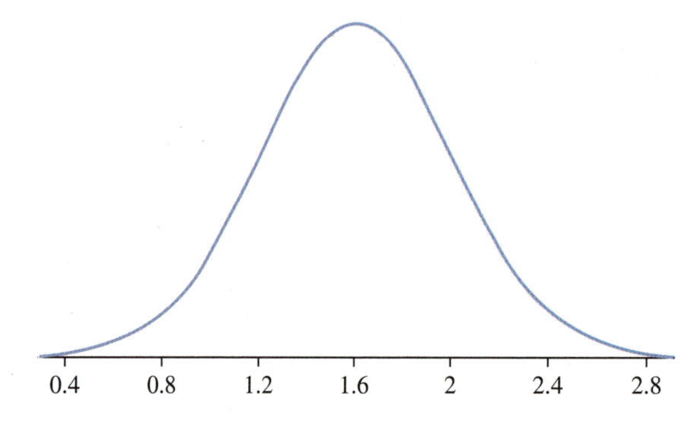

Statement 3: $P(3.2 < x < 3.3)$

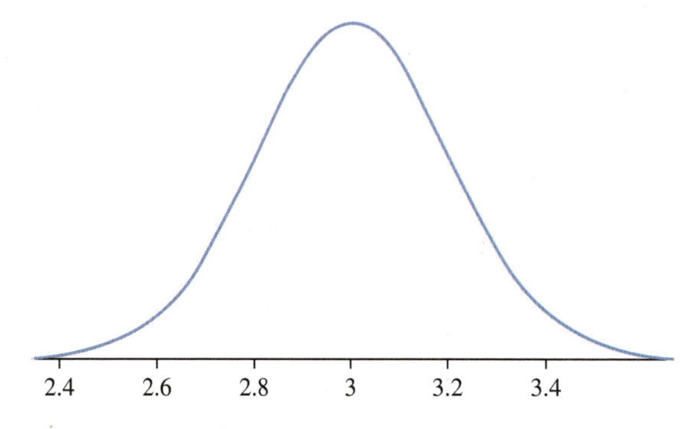

c. For each shaded area, write the probability statement in words and in symbols. Check your answers.

Shaded area 1:

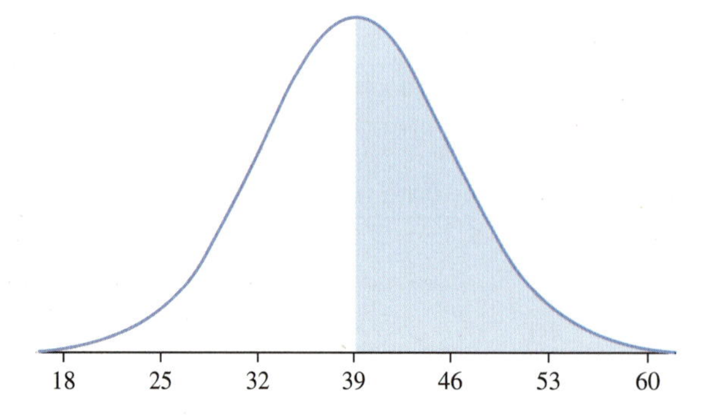

Shaded area 2:

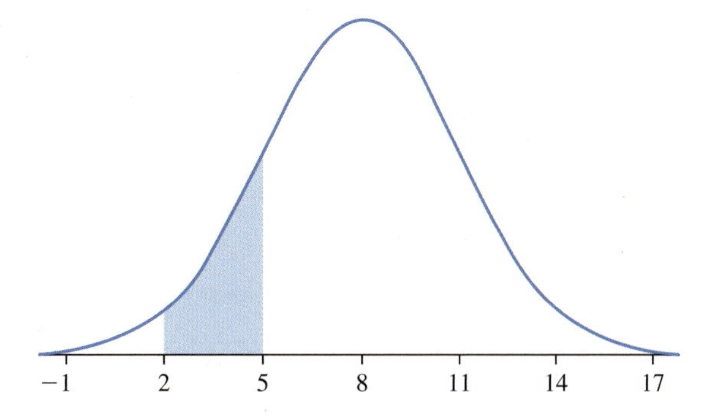

Shaded area 3:

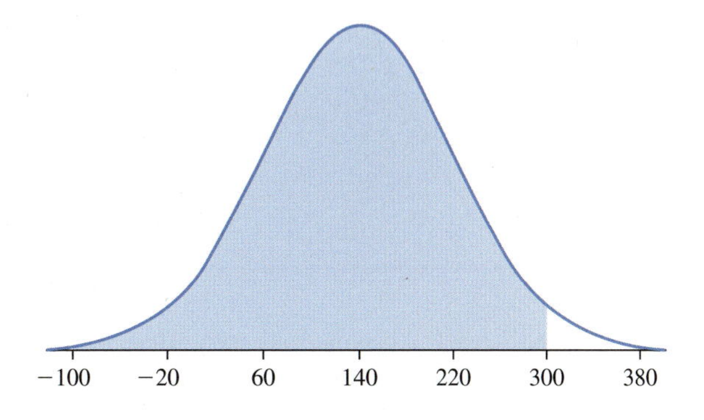

Answers

a.

Statement	Answer
The probability that x takes a value between 23 and 93.	$P(23 < x < 93)$
The probability that x takes a value that is greater than or equal to 1.05.	$P(x \geq 1.05)$
The probability that x takes a value that is at most 97.1.	$P(x \leq 97.1)$

b.

Statement 1

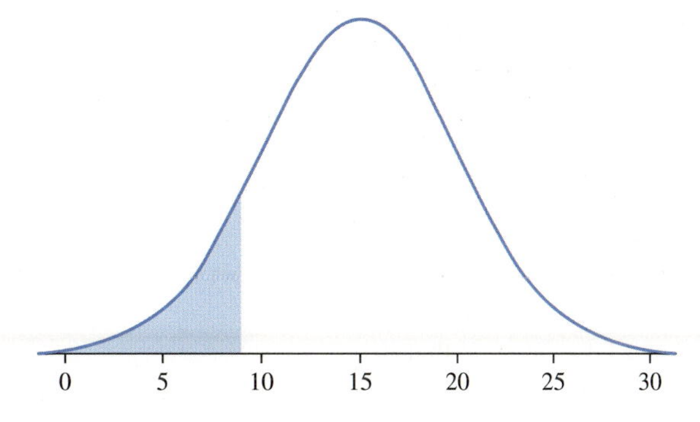

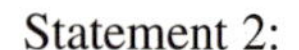

Statement 2:

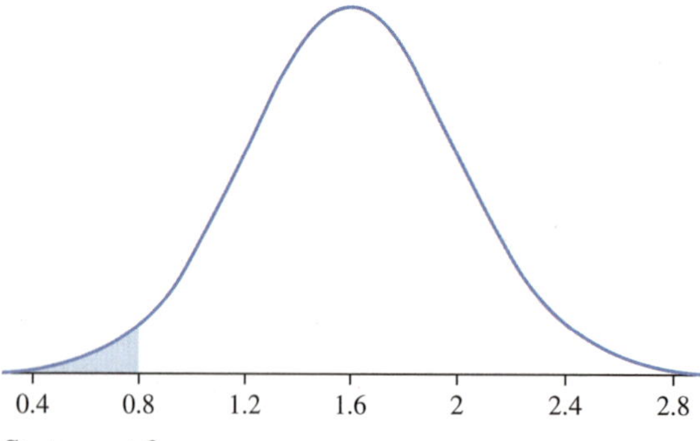

Statement 3:

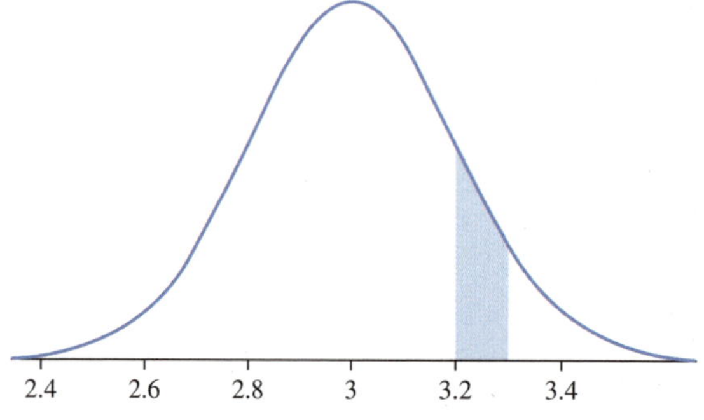

c. Shaded Area 1: Answers may vary. Here is one example: The probability that a variable x is greater than 39. $P(x > 39)$

Shaded Area 2: Answers may vary. Here is one example: The probability that a variable x is between 2 and 5. $P(2 < x < 5)$

Shaded Area 3: Answers may vary. Here is one example: The probability that a variable x is less than 300. $P(x < 300)$

SECTION 6.5 EXERCISES

Write the statements given in Exercises 6.39–6.41 using probability notation.

6.39 The probability that x takes a value between 8.2 and 8.4

6.40 The probability that x takes a value below 48

6.41 The probability that x takes a value that is at least 5.0

For each of the probability statements given in Exercises 6.42–6.44, mark and shade the areas represented by the probability statement.

6.42 Statement: The probability that x takes a value that is at least 52.

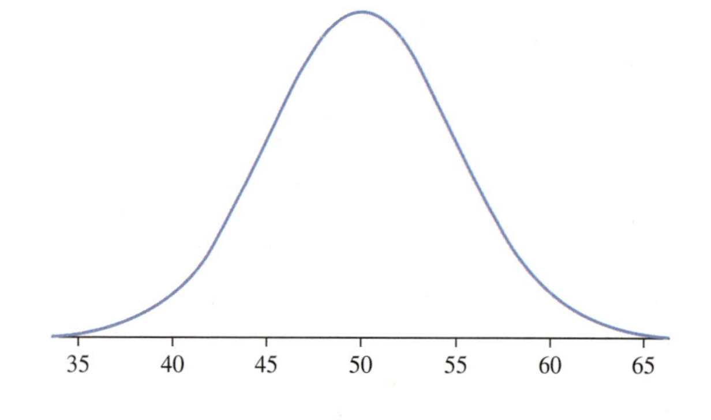

6.43 Statement: $P(x > 16)$

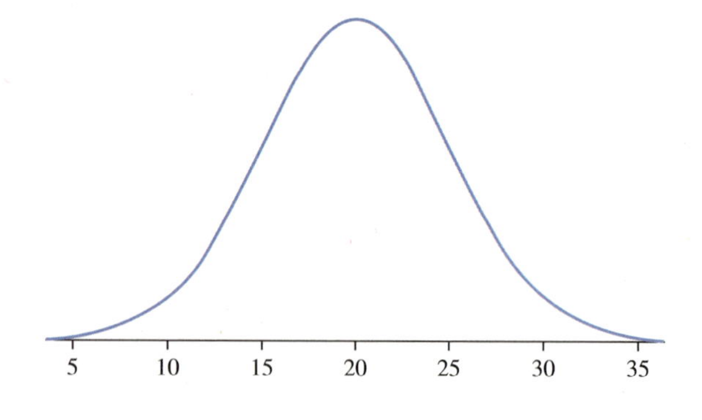

6.44 Statement: The probability that x takes a value between 100 and 900.

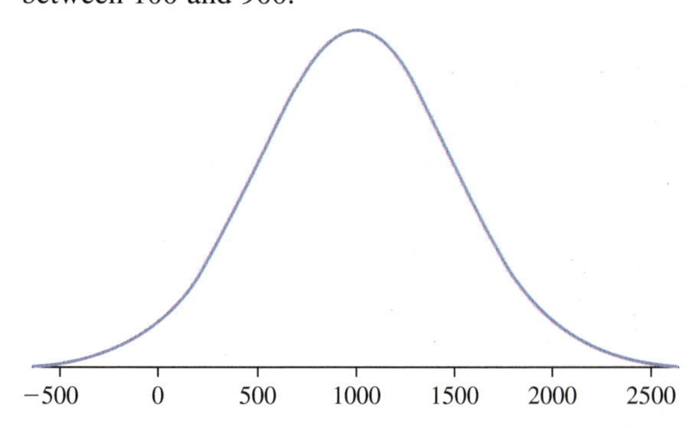

For each shaded area given in Exercises 6.45–6.47, write an appropriate probability statement in words and in symbols.

6.45 Shaded area:

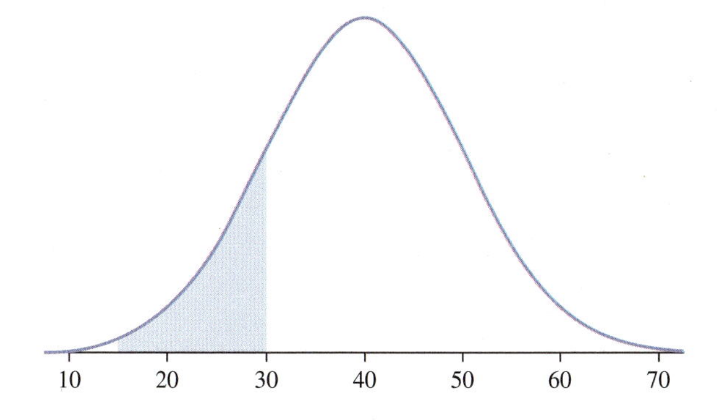

6.46 Shaded area:

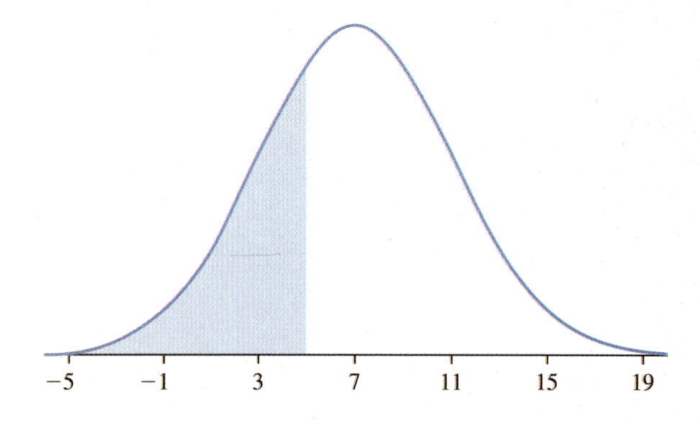

6.47 Shaded area:

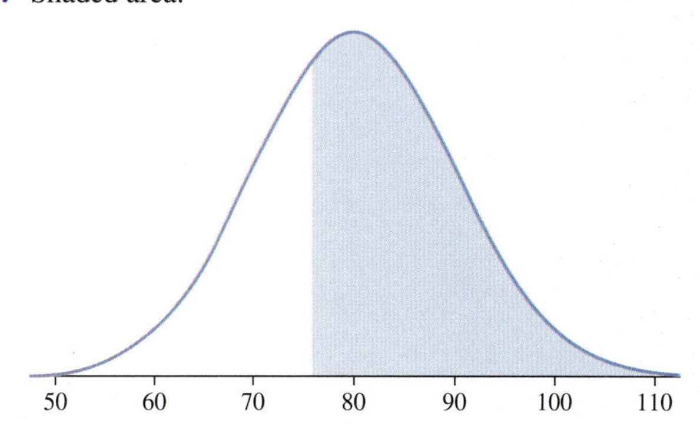

SECTION 6.6 Areas of Rectangles and Areas of Triangles

Calculating probabilities for continuous random variables requires evaluating areas under a curve. For some probability distributions, this may involve calculating the area of a rectangle or a triangle.

To calculate the area of a rectangle, you multiply the length of the rectangle by the height of the rectangle. For example, suppose that you want to find the area above the interval from 6 to 8 under the distribution shown here:

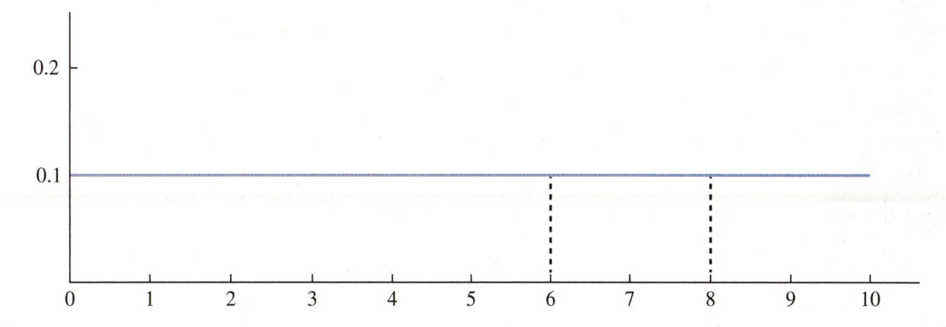

The area above the interval from 6 to 8 is the area of a rectangle, so you will need to determine length of the rectangle (which is the interval width) and the height of the rectangle. Recall that the length of an interval is calculated by subtracting the lower endpoint of the interval from the upper endpoint, so the length of the interval between 6 and 8 is $8 - 6 = 2$. You can determine the height of the rectangle by looking at the vertical axis. Because the vertical axis starts at 0, the height of the rectangle shown is 0.1. With a length of 2 and a height of 0.1, the area of this rectangle is

$$\text{area} = (\text{length})(\text{height}) = (2)(0.1) = 0.2$$

The formula for calculating the area of a triangle is area of a triangle $= \frac{1}{2}$(base)(height). For the triangle shown below, the height is 5 and the base is the length of the interval between 3 and 9: $9 - 3 = 6$. You can now calculate the area of this triangle:

$$\text{area of a triangle} = \frac{1}{2}(\text{base})(\text{height}) = \frac{1}{2}(6)(5) = 15$$

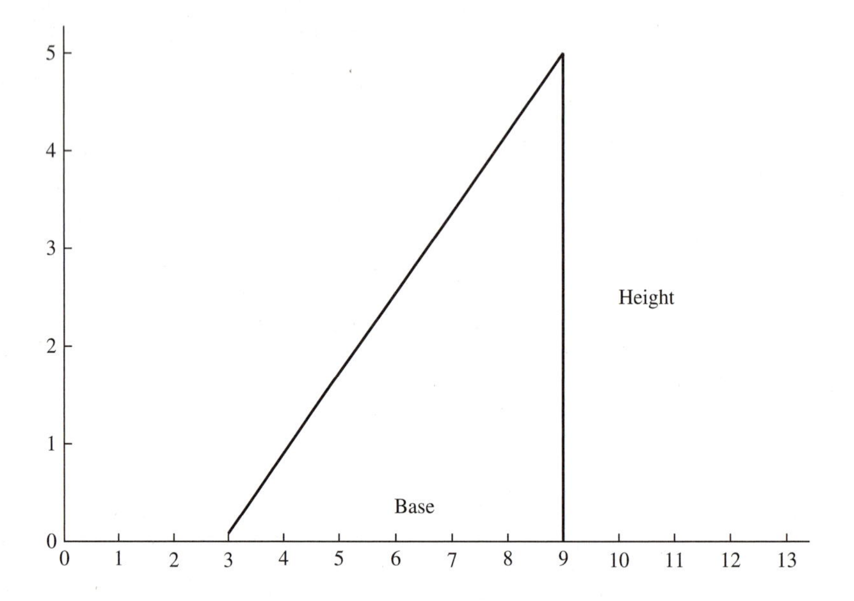

To see why the formula for the area of a triangle is $\frac{1}{2}$(base)(height), look at the figure shown below. Notice that the area of the triangle is one half of the area of the rectangle shown, and the area of the rectangle is equal to the base of the triangle times the height of the triangle.

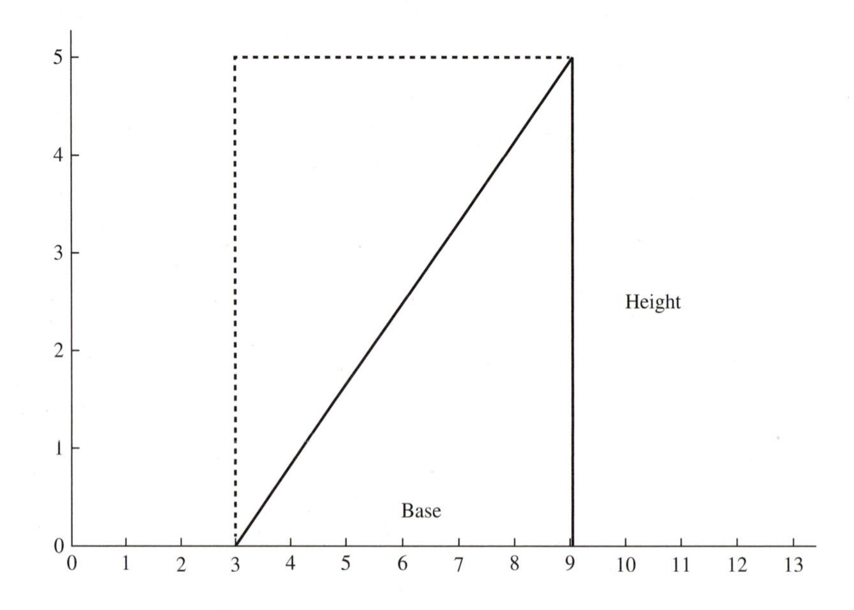

Finding Areas of Rectangles and Triangles

$$\text{area of a rectanagle} = (\text{length})(\text{height})$$

$$\text{area of a triangle} = \frac{1}{2}(\text{base})(\text{height})$$

Check It Out!

a. Find the area above the interval between 0.25 and 0.75 under the distribution shown below, which has height 1.0 over the interval from 0 to 1. Shade the area you are finding on the graph. Check your answer.

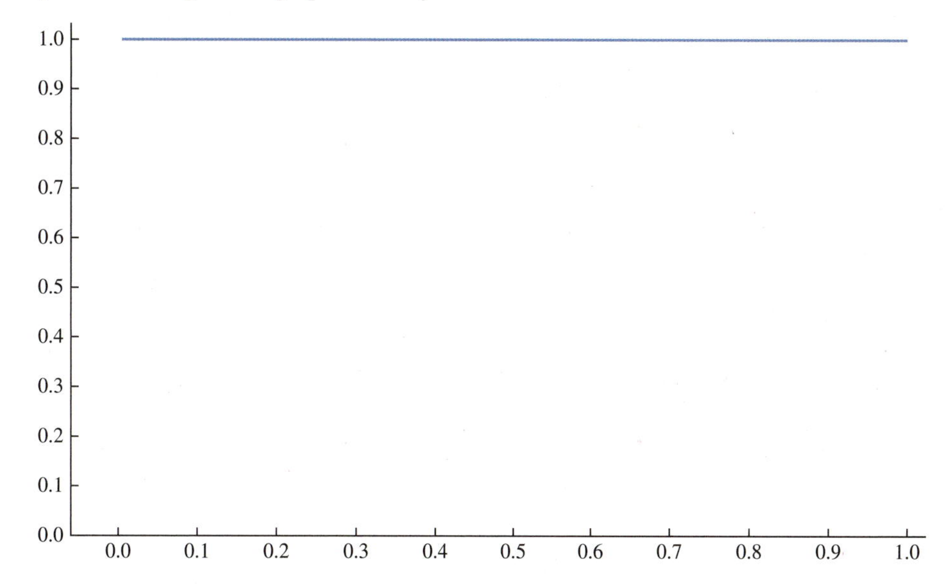

b. Find the area above the interval between 160 and 180 under the distribution below, which has height 0.01 over the interval from 100 to 200. Shade the area you are finding on the graph. Check your answer.

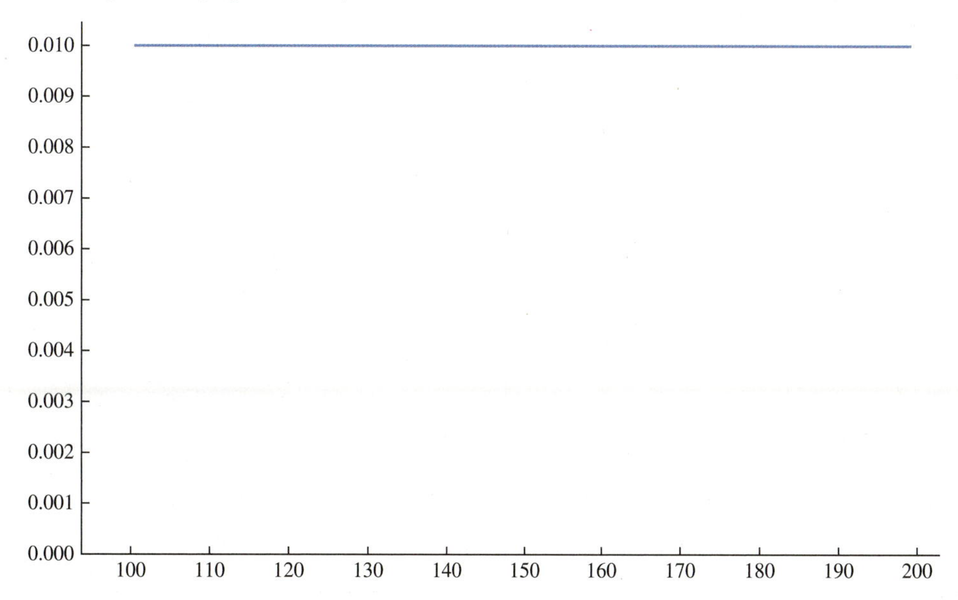

c. Find the area above the interval between 0 and 1 under the distribution below, which is triangular over the interval from 0 to 2. Shade the area you are finding on the graph. Check your answer.

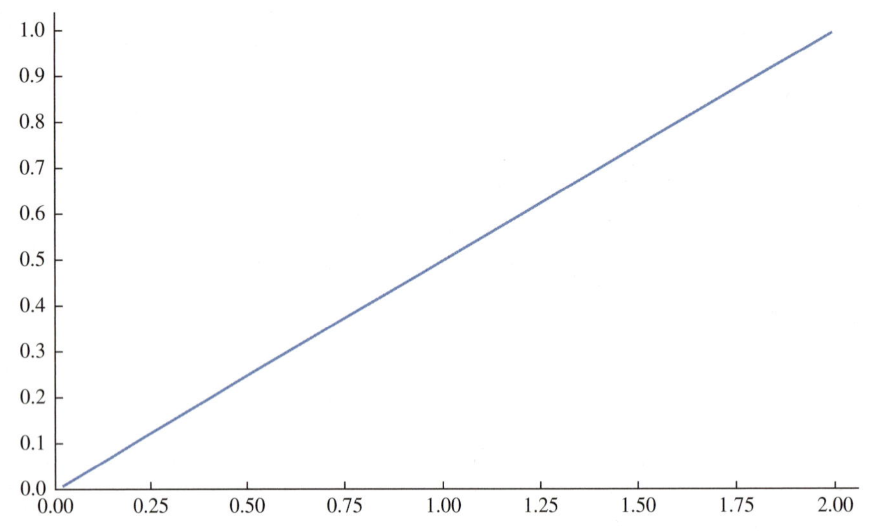

d. Find the area above the interval between 25 and 100 under the distribution below, which is triangular over the interval from 25 to 125. Shade the area you are finding on the graph. Check your answer.

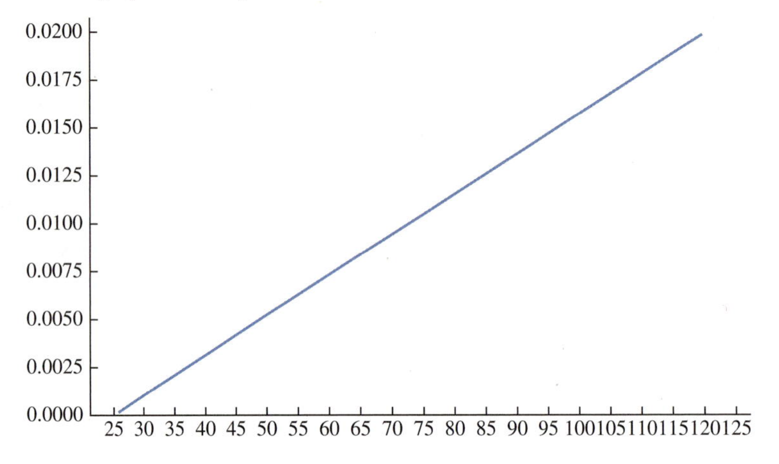

Answers

a. Area = (length)(height) = (0.5)(1) = 0.5

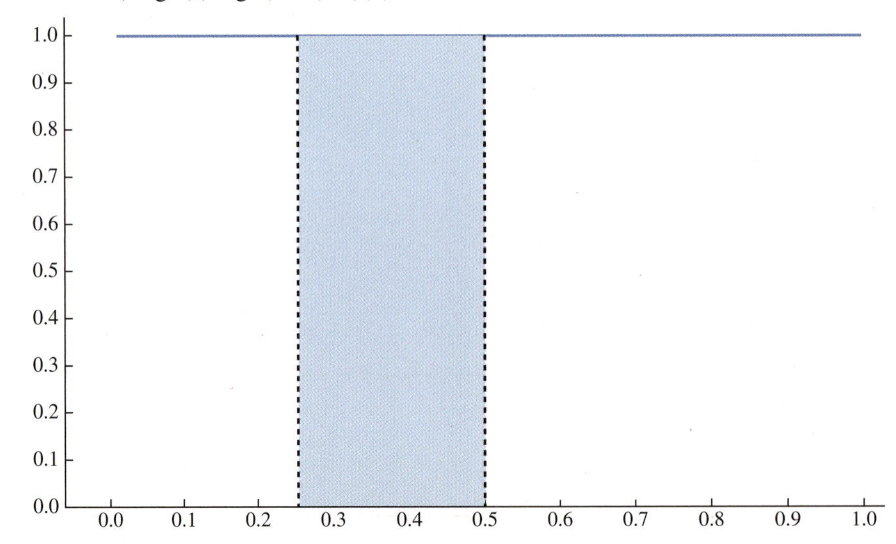

b. Area = (length)(height) = (20)(0.01) = 0.20

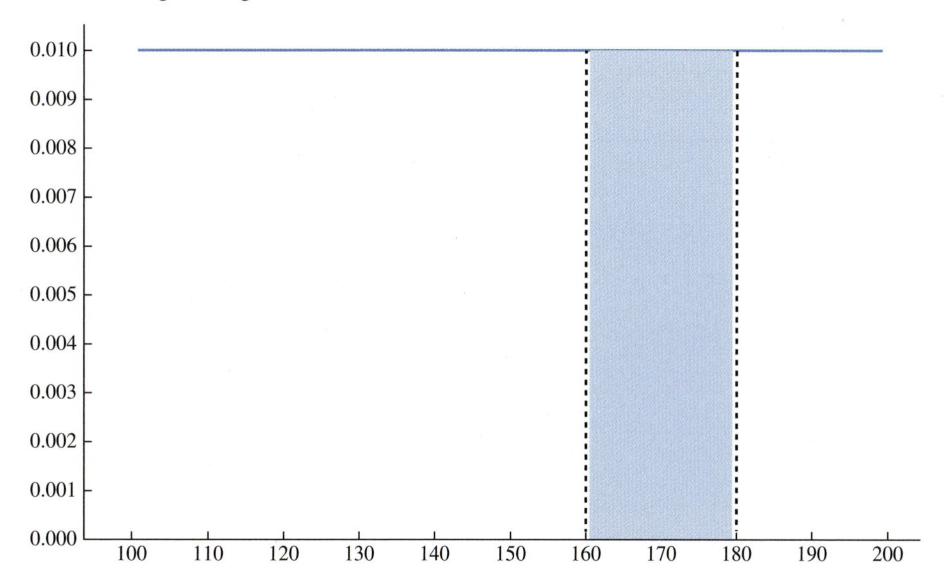

c. Area = ½(base)(height) = ½(1)(0.5) = 0.25

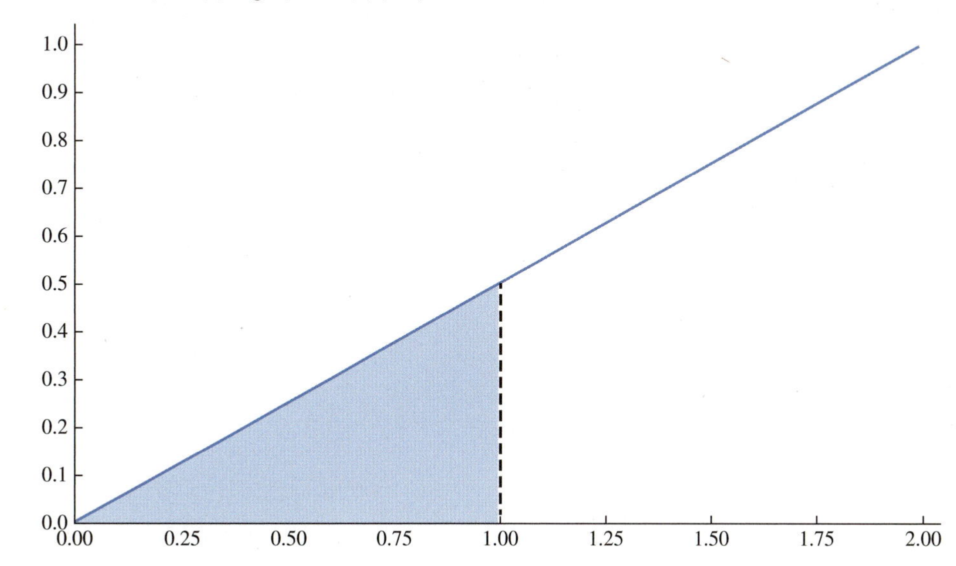

d. Area = ½(base)(height) = ½(75)(0.015) = 0.5625

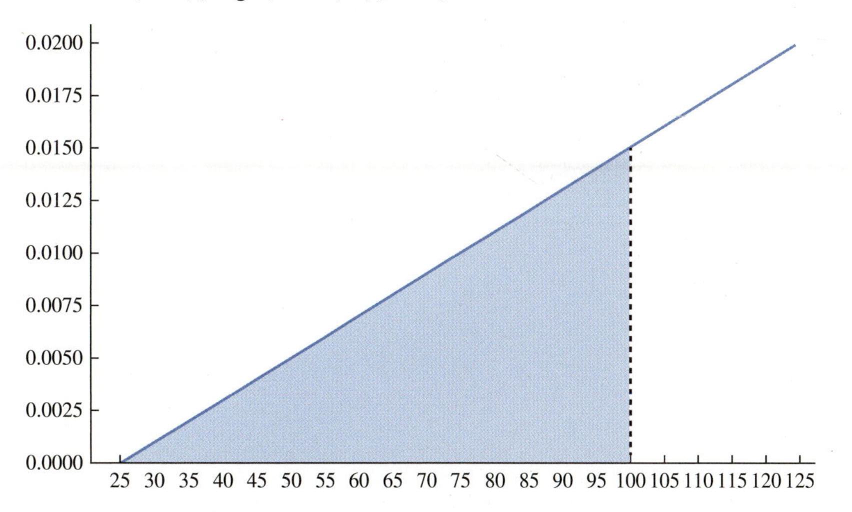

SECTION 6.6 EXERCISES

6.48 Find the area above the interval consisting of all values greater than 3 under the distribution below, which has height 0.25 over the interval from 0 to 4. Shade this area on the graph.

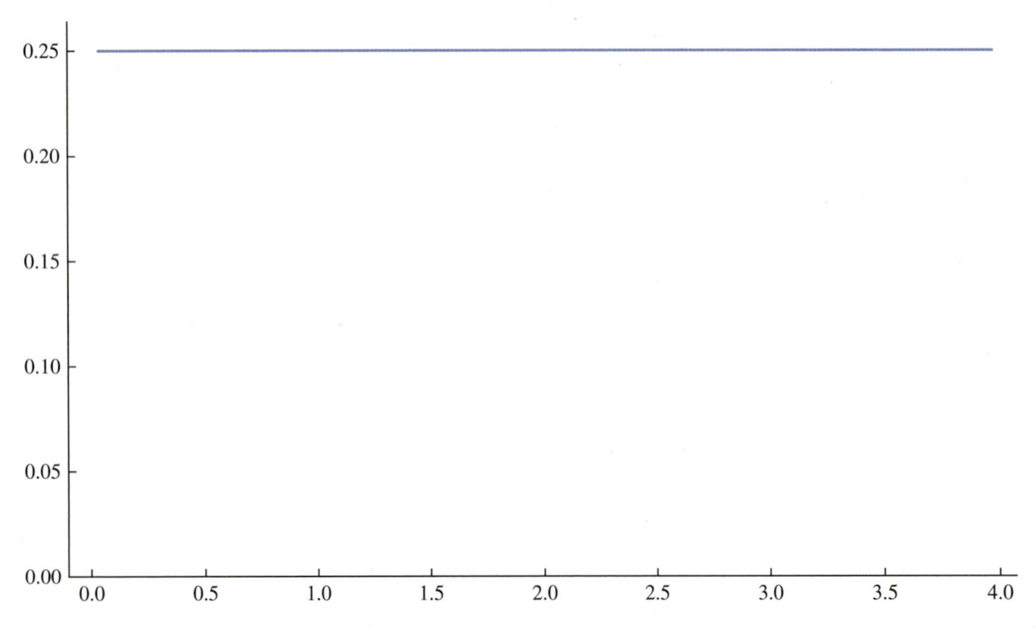

6.49 Find the area above the interval from -0.4 to 0.2 under the distribution below, which has height 0.5 over the interval from -1 to 1. Shade this area on the graph.

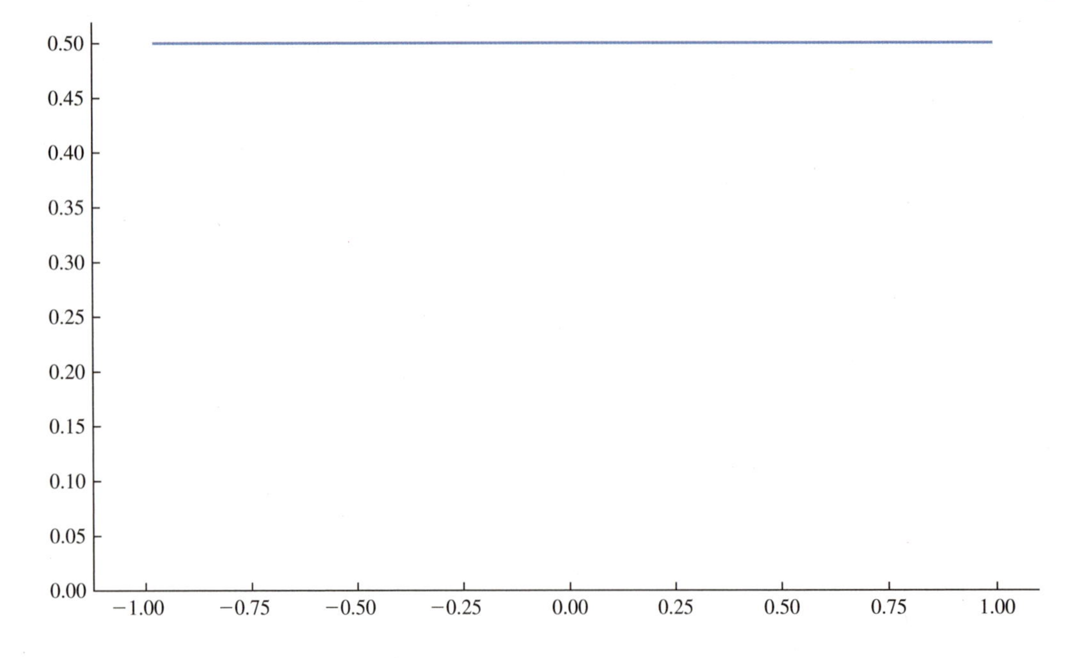

6.50 Find the area above the interval between 0 and 1 under the distribution below, which is triangular over the interval from -1 to 1. Shade this area on the graph.

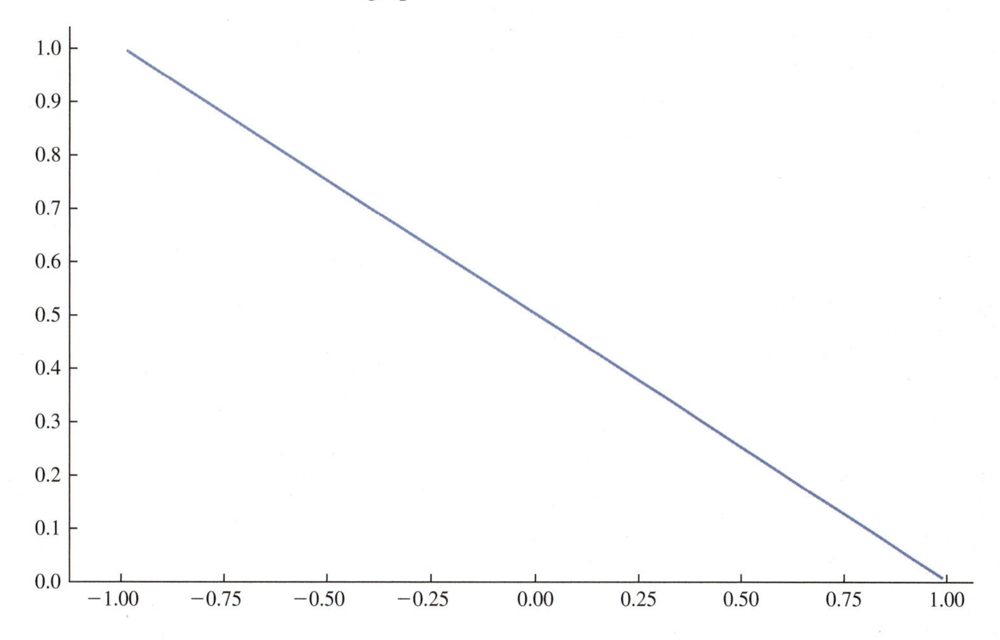

6.51 Find the area above the interval between 7.5 and 10 under the distribution below, which is triangular over the interval from 0 to 10. Shade this area on the graph. Hint: First, find the area of the triangle above the interval from 0 to 7.5, and remember that the area under the distribution for the original triangular distribution is equal to 1.

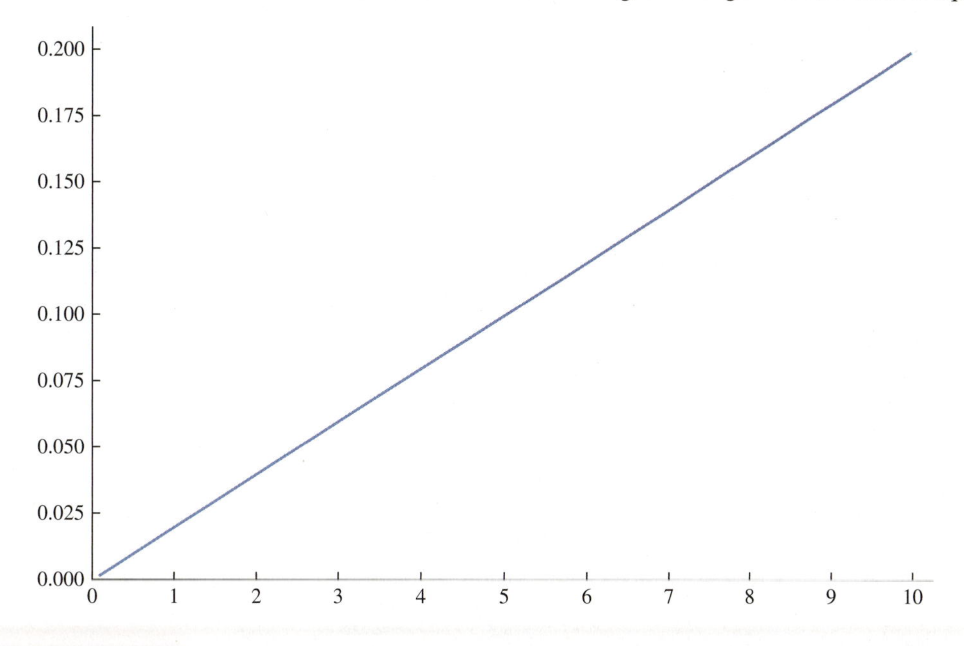

SECTION 6.7 Solving Simple Equations in One Variable

A common strategy for solving a simple linear equation in one variable it to isolate the variable in the equation by itself on one side of the equation. To accomplish this, you can do the following:

1. Add or subtract the same value to both sides of the equation.
2. Multiply or divide both sides of the equation by the same value.

Because an equation is a mathematical statement that specifies that two quantities are equal, adding or subtracting the same value to each side of the equation creates a new equation where the quantities on each side of the equal sign are still equal. Similarly, if you multiply or divide both sides of an equation by the same value, you get a new equation where the quantities of each side of the equal sign are still equal.

In your work with normal distributions, you may need to solve for x in an equation like this one:

$$\frac{x - 8}{3} = 2$$

To solve an equation like this, a good first step is to eliminate the fraction. To do this, you could multiply both sides of the equation by 3 to get

$$(3)\left(\frac{x - 8}{3}\right) = (3)(2)$$

which simplifies to

$$x - 8 = 6$$

It is now possible to isolate the variable x on one side of the equation by adding 8 to both sides of the equation. This results in

$$x - 8 + 8 = 6 + 8$$

which simplifies to

$$x = 14$$

This same strategy works even when the numbers in the equation you are trying to solve aren't integers. For example, to solve the equation

$$\frac{x - 53.6}{14.2} = 1.96$$

for x, you could carry out the steps shown in the following table.

Step	Resulting Equation
Multiply both sides of the equation by 14.2 to eliminate the fraction.	$(14.2)\left(\frac{x - 53.6}{14.2}\right) = (14.2)1.96$
	$x - 53.6 = 27.83$
Add 53.6 to both sides to isolate x.	$x - 53.6 + 53.6 = 27.83 + 53.6$
	$x = 81.43$

Check It Out!

Solve for the unknown variable. Check your answers along the way.

a. Solve for x in the equation:

$$\frac{x}{3.2} = 2$$

b. Solve for y in the equation:

$$\frac{1}{11}(y + 18) = 2.58$$

c. Solve for w in the equation:

$$\frac{w - 100}{15} = 1.65$$

Answers

a. $x = (3.2)(2) = 6.4$

b. $y = (11)(2.58) - 18 = 10.38$

c. $w = (15)(1.65) + 100 = 124.75$

SECTION 6.7 EXERCISES

6.52 Solve for x in the equation: $\frac{-x}{21} = 4.1$

6.53 Solve for u in the equation: $\frac{u - 98.6}{2.3} = 0$

6.54 Solve for v in the equation: $\frac{v - 67.5}{4.6} = -1.0$

SECTION 6.8 Working with Factorials (Optional, for those covering the Binomial Distribution)

If your statistics class covers the Binomial Probability Distribution, you will need to be able to evaluate expressions that include **factorials**. The factorial of a positive integer is the product of that integer and all the positive integers that are smaller. The factorial symbol is !. For example, "3 factorial" is denoted by 3! and is equal to $(3)(2)(1) = 6$. Similarly, 5! is $(5)(4)(3)(2)(1) = 120$.

The only other thing you need to know about factorials is that 0! is defined to be equal to 1.

Factorials

If x is a positive integer, x factorial is denoted by $x!$ and is defined to be the product of x and all the positive integers that are less than x:

$$x! = (x)(x - 1)(x - 2)\cdots(1)$$

In addition,

$$0! = 1$$

Many scientific and graphing calculators will calculate factorials. Look for a key labeled with the factorial symbol, !.

Check It Out!

a. Calculate the following factorials, checking your answers as you go.

Factorial	Result
2!	
6!	

b. Use the factorial function on a calculator to calculate the following factorials, and check your answers.

Factorial	Result
9!	
16!	

Answers

a.

Factorial	Answer
2!	$(2)(1) = 2$
6!	$(6)(5)(4)(3)(2)(1) = 720$

b.

Factorial	Answer
9!	362,880
16!	2.09×10^{13}

Factorials may also appear in arithmetic and algebraic expressions. To evaluate these expressions, you would first calculate the value of the factorial and then substitute that number into the expression. For example, to evaluate the expression

$$7!(10 - 7)!$$

you would first calculate 7! (which is $(7)(6)(5)(4)(3)(2)(1) = 5040$) and $(10 - 7)! = 3!$ (which is $(3)(2)(1) = 6$). Then

$$7!(10 - 7)! = (5040)(6) = 30{,}240$$

To evaluate the expression

$$\frac{5!}{2!3!}$$

you would first calculate the value of each factorial and then use those values to evaluate the expression:

$$\frac{5!}{2!3!} = \frac{120}{(2)(6)} = \frac{120}{12} = 10$$

Check It Out!

Evaluate the expressions, and check your answers.

Expression	Evaluate
$\frac{6!}{3!}$	
$\frac{4!}{2!2!}$	
$\frac{12!}{1!11!}$	

Answers

Expression	Answer
$\frac{6!}{3!}$	$\frac{6!}{3!} = \frac{720}{6} = 120$
$\frac{4!}{2!2!}$	$\frac{4!}{2!2!} = \frac{24}{2 \cdot 2} = 6$
$\frac{12!}{1!11!}$	$\frac{12!}{1!11!} = \frac{12 \cdot 11!}{11!} = 12$

SECTION 6.8 EXERCISES

Calculate the factorials given in Exercises 6.55 and 6.56.

6.55 5!

6.56 7!

Use the factorial function on a calculator to calculate the factorials given in Exercises 6.57 and 6.58.

6.57 10!

6.58 15!

For Exercises 6.59–6.61, evaluate the given expression.

6.59 $\frac{7!}{5!}$

6.60 $\frac{7!}{5!2!}$

6.61 $\frac{10!}{0!10!}$

SECTION 6.9 Evaluating Expressions

Some of the expressions that you will work with in your study of probability distributions involve summation notation, so let's start with a quick review of summation notation. Recall that summation notation is a shorthand notation to represent addition. The summation symbol, $\sum$, is used to represent addition, and when you see it in an expression or a formula it is just telling you to add up the values of whatever follows this symbol. For example, if you see $\sum x$, this means you should add up all the values of the variable x in a data set.

Mean of a Discrete Random Variable

$$\mu_x = \sum xp(x)$$

To calculate the mean of a discrete random variable x, you need to know the possible values that the random variable x can take and the probability that x takes each of these possible values. This is the information provided by the probability distribution of x.

μ_x is the symbol that is used to represent the mean value of the discrete random variable x.

To calculate the value of μ_x, you will need to evaluate the expression $\sum xp(x)$. The summation symbol tells you that you should add up all the values of $xp(x)$, where x is a possible value of the random variable and $p(x)$ is the probability that the variable takes on that value. You would first multiply each possible value by the corresponding probability and then add up those values.

For example, suppose that the random variable x is the number of items purchased by a customer using the express lane checkout at a grocery store. The number of items is limited to 8 for the express checkout, so the possible values of x are 1, 2, 3, 4, 5, 6, 7, and 8. Suppose that the probabilities of observing these values are the ones given in the following table.

x	1	2	3	4	5	6	7	8
p(x)	0.05	0.05	0.15	0.10	0.10	0.15	0.20	0.20

To calculate the mean value of x, μ_x, you first multiply each x value by the corresponding probability:

x	p(x)	xp(x)
1	0.05	1(0.05) = 0.05
2	0.05	2(0.05) = 0.10
3	0.15	3(0.15) = 0.45
4	0.10	4(0.10) = 0.40
5	0.10	5(0.10) = 0.50
6	0.15	6(0.15) = 0.90
7	0.20	7(0.20) = 1.40
8	0.20	8(0.20) = 1.60

Then you add up all the $xp(x)$ values to obtain the value of $\mu_x = \sum xp(x)$:

$$\mu_x = \sum xp(x) = 0.05 + 0.10 + 0.45 + 0.40 + 0.50 + 0.90 + 1.40 + 1.60 = 5.40$$

Check It Out!

a. Calculate the mean of the discrete random variable x, which has the probability distribution given in the accompanying table. Check your calculations.

x	p(x)
10	0.2
20	0.3
30	0.3
40	0.2

b. A male high school athlete decides to practice until he successfully high jumps three times over a bar set at 5 feet. Suppose the discrete random variable y represents the number of tries it will take until the athlete clears 5 feet three times, and assume that y has the probability distribution given in the accompanying table. Use the probability distribution to calculate the mean number of tries needed to clear the bar three times. Check your answer.

y	$p(y)$
3	0.23
4	0.26
5	0.21
6	0.15
7	0.09
8	0.06

Answers

a. $\mu_x = \sum xp(x) = 2.0 + 6.0 + 9.0 + 8.0 = 25.0$

b. $\mu_x = \sum xp(x) = 0.69 + 1.04 + 1.05 + 0.9 + 0.63 + 0.48 = 4.79$

Variance and Standard Deviation of a Discrete Random Variable

$$\sigma_x^2 = \sum (x - \mu_x)^2 p(x)$$
$$\sigma_x = \sqrt{\sigma_x^2}$$

To calculate the variance of a discrete random variable x, you need to know the possible values that the random variable x can take and the probability that x takes each of these possible values. This is the information provided by the probability distribution of x. You also need to know the value of the mean of the random variable, μ_x.

The symbol σ_x^2 is used to represent the variance of the discrete random variable x. To calculate the value of σ_x^2, you will need to evaluate the expression $\sum(x - \mu_x)^2 p(x)$. The summation symbol tells you that you should add up all the values of $(x - \mu_x)^2 p(x)$, where μ_x is the mean of the random variable, x is a possible value of the random variable, and $p(x)$ is the probability that the variable takes on that value.

The easiest way to do this is to follow these steps:

1. If the mean μ_x has not already been calculated, you should calculate the value of μ_x.
2. Subtract the value of μ_x from each of the possible x values to get the values of $(x - \mu_x)$.
3. Square each of the $(x - \mu_x)$ values to get the $(x - \mu_x)^2$ values.
4. Multiply each $(x - \mu_x)^2$ value by the corresponding probability to get the values of $(x - \mu_x)^2 p(x)$.
5. Add up all the $(x - \mu_x)^2 p(x)$ values. This number is the value of $\sigma_x^2 = \sum(x - \mu_x)^2 p(x)$, the variance of the random variable x.

A systematic way to organize these calculations is to use a table with columns labeled as follows:

x	$p(x)$	$(x - \mu_x)$	$(x - \mu_x)^2$	$(x - \mu_x)^2 p(x)$

For example, suppose that the random variable x is the number of items purchased by a customer using the express lane checkout at a grocery store. The number of items is limited to 8 for the express checkout, so the possible values of x are 1, 2, 3, 4, 5, 6, 7, and 8. Suppose that the probabilities of observing these values are the ones given in the following table.

x	1	2	3	4	5	6	7	8
$p(x)$	0.05	0.05	0.15	0.10	0.10	0.15	0.20	0.20

The mean value of x, μ_x, was previously calculated to be 5.4. Start by setting up the following table:

x	$p(x)$	$(x - \mu_x)$	$(x - \mu_x)^2$	$(x - \mu_x)^2p(x)$
1	0.05			
2	0.05			
3	0.15			
4	0.10			
5	0.10			
6	0.15			
7	0.20			
8	0.20			

Fill in the third column of the table by subtracting $\mu_x = 5.4$ from each x value:

x	$p(x)$	$(x - \mu_x)$	$(x - \mu_x)^2$	$(x - \mu_x)^2p(x)$
1	0.05	$1 - 5.4 = -4.4$		
2	0.05	$2 - 5.4 = -3.4$		
3	0.15	−2.4		
4	0.10	−1.4		
5	0.10	−0.4		
6	0.15	0.6		
7	0.20	1.6		
8	0.20	2.6		

The $(x - \mu_x)$ values are now in the third column, so you can complete the fourth column by squaring each of these values to get the $(x - \mu_x)^2$ values:

x	$p(x)$	$(x - \mu_x)$	$(x - \mu_x)^2$	$(x - \mu_x)^2p(x)$
1	0.05	$1 - 5.4 = -4.4$	$(-4.4)^2 = 19.36$	
2	0.05	$2 - 5.4 = -3.4$	$(-3.4)^2 = 11.56$	
3	0.15	−2.4	5.76	
4	0.10	−1.4	1.96	
5	0.10	−0.4	0.16	
6	0.15	0.6	0.36	
7	0.20	1.6	2.56	
8	0.20	2.6	6.76	

To complete the last column, you multiply each $(x - \mu_x)^2$ value by the corresponding probability to get the $(x - \mu_x)^2p(x)$ values:

x	$p(x)$	$(x - \mu_x)$	$(x - \mu_x)^2$	$(x - \mu_x)^2p(x)$
1	0.05	$1 - 5.4 = -4.4$	$(-4.4)^2 = 19.36$	$(19.36)(0.05) = 0.968$
2	0.05	$2 - 5.4 = -3.4$	$(-3.4)^2 = 11.56$	$(11.56)(0.05) = 0.578$
3	0.15	−2.4	5.76	$(5.76)(0.15) = 0.864$
4	0.10	−1.4	1.96	0.196
5	0.10	−0.4	0.16	0.016
6	0.15	0.6	0.36	0.054
7	0.20	1.6	2.56	0.512
8	0.20	2.6	6.76	1.352

Finally, you can now calculate the value of $\sum(x - \mu_x)^2 p(x)$ by adding the numbers in the last column of the table:

$$\sigma_x^2 = \sum(x - \mu_x)^2 p(x) = 0.968 + 0.578 + 0.864 + 0.196 + 0.016 + 0.054 + 0.512 + 1.352 = 4.540$$

The standard deviation of the random variable x is just the square root of the variance, so

$$\sigma_x = \sqrt{\sigma_x^2} = \sqrt{4.540} = 2.131 \text{ (rounded to three decimal places)}$$

Check It Out!

a. Calculate the standard deviation of the discrete random variable x, which has the probability distribution given in the accompanying table. Check your calculations.

x	p(x)
10	0.2
20	0.3
30	0.3
40	0.2

b. A male high school athlete decides to practice until he successfully high jumps three times over a bar set at 5 feet. Suppose the discrete random variable y represents the number of tries it will take until the athlete clears 5 feet three times, and assume that y has the probability distribution given in the accompanying table. Use the probability distribution to calculate the standard deviation of the number of tries needed to clear the bar three times. Check your answer.

y	p(y)
3	0.23
4	0.26
5	0.21
6	0.15
7	0.09
8	0.06

Answers

a. $\mu_x = \sum xp(x) = 10(0.2) + 20(0.3) + 30(0.3) + 40(0.2) = 25$

x	p(x)	$(x - \mu_x)$	$(x - \mu_x)^2$	$(x - \mu_x)^2 p(x)$
10	0.2	−15	225	45
20	0.3	−5	25	7.5
30	0.3	5	25	7.5
40	0.2	15	225	45

$\sigma_x^2 = \sum(x - \mu_x)^2 p(x) = 45 + 7.5 + 7.5 + 45 = 105$

$\sigma_x = \sqrt{\sigma_x^2} = \sqrt{105.0} = 10.25$

b. $\mu_y = \sum yp(y) = 3(0.23) + 4(0.26) + 5(0.21) + 6(0.15) + 7(0.09) + 8(0.06) = 4.79$

y	$p(y)$	$(y - \mu_y)$	$(y - \mu_y)^2$	$(y - \mu_y)^2 p(y)$
3	0.23	−1.79	3.204	0.737
4	0.26	−0.79	0.624	0.162
5	0.21	0.21	0.844	0.009
6	0.15	1.21	1.464	0.220
7	0.09	2.21	4.884	0.440
8	0.06	3.21	10.304	0.618

$$\sigma_y^2 = \sum (y - \mu_y)^2 p(y) = 0.737 + 0.162 + 0.009 + 0.220 + 0.440 + 0.618 = 2.186$$

$$\sigma_y = \sqrt{\sigma_y^2} = \sqrt{2.186} = 1.48$$

Binomial Probabilities

If your statistics course covers the Binomial Probability Distribution, you may need to calculate binomial probabilities. The formula for a binomial probability is

$$p(x) = \frac{n!}{x!(n - x)!} p^x (1 - p)^{n-x}$$

In this formula, n is the number of trials in a binomial experiment and p is the probability of a success on each trial.

If you are not using technology to calculate binomial probabilities, once the values of n and p are known, you can use this formula to calculate them.

Start by substituting the values for n and p into the formula. For example, if $n = 10$ and $p = 0.3$, the formula for calculating binomial probabilities becomes

$$p(x) = \frac{10!}{x!(10 - x)!}(0.3)^x(1 - 0.3)^{10-x} = \frac{10!}{x!(10 - x)!}(0.3)^x(0.7)^{10-x}$$

Notice that there are three terms in the formula for a binomial probability. These three values are multiplied together to obtain the value for $p(x)$. Once a value for x is specified, you can substitute that value for the x's in the formula. Once you have done that, calculate the value of each of the three terms and then multiply them together.

For example, to calculate the value of $p(4)$, substitute 4 into the formula for x to get

$$p(4) = \frac{10!}{4!(10 - 4)!}(0.3)^4(0.7)^{10-4}$$

Next, calculate the value of each term:

$$\frac{10!}{4!(10 - 4)!} = \frac{10!}{4!6!} = \frac{10 \cdot 9 \cdot 8 \cdot 7 \cdot 6 \cdot 5 \cdot 4 \cdot 3 \cdot 2 \cdot 1}{(4 \cdot 3 \cdot 2 \cdot 1)(6 \cdot 5 \cdot 4 \cdot 3 \cdot 2 \cdot 1)} = \frac{3{,}628{,}800}{(24)(720)} = \frac{3{,}628{,}800}{17{,}280} = 210$$

(For a review of factorial notation, see Section 6.8.)

$$(0.3)^4 = (0.3)(0.3)(0.3)(0.3) = 0.0081$$

$$(0.7)^{10-4} = (0.7)^6 = (0.7)(0.7)(0.7)(0.7)(0.7)(0.7) = 0.1177$$

Now you can calculate $p(4)$ by multiplying:

$$p(4) = \frac{10!}{4!(10 - 4)!}(0.3)^4(0.7)^{10-4} = (210)(0.0081)(0.1177) = 0.2002$$

Check It Out!

a. Calculate the probability $p(5)$ for a binomial experiment when $n = 7$ and $p = 0.6$. Check your answer.

b. Calculate the probability $p(x < 2)$ for a binomial experiment when $n = 5$ and $p = 0.5$. Check your answer.

c. Calculate the probability $p(9)$ for a binomial experiment when $n = 10$ and $p = 0.8$. Check your answer.

Answers

a. $p(5) = \frac{7!}{5!(7-2)!}(0.6)^5(1-0.6)^{7-5} = 0.261$

b. $P(x < 2) = p(0) + p(1) = \frac{5!}{0!(5-0)!}(0.5)^0(1-0.5)^{5-0} + \frac{5!}{1!(5-1)!}(0.5)^1(1-0.5)^{5-1}$

$= 0.031 + 0.156 = 0.187$

c. $p(9) = \frac{10!}{9!(10-9)!}(0.8)^9(1-0.8)^{10-9} = 0.268$

SECTION 6.9 EXERCISES

6.62 Use the probability distribution in the table below to calculate the mean and the standard deviation for the discrete random variable, x.

x	$p(x)$
1	0.10
2	0.15
3	0.30
4	0.35
5	0.10

6.63 The random variable y represents the number of paying customers per hour on an online shopping site. The probability distribution for y appears in the table below. Calculate the mean and the standard deviation for y.

y	$p(y)$
0	0.22
1	0.33
2	0.25
3	0.13
4	0.04
5	0.03

6.64 Calculate the probability $p(2)$ for a binomial experiment when $n = 12$ and $p = 0.2$.

6.65 Calculate the probability $p(3)$ for a binomial experiment when $n = 9$ and $p = 0.33$.

7 How to Read a Statistics Problem

SECTION 7.1 Chapter Overview

Now that you have completed Chapters 1–6, you have mastered most of the math that you need to be successful in your statistics course. The math topics covered in those chapters are the ones that are needed for the methods that you will see in the second half of the course, which focuses on drawing conclusions based on data.

With the math foundation now in place, it is time to turn your attention to another important component of solving the types of problems that you will encounter in this part of the course: reading and understanding a statistics problem. At first, this can be challenging because a big part of learning from data is understanding the context of the data. But reading a statistics problem is something that gets easier with practice, especially if you have a systematic strategy.

Section 7.2 provides specific advice for how to approach reading a statistics problem. Section 7.3 will consider several examples of the types of problems that you will see in your statistics class and will guide you through "reading" them using the strategy recommended in Section 7.2. Finally, in Section 7.4, you will have a chance to practice this strategy on several new problems.

SECTION 7.2 A Strategy for Reading a Statistics Problem

Reading a statistics problem is different from reading a newspaper, a magazine, or a novel. It is important to read carefully and to read every word. A common mistake that students make is to just skim the problem looking for numbers without understanding the question that is being asked or the context of the problem.

In statistics, context is critical. Because the goal is to learn from data, you need to understand what data were collected and what the person who collected the data was hoping to learn from the data. Moreover, understanding the context of the problem is the key to being able to communicate the results of any analysis that you do. You will not be able to fully answer the questions posed if you don't understand the context.

One method that has been used successfully by many is to read the problem multiple times, each time focusing on a different aspect of what you need to know to solve the problem. Consider using the following strategy, which involves reading the problem three times.

First Read: Context

First, read for context. Read the entire problem (don't just skim!) without focusing on the numbers. What you are trying to do on this first read is to understand the context and identify any terms (statistical or context-related) that you aren't familiar with.

1. **What is the general context?**
2. **Are there any terms that you don't understand?**

Second Read: Tasks

Second, read to identify what tasks you are being asked to complete. Read the entire question again. As with the first read, don't worry about the numbers yet—just read with a focus on what you are being asked to do or the questions that you are being asked to answer.

1. **What question is being asked? What am I being asked to do?**

Third Read: Details

Third, read to extract the specific information relevant to solving the problem. Now that you understand the context and have identified the tasks that you need to complete, read one more time to identify information that will allow you to solve the problem. This is the read where it is OK to focus on the numbers and other relevant information. Think about the following:

1. **What data were collected? Were the data collected in a reasonable way?**
2. **Are the actual data provided? What is the sample size (or sample sizes if there are more than one sample)?**
3. **Are any summary statistics or additional information given? If so, which statistics and what are the values of these statistics?**

How this strategy can be used to "read" statistics problems is illustrated in the examples of the next section.

SECTION 7.3 Guided Practice Reading Statistics Problems

In this section, you will consider three examples of the types of problems you will encounter as you learn the methods of statistical inference.

Example 1

Consider this problem from a statistics textbook:

> The article "Most Dog Owners Take More Pictures of Their Pet than Their Spouse" (news.fastcompany.com/most-dog-owners-take-more-pictures-of-their-pet-than-their-spouse-4017458, August 22, 2016) indicates that in a sample of 1000 dog owners, 650 said that they take more pictures of their dog than of their significant others or friends. Suppose that it is reasonable to consider this sample as representative of the population of dog owners. Construct and interpret a 90% confidence interval for the proportion of dog owners who take more pictures of their dogs than of their significant others or friends.

First Read: Context

1. **What is the general context?**
   This example describes a survey of dog owners to find out about how often they take pictures of their dog.

2. **Are there any terms that you don't understand?**
   At this point, you probably don't know what a "confidence interval" is because you haven't covered that yet in your statistics course. But if you were further along and this problem was part of your homework assignment and you weren't sure, you should go back to the appropriate section in your statistics text to review the meaning of this term. You might also wonder what "significant other" means. If you aren't familiar with this term, you could look it up in an online dictionary like dictionary.com to find that "significant other" is a spouse, a lover, or a partner.

Now that you understand the context of the question, you are ready for the second read.

Second Read: Tasks

1. **What question is being asked? What am I being asked to do?**
   In this example, you are being asked to construct and interpret a confidence interval estimate.

Now that you understand the context and know what you are being asked to do, you are ready to read the question a third time to extract the specific information that is relevant to completing the work on the problem.

Third Read: Details

1. **What data were collected? Were the data collected in a reasonable way?**
   In this example, the data are from a sample selected from the population of dog owners. The problem states that it is reasonable to consider this sample as representative of the population of dog owners.
2. **Are the actual data provided? What is the sample size (or sample sizes if there are more than one sample)?**
   The actual data (the 1000 individual survey responses) are not given. The sample size is $n = 1000$.
3. **Are any summary statistics or additional information given? If so, which statistics and what are the values of these statistics?**
   In addition to the sample size, the problem states that 650 of the 1000 people surveyed said that they take more pictures of their dogs than their significant others or friends. Using this information, it is possible to calculate the sample proportion $\hat{p} = \frac{650}{1000} = 0.650$.

After this third read, you understand the context, have identified the tasks that need to be completed, and have extracted the specific information that is needed to solve the problem. You are now ready to do the work and communicate the results.

Example 2

Consider this problem from a statistics textbook:

> Students in a representative sample of 65 first-year students selected from a large university in England participated in a study of academic procrastination ("Study Goals and Procrastination Tendencies at Different Stages of the Undergraduate Degree," *Studies in Higher Education* [2016]: 2028–2043). Each student in the sample completed the Tuckman Procrastination Scale (TPS), which measures procrastination tendencies. Scores on this scale can range from 16 to 64, with scores over 40 indicating higher levels of procrastination. For the 65 first-year students in this study, the sample mean TPS score on the procrastination scale was 37.02 and the sample standard deviation was 6.44.

a. Construct a 95% confidence interval for μ, the population mean TPS score for first-year students at this university.

b. Based on your interval, is 40 a plausible value for the population mean TPS score? What does this imply about the population of first-year students?

First Read: Context

1. **What is the general context?**
   This example describes a study on academic procrastination. Study participants were first-year students at a university in England.
2. **Are there any terms that you don't understand?**
   At this point, you probably don't know what a "confidence interval" is because you haven't covered that yet in your statistics course. But if you were further along and this problem was part of your homework assignment and you weren't sure, you should go back to the appropriate section in your statistics text to review the meaning of this term. The symbol μ is used, and if you don't recall what this represents, you should refer to your statistics text, where you would find that μ represents the population mean.

   You might also wonder what "procrastination" means. If you aren't familiar with this term, you could look it up in a dictionary like dictionary.com to find that "procrastination" is defined as the act of putting off or delaying.

Now that you understand the context of the question, you are ready for the second read.

Second Read: Tasks

1. **What question is being asked? What am I being asked to do?**
   In this example, you are being asked to do two things. Part (a) of the question asks you to construct a confidence interval estimate for the population mean TPS score. Part (b) asks you to use the confidence interval that you calculate to determine if 40 is a plausible value for the population mean TPS score for students at this university.

Now that you understand the context and know what you are being asked to do, you are ready to read the question a third time to extract the specific information that is relevant to completing the work on the problem.

Third Read: Details

1. **What data were collected? Were the data collected in a reasonable way?**
   In this example, the data are from a sample selected from the population of first-year students at a university in England. The students completed a scale called the Tuckman Procrastination Scale (TPS), which measures tendency to procrastinate. The problem states that the sample is representative of the population of first-year students at this university.
2. **Are the actual data provided? What is the sample size (or sample sizes if there are more than one sample)?**
   The actual data (the 65 TPS scores) are not given. The sample size is $n = 65$.
3. **Are any summary statistics or additional information given? If so, which statistics and what are the values of these statistics?**
   In addition to the sample size, the problem gives the sample mean TPS score and the sample standard deviation of the TPS scores. These values are $\bar{x} = 37.02$ and $s = 6.44$.

After this third read, you understand the context, have identified what you are being asked to do, and have extracted the specific information that is needed to solve the problem. You are now ready to do the work and communicate the results.

Example 3

Consider this problem from a statistics textbook:

> The paper "Facebook Use and Academic Performance Among College Students: A Mixed-Methods Study with a Multi-Ethnic Sample" (*Computers in Human Behavior* [2015]: 265–272) describes a survey of a sample of 66 male students and a sample of 195 female students at a large university in Southern California. The authors of the paper believed that these samples were representative of male and female

college students in Southern California. For the sample of males, the mean time spent per day on Facebook was 102.31 minutes. For the sample of females, the mean time was 159.61 minutes. The sample standard deviations were not given in the paper, but for purposes of this exercise, suppose that the sample standard deviations were both 100 minutes. Do the data provide convincing evidence that the mean time spent of Facebook is not the same for males and for females? Test the relevant hypotheses using $\alpha = 0.05$.

First Read: Context

1. **What is the general context?**
   This example describes a survey of male and female college students in Southern California to find out how much time they spend per day on Facebook.
2. **Are there any terms that you don't understand?**
   At this point, you probably don't know what "hypotheses" means in this context or what "$\alpha = 0.05$" means because you haven't covered that yet in your statistics course. But if you were further along and this problem was part of your homework assignment and you weren't sure, you should go back to the appropriate section in your statistics text to review the meaning these terms.

Now that you understand the context of the question, you are ready for the second read.

Second Read: Tasks

1. **What question is being asked? What am I being asked to do?**
   In this example, you are being asked to decide if the data provide convincing evidence that the population mean time spent on Facebook is different for males and for females. You are told to reach your decision by testing the relevant hypotheses using $\alpha = 0.05$.

Now that you understand the context and know what you are being asked to do, you are ready to read the question a third time to extract the specific information that is relevant to completing the work on the problem.

Third Read: Details

1. **What data were collected? Were the data collected in a reasonable way?**
   In this example, the data are from two separate samples. One sample was selected from the population of male college students at a university in Southern California, and the second sample was selected from the population of female college students at the same university. The students in both samples were asked about how much time they spend per day using Facebook. The problem states that it is reasonable to consider these samples as representative of the population of male college students in Southern California and the population of female college students in Southern California.
2. **Are the actual data provided? What is the sample size (or sample sizes if there are more than one sample)?**
   The actual data (the 66 Facebook times from the males and the 195 Facebook times from the females) are not given. The sample size for the sample of males (which you can call Sample 1) is $n_1 = 66$. The sample size for the sample of females (Sample 2) is $n_2 = 195$.
3. **Are any summary statistics or additional information given? If so, which statistics and what are the values of these statistics?**
   In addition to the two sample sizes, the problem gives the sample mean and the sample standard deviation for each of the two samples. For the sample of males (Sample 1), $\bar{x}_1 = 102.31$ and $s_1 = 100$. For the sample of females (Sample 2), $\bar{x}_2 = 159.61$ and $s_2 = 100$.

After this third read, you understand the context, have identified what you are being asked to do, and have extracted the specific information that is needed to solve the problem. You are now ready to do the work and communicate the results.

SECTION 7.4 On Your Own

In this section, you will practice using the recommended strategy for reading a statistics problem. Recall that you are focusing on different things on each read. The strategy is summarized below.

First Read: Context

First, read for context. Read the entire problem (don't just skim!) without focusing on the numbers. What you are trying to do on this first read is just to understand the context and identify any terms (statistical or context-related) that you aren't familiar with.

1. **What is the general context?**
2. **Are there any terms that you don't understand?**

Second Read: Tasks

Second, read to identify what specific tasks you are being asked to complete. Read the entire question again. As with the first read, don't worry about the numbers yet—just read with a focus on what you are being asked to do or the question that you are being asked to answer.

1. **What question is being asked? What am I being asked to do?**

Third Read: Details

Third, read to extract specific information relevant to solving the problem. Now that you understand the context and have identified what you need to do, read one more time to identify specific details that will allow you to solve the problem. This is the read where it is OK to focus on the numbers and other relevant information. In particular, think about the following:

1. **What data were collected? Were the data collected in a reasonable way?**
2. **Are the actual data provided? What is the sample size (or sample sizes if there are more than one sample)?**
3. **Are any summary statistics or additional information given? If so, which statistics and what are the values of these statistics?**

Check It Out!

For each of the following study contexts, use the three-read strategy and answer the questions associated with each read. Then check your answers with those given at the end of this section.

Study Context 1

Consider the following problem from a statistics textbook:

> People in a random sample of 236 students enrolled at a liberal arts college were asked questions about how much sleep they get each night ("Alcohol Consumption, Sleep, and Academic Performance Among College Students," *Journal of Studies on Alcohol and Drugs* [2009]: 355–363). The sample mean sleep duration (average hours of daily sleep) was 7.71 hours and the sample standard deviation was 1.03 hours. The recommended number of hours of sleep for college-age students is 8.4 hours per day. Is there convincing evidence that the population mean daily sleep duration for students at this college is less than the recommended number of 8.4 hours? Test the relevant hypotheses using $\alpha = 0.01$.

Answer the questions associated with each read. Then check your answers with those given at the end of the section.

Study Context 2

Consider the following problem from a statistics textbook:

> The authors of the paper "Influence of Biofeedback Weight Bearing Training in Sit to Stand to Sit and the Limits of Stability on Stroke Patients" (*The Journal of Physical Therapy Science* [2016]: 3011–2014) randomly selected two samples of patients admitted to the hospital after suffering a stroke. One sample was selected from patients who received biofeedback weight training for 8 weeks and the other sample was selected from patients who did not receive this training. At the end of 8 weeks, the time it took (in seconds) to stand from a sitting position and then to sit down again (called sit-stand-sit time) was measured for the people in each sample. Data consistent with summary quantities given in the paper are given below. For purposes of this exercise, you can assume that the samples are representative of the population of stroke patients who receive the biofeedback training and the population of stroke patients who do not receive this training. Use the given data to construct and interpret a 95% confidence interval for the difference in mean sit-stand-sit time for these two populations.
>
> **Biofeedback Group**
>
> 1.9 2.6 4.3 2.1 2.7 4.1 3.2 4.0 3.2 3.5 2.8 3.5 3.5 2.3 3.1
>
> **No Biofeedback Group**
>
> 5.1 4.7 3.9 4.2 4.7 4.3 4.2 5.1 3.4 4.2 5.1 4.4 4.0 3.4 3.9

Answer the questions associated with each read. Then check your answers with those given at the end of the section.

Study Context 3

Consider the following problem from a statistics textbook:

> The article "Footwear, Traction, and the Risk of Athletic Injury" (January 2016, lermagazine.com/article/footwear-traction-and-the-risk-of-athletic-injury, retrieved December 15, 2016) describes a study in which high school football players were given either a conventional football cleat or a swivel disc shoe. Of 2373 players who wore the conventional cleat, 372 experienced an injury during the study period. Of the 466 players who wore the swivel disc shoe, 24 experienced an injury. The question of interest is whether there is evidence that the injury proportion is smaller for the swivel disc shoe than it is for conventional cleats.

a. What are the two treatments in this experiment?

b. The article didn't state how the players in the study were assigned to the two groups. Explain why it is important to know if they were assigned to the groups at random.

c. For purposes of this example, assume that the players were randomly assigned to the two treatment groups. Carry out a hypothesis test to determine if there is evidence that the injury proportion is smaller for the swivel disc shoe than it is for conventional cleats. Use a significance level of 0.05.

Answer the questions associated with each read. Then check your answers with those given at the end of the section.

Answers

Study Context 1 Answers

First Read: Context.

1. What is the general context?

The problem describes a study to investigate how long college students sleep each night, although this appears to be part of a larger study.

2. Are there any terms that you don't understand?

You may not yet have encountered statistical terms such as "test of hypotheses," or "standard deviation," or α, the Greek letter "alpha." If this is the case, you would turn to your statistics text to review the meaning of these terms and symbols before proceeding.

Second Read: Tasks

1. **What question is being asked? What am I being asked to do?**
   You are being asked to carry out a hypothesis test to determine if "the population mean daily sleep duration for students at this college is less than the recommended number of 8.4 hours."

Third Read: Details

1. **What data were collected? Were the data collected in a reasonable way?**
   Average hours of daily sleep was recorded for each person in a random sample of students at a liberal arts college. Even though a random sample was selected, the sample was selected from the population of students at one particular liberal arts college, so it is not reasonable to generalize to all college students.
2. **Are the actual data provided? What is the sample size (or sample sizes if there are more than one sample)?**
   The actual data are not included—only summary statistics are given. The sample size was $n = 236$ college students.
3. **Are any summary statistics or additional information given? If so, which statistics and what are the values of these statistics?**
   The sample mean sleep duration is reported to be 7.71 hours, and the sample standard deviation is reported to be 1.03 hours.

Study Context 2 Answers

First Read: Context

1. **What is the general context?**
   Stroke patients who received biofeedback weight training were compared to stroke patients who did not receive biofeedback weight training to investigate whether the biofeedback weight training has an effect of stability. Stability was measured using sit-stand-sit time.
2. **Are there any terms that you don't understand?**
   "Confidence interval" and "biofeedback" are terms that you may not have encountered before reading a description like this. You could turn to a dictionary or look online to find the meaning of biofeedback. If you were trying to complete this problem and didn't recall the definition of confidence interval, you should review the appropriate section of your statistics text before proceeding.

Second Read: Tasks

1. **What question is being asked? What am I being asked to do?**
   The question asks you to "construct and interpret a 95% confidence interval for the difference in mean sit-stand-sit time for these two populations."

Third Read: Details

1. **What data were collected? Were the data collected in a reasonable way?**
   In this example, the data were collected by selecting a sample of stroke patients from those who received biofeedback weight training, and by selecting a sample of stroke patients from those who did not receive biofeedback weight training. Sit-stand-sit times were recorded for each patient participating in the study. The statement of the problem indicates that you may assume that the data are representative of the populations of all stroke patients who receive biofeedback weight training and all stroke patients who do not receive biofeedback weight training.
2. **Are the actual data provided? What is the sample size (or sample sizes if there are more than one sample)?**
   Yes, the actual data on sit-stand-sit time are provided. There were 15 stroke patients in each of the two samples.

3. **Are any summary statistics or additional information given? If so, which statistics and what are the values of these statistics?**
   Summary statistics are not provided, and you will have to use the data provided to calculate them or use an appropriate technology to compute them.

Study Context 3 Answers

First Read: Context

1. **What is the general context?**
   A new swivel disc shoe is being tested to determine if it can reduce the proportion of high school football players who are injured.
2. **Are there any terms that you don't understand?**
   You may not yet know the terms "hypothesis test," and "significance level." If you were working on this problem as part of a homework assignment and were unclear on the meaning of these terms, you should review the appropriate section of you statistics text.

Second Read: Tasks

1. **What question is being asked? What am I being asked to do?**
   Part (a) asks you to identify the two treatments in the experiment. Part (b) asks you to explain why it would have been important to randomly assign players into the two treatment groups. In part (c), you are asked to "carry out a hypothesis test to determine if there is evidence that the injury proportion is smaller for the swivel disc shoe than it is for conventional cleats."

Third Read: Details

1. **What data were collected? Were the data collected in a reasonable way?**
   This study is an experiment, and football players are assumed to have been randomly assigned into two treatment groups. For each player who participated in the study, whether or not an injury occurred was determined. The data are the responses for the 2373 players who wore conventional shoes and the 466 players who wore swivel disc shoes. The article doesn't state whether random assignment was used or not, although you are asked to explain why this would be important and are also told to assume that random assignment was in fact used.
2. **Are the actual data provided? What is the sample size (or sample sizes if there are more than one sample)?**
   The actual data for the individual players are not provided. Instead, counts of how many players were in each treatment group and how many in each group experienced injuries are provided. One sample of 2839 high school football players was studied, and you are told to assume that they were randomly assigned to two groups, one containing 2373 players and the other containing 466 players.
3. **Are any summary statistics or additional information given? If so, which statistics and what are the values of these statistics?**
   The counts are provided, and these can be used to calculate sample proportions.

SECTION 7.4 EXERCISES

7.1 Consider the following problem from a statistics textbook. Use the three-read strategy and answer the questions associated with each read.

The authors of the paper "Driving Performance While Using a Mobile Phone: A Simulation Study of Greek Professional Drivers" (*Transportation Research Part F* [2016]: 164–170) describe a study to evaluate the effect of mobile phone use by taxi drivers in Greece. Fifty taxi drivers drove in a driving simulator where they were following a lead car. The drivers were asked to carry out a conversation on a mobile phone while driving, and the following distance (the distance between the taxi and the lead car) was recorded. The sample mean following distance was 3.20 meters and the sample standard deviation was 1.11 meters.

a. Construct and interpret a 95% confidence interval for μ, the population mean following distance while talking on a mobile phone for the population of taxi drivers.

b. What assumption must be made to generalize this confidence interval to the population of all taxi drivers in Greece?

7.2 Consider the following problem from a statistics textbook. Use the three-read strategy and answer the questions associated with each read.

According to a large national survey conducted by the Pew Research Center ("What Americans Think About NSA Surveillance, National Security and Privacy," May 2, 2015, pewresearch.org, retrieved December 1, 2016), 54% of adult Americans disapprove of the National Security Agency collecting records of phone and Internet data. Suppose that this estimate was based on a random sample of 1000 adult Americans.

a. Is there convincing evidence that a majority of adult Americans feel this way? Test the relevant hypotheses using a 0.05 significance level.
b. The actual sample size was much larger than 1000. If you had used the actual sample size when doing the calculations for the test in Part (a), would the *P*-value have been larger than, the same as, or smaller than the *P*-value you obtained in Part (a)? Provide a justification for your answer.

7.3 Consider the following problem from a statistics textbook. Use the three-read strategy and answer the questions associated with each read.

A Harris Poll press release dated November 1, 2016 summarized results of a survey of 2463 adults and 510 teens age 13 to 17 ("American Teens No Longer More Likely Than Adults to Believe in God, Miracles, Heaven, Jesus, Angels, or the Devil," theharrispoll .com, retrieved December 12, 2016). It was reported that 19% of the teens surveyed and 26% of the adults surveyed indicated that they believe in reincarnation. The samples were selected to be representative of American adults and teens. Use the data from this survey to construct a confidence interval to estimate the difference in the proportion of teens who believe in reincarnation and the proportion of adults who believe in reincarnation. Be sure to interpret your interval in context.

7.4 Consider the following problem from a statistics textbook. Use the three-read strategy and answer the questions associated with each read.

The article "Activity Trackers May Undermine Weight Loss Efforts" (*The New York Times*, September 20, 2016) describes a study published in the *Journal of the American Medical Association* ("The Effect of Wearable Technology Combined with a Lifestyle Intervention on Long-Term Weight Loss" [2016]: 1161–1171). In this study, subjects followed a low-calorie diet and exercise program for 6 months. After 6 months, the subjects were randomly assigned to one of two groups. The people in one group were provided with a website they could use to self-monitor diet and physical activity. The people in the second group were provided with a wearable fitness tracker with an accompanying web interface to monitor diet and physical activity.

The researchers were interested in learning if the mean weight loss (in kilograms) at the end of two years was different for the two treatments (self-monitoring and fitness tracker monitoring). Data from this experiment are summarized in the accompanying table.

Group	Sample Size	Mean Weight Loss	Standard Deviation
Self-Monitoring	170	5.9 kg	6.8 kg
Fitness Tracker Monitoring	181	3.5 kg	6.3 kg

Do the data from this experiment provide evidence that the mean weight loss differs for the two treatments? Test the relevant hypotheses using a significance level of 0.01.

8 Estimating a Population Proportion—The Math You Need to Know

SECTION 8.1 Evaluating Expressions

There are several expressions that you will encounter as you learn about methods for estimating a population proportion.

Note: Depending on your access to technology, such as a graphing calculator or a statistics software package, you may not need to evaluate these expressions by hand. But just in case your statistics class is not relying on technology to do these calculations, this section looks at how to evaluate these expressions.

Margin of Error When Estimating a Population Proportion (using a 95% confidence level)

$$\text{margin of error} = 1.96\sqrt{\frac{\hat{p}(1-\hat{p})}{n}}$$

To calculate the margin of error associated with an estimate of a population proportion, you need to verify that the sample size is large enough to calculate margin of error using the given formula. You also need to confirm that the sample is a random sample or that it was selected in a way that makes it reasonable to think that the sample is representative of the population.

The sample size is large enough to use the margin of error formula if $n\hat{p}$ and $n(1-\hat{p})$ are both greater than or equal to 10. For example, if $n = 40$ and $\hat{p} = 0.4$, you would use these values to check the sample size condition as follows:

$$n\hat{p} = (40)(0.4) = 16$$
$$n(1-\hat{p}) = (40)(1-0.4) = (40)(0.6) = 24$$

Because 16 is greater than or equal to 10 and 24 is greater than or equal to 10, the sample size of 40 is large enough to use the formula for margin of error.

Next, substitute the values of $\hat{p}$ and n into the formula for margin of error to obtain

$$\text{margin of error} = 1.96\sqrt{\frac{\hat{p}(1-\hat{p})}{n}} = 1.96\sqrt{\frac{(0.4)(1-0.4)}{40}}$$

To evaluate this expression, start by evaluating the part of the expression that appears under the square root. First calculate $(1 - 0.4) = 0.6$. Then multiply the two numbers in the numerator, and finally divide that number by 40. This is the sequence of steps shown here:

$$\frac{(0.4)(1-0.4)}{40} = \frac{(0.4)(0.6)}{40} = \frac{0.24}{40} = 0.006$$

To complete the calculation of margin of error, you would find the square root of this number and then multiply that number by 1.96 (for a 95% confidence level), as shown here:

$$\text{margin of error} = 1.96\sqrt{0.006} = 1.96(0.077) = 0.151$$

The margin of error can be interpreted as the maximum likely estimation error. If 0.4 is the estimate of the population proportion based on this sample of size 40, it is unlikely that this estimate differs from the actual value of the population proportion by more than 0.151.

Suppose that the sample size had been $n = 20$ rather than $n = 40$. If $\hat{p} = 0.4$, when you check the sample size condition you would get

$$n\hat{p} = (20)(0.4) = 8$$
$$n(1-\hat{p}) = (20)(1-0.4) = (20)(0.6) = 12$$

Because the value of $n\hat{p}$ is 8, which is not greater than or equal to 10, the sample size condition is not met and it would not be appropriate to use the given formula to calculate a margin of error.

Sometimes the margin of error is calculated using 2 in place of 1.96 in the expression for margin of error. This simplifies the calculations and doesn't usually change the calculated value by much (it results in a slightly larger value for the margin of error).

Check It Out!

a. Suppose that you want to estimate a population proportion and calculate a margin of error using data from a random sample. The sample size is $n = 12$ and $\hat{p} = 0.9$. Check the necessary sample size condition, and comment on whether the parts are both satisfied. If the sample size condition is satisfied, then calculate the margin of error for a 95% confidence level, and interpret its value.

b. Suppose that you want to estimate a population proportion and calculate a margin of error using data from a random sample. The sample size is $n = 96$ and $\hat{p} = 0.63$. Check the necessary sample size condition, and comment on whether the parts are both satisfied. If the sample size condition is satisfied, then calculate the margin of error for a 95% confidence level, and interpret its value.

c. Suppose that you want to estimate a population proportion and calculate a margin of error using data from a random sample. The sample size is $n = 42$ and $\hat{p} = 0.02$. Check the necessary sample size condition, and comment on whether the parts are both satisfied. If the sample size condition is satisfied, then calculate the margin of error for a 95% confidence level, and interpret its value.

Answers

a.

$$n\hat{p} = (12)(0.9) = 10.8 \geq 10$$
$$n(1-\hat{p}) = (12)(1-0.9) = (12)(0.1) = 1.2 < 10$$

Because the value of $n(1 - \hat{p}) = 1.2$ is less than 10, the sample size condition is not satisfied. You would not use the given formula to calculate the margin of error, because the sample size condition is not satisfied.

b.

$$n\hat{p} = (96)(0.63) = 60.48 \geq 10$$
$$n(1-\hat{p}) = (96)(1-0.63) = (96)(0.37) = 35.52 \geq 10$$

Both parts of the sample size condition are satisfied, so it is appropriate to calculate and interpret the margin of error.

$$\text{margin of error} = 1.96\sqrt{\frac{\hat{p}(1-\hat{p})}{n}} = 1.96\sqrt{\frac{(0.63)(1-0.63)}{96}}$$

Start by evaluating the part of the expression that appears under the square root. First calculate $(1 - 0.63) = 0.37$. Then multiply the two numbers in the numerator, and finally divide that number by 96. This is the sequence of steps shown here:

$$\frac{(0.63)(1-0.63)}{96} = \frac{(0.63)(0.37)}{96} = \frac{0.2331}{96} = 0.0024$$

To complete the calculation of margin of error using a 95% confidence level, you would find the square root of this number and then multiply that number by 1.96, as shown here:

$$\text{margin of error} = 1.96\sqrt{0.0024} = 1.96(0.049) = 0.096$$

The margin of error can be interpreted as the maximum likely estimation error. If 0.63 is the estimate of the population proportion based on this sample of size 96, it is unlikely that this estimate differs from the actual value of the population proportion by more than 0.096.

c.

$$n\hat{p} = (42)(0.02) = 0.84 < 10$$
$$n(1-\hat{p}) = (42)(1-0.02) = (42)(0.98) = 41.16 \geq 10$$

Because the value of $n\hat{p} = 0.84$ is less than 10, the sample size condition is not satisfied. You would not use the given formula to calculate the margin of error, because the sample size condition is not satisfied.

Confidence Interval for a Population Proportion

$$\hat{p} \pm (z \text{ critical value})\sqrt{\frac{\hat{p}(1-\hat{p})}{n}}$$

To calculate a confidence interval for a population proportion, you need to know the value of the sample proportion, $\hat{p}$, and the sample size, n. You also need to know the confidence level, which determines the z critical value that is used to calculate the confidence interval. Before using the given expression to calculate a confidence interval, you need to verify that the sample size is large enough to calculate a confidence interval using the given formula. You also need to confirm that the sample is a random sample or that it was selected in a way that makes it reasonable to think that the sample is representative of the population.

The sample size is large enough to use the confidence interval formula if $n\hat{p}$ and $n(1 - \hat{p})$ are both greater than or equal to 10. For examples of how the sample size condition is checked, see the above discussion of margin of error.

Recall that an interval is defined by its endpoints—it has a lower endpoint and an upper endpoint. Notice that the expression for the confidence interval includes the symbol $\pm$, which is read as "plus and minus." This means that you will do two calculations, once using $+$ and once using $-$. Each one of these calculations results in one of the two interval endpoints.

To calculate the endpoints of the confidence interval, you can start by determining the appropriate z critical value for the specified confidence level. Your statistics textbook should have instructions for how this is done. The z critical values for commonly used confidence levels are given in the following table:

Confidence Level	z Critical Value
90%	1.645
95%	1.96
99%	2.58

Once you have determined the appropriate z critical value, you can substitute the values of n and $\hat{p}$ into the expression for the confidence interval. For example, suppose you want to calculate a 95% confidence interval for a population proportion and that the sample size is $n = 50$ and the sample proportion is $\hat{p} = 0.29$. For a 95% confidence level, the z critical value is 1.96. Substituting these values into the expression for the confidence interval results in the following:

$$\hat{p} \pm (z \text{ critical value})\sqrt{\frac{\hat{p}(1-\hat{p})}{n}} = 0.29 \pm (1.96)\sqrt{\frac{(0.29)(1-0.29)}{50}}$$

To evaluate this expression, start by evaluating the part of the expression that appears under the square root. First calculate $(1 - 0.29) = 0.71$. Then multiply the two numbers in the numerator, and finally divide that number by 50. This is the sequence of steps shown here:

$$\frac{(0.29)(1-0.29)}{50} = \frac{(0.29)(0.71)}{50} = \frac{0.2059}{50} = 0.0041$$

Next, you would find the square root of this number and then multiply that number by 1.96, as shown here:

$$1.96\sqrt{0.0041} = 1.96(0.064) = 0.125$$

This number is the "plus and minus" part of the confidence interval. To calculate the endpoints of the interval, you would evaluate

$$0.29 \pm 0.125$$

The lower endpoint of the interval is $0.29 - 0.125 = 0.165$. The upper endpoint of the interval is $0.29 + 0.125 = 0.415$.

The 95% confidence interval estimate of the population proportion is (0.165, 0.415). You would interpret this interval by saying that based on the data from the sample, you can be 95% confident that the actual value of the population proportion is somewhere between 0.165 and 0.415.

Check It Out!

a. Suppose that you want to use data from a random sample to estimate a population proportion with a confidence interval. The sample size is $n = 46$ and $\hat{p} = 0.57$. Check the necessary sample size condition, and comment on whether both parts are satisfied. If the sample size condition is satisfied, then calculate the endpoints of a 95% confidence interval, and interpret the confidence interval.

b. Suppose that you want to use data from a random sample to estimate a population proportion with a confidence interval. The sample size is $n = 30$ and $\hat{p} = 0.47$. Check the necessary sample size condition, and comment on whether both parts are satisfied. If the sample size condition is satisfied, then calculate the endpoints of a 99% confidence interval, and interpret the confidence interval.

c. Suppose that you want to use data from a random sample to estimate a population proportion with a confidence interval. The sample size is $n = 15$ and $\hat{p} = 0.73$. Check the necessary sample size condition, and comment on whether both parts are satisfied. If the sample size condition is satisfied, then calculate the endpoints of a 90% confidence interval, and interpret the confidence interval.

Answers

a.

$$n\hat{p} = (46)(0.57) = 26.22 \geq 10$$
$$n(1-\hat{p}) = (46)(1-0.57) = (46)(0.43) = 19.78 \geq 10$$

Both parts of the sample size condition are satisfied and the sample is a random sample, so it is appropriate to calculate and interpret the 95% confidence interval.

For a 95% confidence level, the z critical value is 1.96. Substituting the values into the expression for the confidence interval results in the following:

$$\hat{p} \pm (z \text{ critical value})\sqrt{\frac{\hat{p}(1-\hat{p})}{n}} = 0.57 \pm (1.96)\sqrt{\frac{(0.57)(1-0.57)}{46}}$$

To evaluate this expression, start by evaluating the part of the expression that appears under the square root. First calculate $(1 - 0.57) = 0.43$. Then multiply the two numbers in the numerator, and finally divide that number by 46. This is the sequence of steps shown here:

$$\frac{(0.57)(1-0.57)}{46} = \frac{(0.57)(0.43)}{46} = \frac{0.2451}{46} = 0.0053$$

Next, you would find the square root of this number and then multiply that number by 1.96, as shown here:

$$1.96\sqrt{0.0053} = 1.96(0.073) = 0.143$$

This number is the "plus and minus" part of the confidence interval. To calculate the endpoints of the interval, you would evaluate

$$0.57 \pm 0.143$$

The lower endpoint of the interval is $0.57 - 0.143 = 0.427$. The upper endpoint of the interval is $0.57 + 0.143 = 0.713$.

The 95% confidence interval estimate of the population proportion is (0.427, 0.713). You would interpret this interval by saying that based on the data from the sample, you can be 95% confident that the actual value of the population proportion is somewhere between 0.427 and 0.713.

b.

$$n\hat{p} = (30)(0.47) = 14.1 \geq 10$$
$$n(1-\hat{p}) = (30)(1-0.47) = (30)(0.53) = 15.9 \geq 10$$

Both of the sample size conditions are satisfied and the sample is a random sample, so it is appropriate to calculate and interpret the 99% confidence interval.

For a 99% confidence level, the z critical value is 2.58. Substituting the values into the expression for the confidence interval results in the following:

$$\hat{p} \pm (z \text{ critical value})\sqrt{\frac{\hat{p}(1-\hat{p})}{n}} = 0.47 \pm (2.58)\sqrt{\frac{(0.47)(1-0.47)}{30}}$$

To evaluate this expression, start by evaluating the part of the expression that appears under the square root. First calculate $(1 - 0.47) = 0.53$. Then multiply the two numbers in the numerator, and finally divide that number by 30. This is the sequence of steps shown here:

$$\frac{(0.47)(1-0.47)}{30} = \frac{(0.47)(0.53)}{30} = \frac{0.2491}{30} = 0.0083$$

Next, you would find the square root of this number and then multiply that number by 2.58, as shown here:

$$2.58\sqrt{0.0083} = 2.58(0.091) = 0.235$$

This number is the "plus and minus" part of the confidence interval. To calculate the endpoints of the interval, you would evaluate

$$0.47 \pm 0.235$$

The lower endpoint of the interval is $0.47 - 0.235 = 0.235$. The upper endpoint of the interval is $0.47 + 0.235 = 0.705$.

The 99% confidence interval estimate of the population proportion is (0.235, 0.705). You would interpret this interval by saying that based on the data from the sample, you can be 99% confident that the actual value of the population proportion is somewhere between 0.235 and 0.705.

c.
$$n\hat{p} = (15)(0.73) = 10.95 \geq 10$$
$$n(1 - \hat{p}) = (15)(1 - 0.73) = (15)(0.27) = 4.05 < 10$$

Because the value of $n(1 - \hat{p}) = 4.05$ is less than 10, the sample size conditions are not satisfied. You would not use the given formula to calculate the confidence interval, because the sample size conditions are not satisfied.

Determining Sample Size

$$n = p(1 - p)\left(\frac{1.96}{M}\right)^2$$

When estimating a population proportion, the larger the sample size, the smaller the margin of error will be. This makes sense because you would expect that an estimate based on a larger sample size would tend to be closer to the actual value of the population proportion than an estimate based on a smaller sample. Before collecting any data, you might want to determine how large the sample size should be to achieve a specified margin of error. When this is the case, the expression given above is used to determine sample size.

To use this expression to calculate a sample size, you need to know the desired value for the margin of error, M. The other value you need to evaluate the expression for sample size is p, the value of the population proportion. This value will not be known (it is what you hope to estimate). Sometimes, you will have prior information about the value of p (for example, from a preliminary study), but most likely you will use $p = 0.5$ in the expression for sample size. The reason that 0.5 is used if you don't have any information about the value of p ahead of time is that this value for p is the one that results in the largest sample size compared to any other choice for the value of p.

Begin by substituting the values for M and p (in many cases this will be 0.5) into the expression. For example, suppose you want to determine the sample size necessary to estimate a population proportion with a margin of error of $M = 0.03$. Substituting $M = 0.03$ and $p = 0.5$ into the expression results in

$$n = p(1 - p)\left(\frac{1.96}{M}\right)^2 = (0.5)(1 - 0.5)\left(\frac{1.96}{0.03}\right)^2$$

To evaluate this expression, you would first calculate the value of $\left(\frac{1.96}{0.03}\right)$. You would then square this number, and finally multiply the result by $(0.5)(1 - 0.5)$. This sequence of steps illustrated here:

$$(0.5)(1 - 0.5)\left(\frac{1.96}{0.03}\right)^2 = (0.5)(1 - 0.5)(65.333)^2 = (0.5)(0.5)(4268.401) = 1067.100$$

In sample size calculations, you always round up to the next whole number to obtain the necessary sample size. For this example, a sample size of 1068 would be needed to achieve a margin of error of $M = 0.03$.

Check It Out!

a. Determine the sample size necessary to estimate a population proportion with a margin of error of $M = 0.05$, using a 95% confidence level.

b. Determine the sample size necessary to estimate a population proportion with a margin of error of $M = 0.01$ and a 95% confidence level. Suppose that there is evidence that the value of the population proportion is 0.9, or near 0.9.

Answers

a. Substituting $M = 0.05$ and $p = 0.5$ into the expression results in

$$n = p(1 - p)\left(\frac{1.96}{M}\right)^2 = (0.5)(1 - 0.5)\left(\frac{1.96}{0.05}\right)^2$$

To evaluate this expression, you would first calculate the value of $\left(\frac{1.96}{0.05}\right)$. You would then square this number, and finally multiply the result by $(0.5)(1 - 0.5)$. This sequence of steps illustrated here:

$$(0.5)(1 - 0.5)\left(\frac{1.96}{0.05}\right)^2 = (0.5)(1 - 0.5)(39.2)^2 = (0.5)(0.5)(1536.64) = 384.16$$

In sample size calculations, you always round up to the next whole number to obtain the necessary sample size. For this example, a sample size of at least 385 would be needed to achieve a margin of error no greater than $M = 0.05$.

b. Substituting $M = 0.01$ and $p = 0.9$ into the expression results in

$$n = p(1 - p)\left(\frac{1.96}{M}\right)^2 = (0.9)(1 - 0.9)\left(\frac{1.96}{0.01}\right)^2$$

To evaluate this expression, you would first calculate the value of $\left(\frac{1.96}{0.01}\right)$. You would then square this number, and finally multiply the result by $(0.9)(1 - 0.9)$. This sequence of steps illustrated here:

$$(0.9)(1 - 0.9)\left(\frac{1.96}{0.01}\right)^2 = (0.9)(1 - 0.9)(196)^2 = (0.9)(0.1)(38{,}416) = 3457.44$$

In sample size calculations, you always round up to the next whole number to obtain the necessary sample size. For this example, a sample size of at least 3458 would be needed to achieve a margin of error no greater than $M = 0.01$.

SECTION 8.1 EXERCISES

8.1 Suppose that you want to estimate a population proportion and calculate a margin of error using data from a random sample. The sample size is $n = 73$ and $\hat{p} = 0.48$. Check the necessary sample size condition, and comment on whether both parts are satisfied. If the sample size condition is satisfied, then calculate the margin of error for a 95% confidence level, and interpret its value.

8.2 Suppose that you want to estimate a population proportion and calculate a margin of error using data from a random sample. The sample size is $n = 50$ and $\hat{p} = 0.64$. Check the necessary sample size condition, and comment on whether both parts are satisfied. If the sample size condition is satisfied, then calculate the margin of error for a 95% confidence level, and interpret its value.

8.3 Suppose that you want to use data from a random sample to estimate a population proportion with a confidence interval. The sample size is $n = 42$ and $\hat{p} = 0.38$. Check the necessary sample size condition, and comment on whether both parts are satisfied. If the sample size condition is satisfied, then calculate the endpoints of a 90% confidence interval, and interpret the confidence interval.

8.4 Suppose that you want to use data from a random sample to estimate a population proportion with a confidence interval. The sample size is $n = 31$ and $\hat{p} = 0.58$. Check the necessary sample size condition, and comment on whether both parts are satisfied. If the sample size condition is satisfied, then calculate the endpoints of a 95% confidence interval, and interpret the confidence interval.

8.5 Determine the sample size necessary to estimate a population proportion with a margin of error of $M = 0.04$, using a 95% confidence level. Suppose that there is no available information about the value of the population proportion.

8.6 Determine the sample size necessary to estimate a population proportion with a margin of error of $M = 0.03$, using a 90% confidence level. Suppose that you have reason to believe that the value of the population proportion is 0.3 (or near 0.3).

SECTION 8.2 Guided Practice—Margin of Error

In this section, you will consider a typical situation that involves finding a margin of error. Consider this problem from a statistics textbook:

> The paper "Sleeping with Technology: Cognitive, Affective and Technology Usage Predictors of Sleep Problems Among College Students" (*Sleep Health* [2016]: 49–56) summarized data from a survey of a sample of college students. Of the 734 students surveyed, 125 reported that they sleep with their cell phones near the bed and check their phones for something other than the time at least twice during the night. For purposes of this exercise, assume that this sample is representative of college students in the U.S.

a. Use the given information to estimate the proportion of college students who check their phones for something other than the time at least twice during the night.
b. Verify that the sample size condition needed for the margin of error formula to be appropriate is met.
c. Calculate the margin of error, using a 95% confidence level.
d. Interpret the margin of error in context.

If you were to encounter a problem like this, you should start by reading the problem carefully. You can use the three-read strategy introduced in Chapter 7.

First Read: Context

1. **What is the general context?**
   A study was conducted to learn about the proportion of people who check their cell phones during the night. A sample of college students was surveyed.

2. **Are there any terms that you don't understand?**
   This question is about margin of error, so if you don't recall the definition of margin of error, you would want to go to the appropriate section in a textbook to review the definition.

Second Read: Tasks

1. **What question is being asked? What am I being asked to do?**
   This problem asks for four things:
   1. Calculate an estimate of the proportion of college students who check their phones for something other than the time at least twice during the night.
   2. Verify that the sample size is large enough to use the margin of error formula.
   3. Calculate the margin of error.
   4. Interpret the margin of error in context.

Third Read: Details

1. **What data were collected. Were the data collected in a reasonable way?**
   Data were collected from a sample of 734 college students. The problem states that you can assume that the sample is representative of college students in the U.S.

2. **Are the actual data provided? What is the sample size (or sample sizes if there are more than one sample)?**
   The sample size was $n = 734$ (the 734 students surveyed). The actual data (the 734 individual responses) are not given, but you know that 125 of the responses were "yes," indicating that they do check their phones at least twice during the night for something other than the time.

3. **Are any summary statistics or additional information given? If so, which statistics and what are the values of these statistics?**
   The sample proportion is not given, but you know that 125 of the 734 responses were "yes," and this is the information needed to calculate the sample proportion.

Now you are ready to tackle the tasks required to complete this exercise.

Use the given information to estimate the proportion of college students who check their phones for something other than the time at least twice during the night.

This is asking you to estimate a population proportion. The population is college students, and you are asked to estimate the proportion who check their phones for something other than the time at least twice during the night for this population. If you have a random sample or a sample that you think is representative of the population, the usual estimate of a population proportion is the sample proportion. You know that 125 of the 734 students in the sample indicated that they check their phones for something other than the time at least twice during the night. You can now calculate the sample proportion:

$$\hat{p} = \frac{125}{734} = 0.170$$

This tells you that the sample proportion was 0.170, or 17.0%. Based on this sample, you would estimate the population proportion of college students who check their phones for something other than the time at least twice during the night to be 0.170, or 17.0%.

Verify that the sample size condition needed for the margin of error formula to be used is met.

The sample size is large enough to use the margin of error formula if $n\hat{p}$ and $n(1 - \hat{p})$ are each greater than or equal to 10. Here, $n = 734$ and $\hat{p} = 0.170$; you would use these values to check the sample size condition as follows:

$$n\hat{p} = (734)(0.170) = 124.780$$
$$n(1 - \hat{p}) = (734)(1 - 0.170) = (734)(0.830) = 609.220$$

Because 124.780 is greater than or equal to 10 and 609.220 is greater than or equal to 10, the sample size of 734 is large enough to use the formula for margin of error.

Calculate the margin of error.

To calculate the value of the margin of error, substitute the values of $\hat{p}$ and n into the formula for margin of error to obtain

$$\text{margin of error} = 1.96\sqrt{\frac{\hat{p}(1 - \hat{p})}{n}} = 1.96\sqrt{\frac{(0.170)(1 - 0.170)}{734}}$$

To evaluate this expression, start by evaluating the part of the expression that appears under the square root. First calculate $(1 - 0.170) = 0.830$. Then multiply the two numbers in the numerator, and finally divide that number by 734. This is the sequence of steps shown here:

$$\frac{(0.170)(1 - 0.170)}{734} = \frac{(0.170)(0.830)}{734} = \frac{0.1411}{734} = 0.0002$$

To complete the calculation of margin of error, you would find the square root of this number and then multiply that number by 1.96, as shown here:

$$\text{margin of error} = 1.96\sqrt{0.0002} = 1.96(0.014) = 0.027$$

Interpret the margin of error in context.

The margin of error can be interpreted as the maximum likely estimation error. Your estimate of the population proportion of college students who check their phones at least twice for something other than the time during the night is 0.170. It is unlikely that this estimate differs from the actual value of the population proportion by more than 0.027.

SECTION 8.2 EXERCISES

8.7 Use the three-read strategy to understand the following exercise from a statistics textbook, and then complete the exercise.

The authors of the paper "U.S. College Students' Internet Use: Race, Gender and Digital Divides" (*Journal of Computer-Mediated Communication* [2009]: 244–264) described a survey of 7421 students from 40 colleges and universities. Of the students surveyed, 2998 reported Internet use of more than 3 hours per day. For purposes of this exercise, assume that this sample is representative of students in the U.S.

a. Use the given information to estimate the proportion of college students who use the Internet more than 3 hours per day.
b. Verify that the sample size condition needed for the margin of error formula to be appropriate is met.
c. Calculate the margin of error for a 90% confidence level.
d. Interpret the margin of error in context.

8.8 Use the three-read strategy to understand the following exercise from a statistics textbook, and then complete the exercise.

The article "Effects of Fast-Food Consumption on Energy Intake and Diet Quality Among Children" (*Pediatrics* [2004]: 112–118) reported that 1720 of those in a random sample of 6212 American children indicated that they ate fast food on a typical day.

a. Use the given information to estimate the proportion of American children who eat fast food on a typical day.
b. Verify that the sample size condition needed for the margin of error formula to be appropriate is met.
c. Calculate the margin of error for a 95% confidence level.
d. Interpret the margin of error in context.

SECTION 8.3 Guided Practice—Large Sample Confidence Interval for a Population Proportion

Consider this problem from a statistics textbook:

> In a survey of 800 undergraduate students in the U.S., 576 indicated that they believe that a student or faculty member on campus who uses language considered racist, sexist, homophobic, or offensive should be subject to disciplinary action ("Listening to Dissenting Views Part of Civil Debate," *USA TODAY*, November 17, 2015). Assuming that the sample is representative of college students in the U.S., construct and interpret a 95% confidence interval for the proportion of undergraduate students who have this belief.

If you were to encounter a problem like this, you should start by reading the problem carefully. You can use the three-read strategy introduced in Chapter 7.

First Read: Context

1. **What is the general context?**
A survey was conducted to learn about the proportion of undergraduate students who believe that a student or faculty member on campus who uses language considered racist, sexist, homophobic, or offensive should be subject to disciplinary action.

2. **Are there any terms that you don't understand?**
This question is about constructing a confidence interval, so if you aren't comfortable with the meaning of this term you would want to go to the appropriate section in your textbook to review the definition. If any of the other non-statistical terms are unfamiliar (such as "homophobic"), you can look up the meaning of those terms in a dictionary to make sure that you fully understand the context.

Second Read: Tasks

1. **What question is being asked? What am I being asked to do?**
This problem asks you to construct and interpret a 95% confidence interval for the proportion of undergraduate students who believe that a student or faculty member on campus who uses language considered racist, sexist, homophobic, or offensive should be subject to disciplinary action.

Third Read: Details

1. **What data were collected? Were the data collected in a reasonable way?**
   Data were collected from a sample of 800 college students. Each student indicated if they believe that a student or faculty member on campus who uses language considered racist, sexist, homophobic, or offensive should be subject to disciplinary action. The problem states that you can assume that the sample is representative of college students in the U.S.

2. **Are the actual data provided? What is the sample size (or the sample sizes if there are more than one sample)?**
   The sample size was $n = 800$. The actual data (the 800 individual responses) are not provided, but you are told that 576 of the responses indicated belief in the statement.

3. **Are any summary statistics or additional information given? If so, which statistics and what are the values of these statistics?**
   The sample proportion is not given, but you know that 576 of the 800 responses indicated belief in the statement.

Now you are ready to tackle the tasks required to complete this exercise. This problem asks you to construct and interpret a 95% confidence interval for the proportion of undergraduate students who believe that a student or faculty member on campus who uses language considered racist, sexist, homophobic, or offensive should be subject to disciplinary action.

The expression used to calculate a confidence interval for a population proportion is

$$\hat{p} \pm (z \text{ critical value})\sqrt{\frac{\hat{p}(1-\hat{p})}{n}}$$

To calculate a confidence interval for a population proportion, you need to know the value of the sample proportion $\hat{p}$, the sample size n, and the confidence level, which determines the z critical value that is used to calculate the confidence interval. The sample size $n = 800$ and the confidence level of 95% are given in the problem. The value of the sample proportion is not given, but you can calculate it because you know that 576 of the 800 students indicated belief in the statement:

$$\hat{p} = \frac{576}{800} = 0.72$$

Before using the confidence interval formula, you need to verify that the sample size is large enough to calculate the confidence interval using the given formula. You also need to confirm that the sample is a random sample or that it was selected in a way that makes it reasonable to think that the sample is representative of the population.

The problem states that you can assume that the sample is representative of the population of college students in the U.S. The sample size is large enough to use this formula if $n\hat{p}$ and $n(1 - \hat{p})$ are each greater than or equal to 10. Here, $n = 800$ and $\hat{p} = 0.72$. These values are used to check the sample size condition as follows:

$$n\hat{p} = (800)(0.72) = 576$$
$$n(1-\hat{p}) = (800)(1-0.72) = (800)(0.28) = 224$$

Because 576 is greater than or equal to 10 and 224 is greater than or equal to 10, the sample size of 800 is large enough to use the given expression to calculate a confidence interval for the population proportion.

To calculate the endpoints of the confidence interval, you can start by determining the appropriate z critical value for the specified confidence level. For a 95% confidence level, the corresponding z critical value is 1.96. (Your statistics text should have instructions for how to find this value.)

Next, substitute the values of n, $\hat{p}$, and the z critical value into the expression for the confidence interval. In this problem, the sample size is $n = 800$ and the sample proportion is $\hat{p} = 0.72$. For a 95% confidence level, the z critical value is 1.96. Substituting these values into the expression for the confidence interval results in the following:

$$\hat{p} \pm (z \text{ critical value})\sqrt{\frac{\hat{p}(1-\hat{p})}{n}} = 0.72 \pm (1.96)\sqrt{\frac{(0.72)(1-0.72)}{800}}$$

To evaluate this expression, start by evaluating the part of the expression that appears under the square root. First calculate $(1 - 0.72) = 0.28$. Then multiply the two numbers in the numerator, and finally divide that number by 800. This is the sequence of steps shown here:

$$\frac{(0.72)(1-0.72)}{800} = \frac{(0.72)(0.28)}{800} = \frac{0.2016}{800} = 0.0003$$

Next, you would find the square root of this number and then multiply that number by 1.96, as shown here:

$$1.96\sqrt{0.0003} = 1.96(0.017) = 0.033$$

This number is the "plus and minus" part of the confidence interval. To calculate the endpoints of the interval, you would evaluate

$$0.72 \pm 0.033$$

The lower endpoint of the interval is $0.72 - 0.033 = 0.687$. The upper endpoint of the interval is $0.72 + 0.033 = 0.753$. The 95% confidence interval estimate of the population proportion is (0.687, 0.753).

You should also interpret the confidence interval in context. You would interpret this interval by saying that based on the data from the sample, you can be 95% confident that the actual value of the population proportion of college students who believe that a student or faculty member on campus who uses language considered racist, sexist, homophobic, or offensive should be subject to disciplinary action is somewhere between 0.687 and 0.753. You could also interpret the interval in terms of percentages, saying that you are 95% confident that the percentage of college students who believe that a student or faculty member on campus who uses language considered racist, sexist, homophobic, or offensive should be subject to disciplinary action is between 68.7% and 75.3%.

SECTION 8.3 EXERCISES

8.9 Use the three-read strategy to understand the following exercise from a statistics textbook, and then proceed to complete the exercise.

The authors of the article "CSI Effect Has Juries Wanting More Evidence" (*USA TODAY*, August 5, 2004) examined how the popularity of crime-scene investigation television shows is influencing jurors' expectations of what evidence should be produced in a trial. In a survey of 500 potential jurors, one study found that 350 were regular watchers of at least one crime-scene investigation television series. Assuming that the sample was representative of the population of potential jurors, construct and interpret a 90% confidence interval for the proportion of potential jurors who are regular watchers of at least one crime-scene investigation television series.

8.10 Use the three-read strategy to understand the following exercise from a statistics textbook and then proceed to complete the exercise.

In a poll about sports (Associated Press, December 18, 2005), 394 of 1000 randomly selected U.S. adults indicated that they consider themselves to be baseball fans. Construct and interpret a 95% confidence interval for the proportion of U.S. adults who consider themselves to be baseball fans.

SECTION 8.4 Guided Practice—Determining Sample Size

Consider this problem from a statistics textbook:

> A discussion of digital ethics appears in the article "Academic Cheating, Aided by Cell Phones or Web, Shown to be Common" (*Los Angeles Times*, June 17, 2009). One question posed in the article is: What proportion of college students have used cell

phones to cheat on an exam? Suppose you have been asked to estimate this proportion for students enrolled at a large college. How many students should you include in a random sample of students from this college if you want to estimate this proportion with a margin of error of 0.02, using a 95% confidence level?

If you were to encounter a problem like this, you should start by reading the problem carefully. You can use the three-read strategy introduced in Chapter 7.

First Read: Context

1. **What is the general context?**
   You are planning to collect data to estimate the proportion of college students enrolled at a large college who have used cell phones to cheat on an exam. You want to know how large a sample should be selected.

2. **Are there any terms that you don't understand?**
   The term margin of error is used. If you don't recall the definition of this term, you should go to the appropriate section of a statistics textbook to review the meaning of this term.

Second Read: Tasks

1. **What question is being asked? What am I being asked to do?**
   You are asked to determine how many students should be included in the sample to achieve a given margin of error.

Third Read: Details

1. **What data were collected? Were the data collected in a reasonable way?**
   No data have been collected yet—you are just in the process of planning for data collection.

2. **Are the actual data provided? What is the sample size (or the sample sizes if there are more than one sample)?**
   A sample has not been selected yet—you want to determine the size of the sample that should be selected.

3. **Are any summary statistics or additional information given? If so, which statistics and what are the values of these statistics?**
   You know that you want to choose a sample size that would result in a margin of error of 0.02.

Now you are ready to tackle the tasks required to complete this exercise. This problem asks you to determine how many students should be included in a random sample of students from this college if you want to estimate the proportion with a margin of error of 0.02. To answer this question, you can use the formula for sample size:

$$n = p(1 - p)\left(\frac{1.96}{M}\right)^2$$

To use this expression to calculate a sample size, you need to know the desired value for the margin of error, M. In this problem, you want $M = 0.02$.

The other value you need to evaluate the expression for sample size is p, the value of the population proportion. This value is not known, so you should use $p = 0.5$ in the expression for sample size. This value for p is the one that results in the largest sample size compared to any other choice for the value of p.

Begin by substituting the values for M and p (0.5 for this problem) into the sample size expression. Substituting $M = 0.02$ and $p = 0.5$ into the expression, and using a 95% confidence level, results in

$$n = p(1 - p)\left(\frac{1.96}{M}\right)^2 = (0.5)(1 - 0.5)\left(\frac{1.96}{0.02}\right)^2$$

To evaluate this expression, you would first calculate the value of $\left(\frac{1.96}{0.02}\right)$. You would then square this number, and finally multiply the result by (0.5)(1 − 0.5). This sequence of steps illustrated here:

$$(0.5)(1-0.5)\left(\frac{1.96}{0.02}\right)^2 = (0.5)(1-0.5)(98)^2 = (0.5)(1-0.5)(9604) = 2401$$

In this example, the sample size calculation resulted in a whole number. If this had not been the case, you would have rounded the result up to the next whole number to obtain the necessary sample size.

Based on these calculations, you would use a sample size of 2401 when collecting data to estimate the proportion of students at the college who have used cell phones to cheat on an exam. This sample size should result in a margin of error of $M = 0.02$ or less, using a 95% confidence level.

SECTION 8.4 EXERCISES

8.11 Use the three-read strategy to understand the following exercise from a statistics textbook, and then proceed to complete the exercise.

The authors of the article "Consumers Show Increased Liking for Diesel Autos" (*USA TODAY*, January 29, 2003) reported that 27% of U.S. consumers would opt for a diesel car if it ran as cleanly and performed as well as a car with a gas engine. Suppose that you think that the proportion might be different in your area, and you will conduct a survey to estimate this proportion for the adult residents of your town. What is the required sample size if you wish to estimate the proportion with a margin of error of 0.01 using a confidence level of 95%?

8.12 Use the three-read strategy to understand the following exercise from a statistics textbook, and then proceed to complete the exercise.

The authors of the article "Hospitals Dispute Medtronic Data on Wires" (*The Wall Street Journal*, February 4, 2010) described a study of 89 patients with heart defibrillators. Eighteen of these patients experienced a defibrillator failure within the first 2 years. Suppose you have been asked to estimate the population proportion of patients with heart defribulators who experience defibrillator failure within 2 years, with a margin of error of 0.03 and using a 95% confidence level. How many patients with heart defibrillators should you include in a random sample?

9 Testing Hypotheses About a Population Proportion—The Math You Need to Know

SECTION 9.1 Evaluating Expressions

To determine if there is convincing evidence against a claim about a population proportion, you may calculate the value of a test statistic as part of the process for carrying out a test.

Note: Depending on your access to technology, such as a graphing calculator or a statistics software package, you may not need to evaluate the expression for the test statistic by hand. But just in case your statistics class is not relying on technology to do the calculation, this section looks at how to evaluate this expression.

Test Statistic for Testing Hypotheses About a Population Proportion

$$z = \frac{\hat{p} - p_0}{\sqrt{\frac{p_0(1 - p_0)}{n}}}$$

A hypothesis test about a population proportion uses data from a sample to decide between two competing hypotheses—a null hypothesis (denoted by H_0) and an alternative hypothesis (denoted by H_a). The null hypothesis specifies a particular hypothesized value of the population proportion, denoted by p_0. The null hypothesis will be of the form $H_0: p = p_0$. For example, you might want to carry out a hypothesis test where the null hypothesis states that population proportion is 0.6. The null hypothesis would then be $H_0: p = 0.6$. For purposes of calculating the value of the test statistic, the value of p_0 for this null hypothesis is $p_0 = 0.6$.

Before calculating the value of the test statistic, you need to verify that the sample size is large enough for the large-sample z test to be appropriate. You also need to confirm that the sample is a random sample or that it was selected in a way that makes it reasonable to think that the sample is representative of the population.

The sample size is large enough to use the large-sample z test for a population proportion if np_0 and $n(1 - p_0)$ are both greater than or equal to 10. For example, suppose you wanted to use data from a random sample of size $n = 40$ to test the null hypothesis

$H_0: p = 0.6$. Then $p_0 = 0.6$, and you would use this value to check the sample size condition as follows:

$$np_0 = (40)(0.6) = 24$$
$$n(1 - p_0) = (40)(1 - 0.6) = (40)(0.4) = 16$$

Because 24 is greater than or equal to 10 and 16 is also greater than or equal to 10, the sample size of 40 is large enough to use the large-sample z test for a population proportion.

Suppose that the sample contained 22 "successes," resulting in a sample proportion of $\hat{p} = \frac{22}{40} = 0.55$. To calculate the value of the test statistic, substitute the values of $\hat{p}$ (the value of the sample proportion), p_0 (the hypothesized value from the null hypothesis), and n into the formula for the test statistic to obtain:

$$z = \frac{\hat{p} - p_0}{\sqrt{\frac{p_0(1 - p_0)}{n}}} = \frac{0.55 - 0.6}{\sqrt{\frac{(0.6)(1 - 0.6)}{40}}}$$

Begin by evaluating the part of the expression in the numerator of the test statistic: $0.55 - 0.6 = -0.05$. The next step is to evaluate the part of the expression that appears under the square root. First calculate $(1 - 0.6) = 0.4$. Then multiply the two numbers in the numerator, and finally divide that number by 40. This is the sequence of steps shown here:

$$\frac{(0.6)(1 - 0.6)}{40} = \frac{(0.6)(0.4)}{40} = \frac{0.24}{40} = 0.006$$

To complete the calculation of the test statistic, find the square root of this number, which is $\sqrt{0.006} = 0.077$. Complete the calculation by dividing the -0.05 in the numerator by the value in the denominator (0.077) to obtain

$$z = \frac{\hat{p} - p_0}{\sqrt{\frac{p_0(1 - p_0)}{n}}} = \frac{0.55 - 0.6}{\sqrt{\frac{(0.6)(1 - 0.6)}{40}}} = \frac{-0.05}{\sqrt{0.006}} = \frac{-0.05}{0.077} = -0.65$$

(rounded to two decimal places)

The value of the test statistic for this example is $z = -0.65$.

Now suppose that you are carrying out a hypothesis test where the null hypothesis is $H_0: p = 0.35$ and that a random sample of size $n = 50$ resulted in a sample proportion of $\hat{p} = 0.54$. When carrying out the test, you will need to verify that the sample size is large enough for the large-sample z test to be appropriate, and if it is appropriate calculate the value of the z test statistic.

You would check the sample size condition as follows:

$$np_0 = (50)(0.35) = 17.5$$
$$n(1 - p_0) = (50)(1 - 0.35) = (50)(0.65) = 32.5$$

Because 17.5 is greater than or equal to 10 and 32.5 is also greater than or equal to 10, the sample size of 50 is large enough to use the large-sample z test for a population proportion.

To calculate the value of the test statistic, substitute the values of $\hat{p}$ (the value of the sample proportion), p_0 (the hypothesized value from the null hypothesis), and n into the formula for the test statistic to obtain:

$$z = \frac{\hat{p} - p_0}{\sqrt{\frac{p_0(1 - p_0)}{n}}} = \frac{0.54 - 0.35}{\sqrt{\frac{(0.35)(1 - 0.35)}{50}}}$$

You can evaluation this expression by first evaluating the numerator of the test statistic: $0.54 - 0.35 = 0.19$. The next step is to evaluate the part of the expression that appears under the square root. First calculate $(1 - 0.35) = 0.65$. Then multiply the two numbers in the numerator, and finally divide that number by 50. This is the sequence of steps shown here:

$$\frac{(0.35)(1-0.35)}{50}=\frac{(0.35)(0.65)}{50}=\frac{0.2275}{50}=0.0046$$

To complete the calculation of the test statistic, find the square root of this number, which is $\sqrt{0.0046}=0.068$. Complete the calculation by dividing the 0.19 in the numerator by the value in the denominator (0.068) to obtain

$$z=\frac{\hat{p}-p_0}{\sqrt{\frac{p_0(1-p_0)}{n}}}=\frac{0.54-0.35}{\sqrt{\frac{(0.35)(1-0.35)}{50}}}=\frac{0.19}{\sqrt{0.0046}}=\frac{0.19}{0.068}=2.79$$

(rounded to two decimal places)

(The numbers shown here are based on rounding intermediate calculations to three decimal places. If you keep more decimal places in the calculations, you may get a slightly different answer.)

The value of the test statistic for this example is $z=2.79$.

Check It Out!

a. Consider the null hypothesis $H_0: p=0.75$. A random sample of size $n=96$ was selected and contained 77 "successes." First calculate the value of the sample proportion, $\hat{p}$, and then calculate the value of the test statistic, z. Then, check the necessary conditions to determine if the large-sample z test is appropriate.

b. Consider the null hypothesis $H_0: p=0.4$. A random sample of size $n=76$ was selected, and contained 28 "successes." First calculate the value of the sample proportion, $\hat{p}$, and then calculate the value of the test statistic, z. Then, check the necessary conditions to determine if the large-sample z test is appropriate.

Answers

a. The sample of $n=96$ contained 77 "successes," so $\hat{p}=\frac{77}{96}=0.802$. Calculate the value of the test statistic by substituting the values of $\hat{p}$, p_0, and n into the formula to obtain

$$z=\frac{\hat{p}-p_0}{\sqrt{\frac{p_0(1-p_0)}{n}}}=\frac{0.802-0.75}{\sqrt{\frac{(0.75)(1-0.75)}{96}}}$$

The numerator of the test statistic is $0.802-0.75=0.052$.

For the denominator, evaluate

$$\frac{(0.75)(1-0.75)}{96}=\frac{(0.75)(0.25)}{96}=\frac{0.1875}{96}=0.0020$$

Calculate the square root of the value 0.0020 to obtain $\sqrt{0.0020}=0.045$.

Then the value of the test statistic is $z=1.11$:

$$z=\frac{\hat{p}-p_0}{\sqrt{\frac{p_0(1-p_0)}{n}}}=\frac{0.802-0.75}{\sqrt{\frac{(0.75)(1-0.75)}{96}}}=\frac{0.052}{\sqrt{0.0020}}=\frac{0.052}{0.045}=1.16$$

One condition is satisfied because you were told that the sample of $n=96$ was a random sample from the population of interest.

Check the sample size condition as follows:

$$np_0 = (96)(0.75) = 72$$
$$n(1 - p_0) = (96)(0.25) = 24$$

Because 72 is greater than or equal to 10 and 24 is also greater than or equal to 10, the sample size of 96 is large enough to use the large-sample z test for a population proportion.

b. The sample of $n = 76$ contained 28 "successes," so $\hat{p} = \frac{28}{76} = 0.368$. Calculate the value of the test statistic by substituting the values of $\hat{p}$, p_0, and n into the formula for the test statistic to obtain

$$z = \frac{\hat{p} - p_0}{\sqrt{\frac{p_0(1 - p_0)}{n}}} = \frac{0.368 - 0.4}{\sqrt{\frac{(0.4)(1 - 0.4)}{76}}}$$

The numerator of the test statistics is $0.368 - 0.4 = -0.032$.
For the denominator, evaluate

$$\frac{(0.4)(1 - 0.4)}{76} = \frac{(0.4)(0.6)}{76} = \frac{0.24}{76} = 0.0032$$

Calculate the square root of the value 0.0032 to obtain $\sqrt{0.0032} = 0.057$. Then the value of the test statistics is $z = -0.56$:

$$z = \frac{\hat{p} - p_0}{\sqrt{\frac{p_0(1 - p_0)}{n}}} = \frac{0.368 - 0.4}{\sqrt{\frac{(0.4)(1 - 0.4)}{76}}} = \frac{-0.032}{\sqrt{0.0032}} = \frac{-0.032}{0.057} = -0.56$$

One condition is satisfied because you were told that the sample of $n = 76$ was a random sample from the population of interest.
Check the sample size condition as follows:

$$np_0 = (76)(0.4) = 30.4$$
$$n(1 - p_0) = (76)(0.6) = 45.6$$

Because 30.4 is greater than or equal to 10 and 45.6 is also greater than or equal to 10, the sample size of 76 is large enough to use the large-sample z test for a population proportion.

SECTION 9.1 EXERCISES

9.1 Consider the null hypothesis $H_0: p = 0.45$. A random sample of size $n = 91$ was selected from a population of "successes" and "failures," and the sample contained 34 "successes." First calculate the value of the sample proportion, $\hat{p}$, and then calculate the value of the test statistic, z. Finally, check the necessary conditions to determine if the large-sample z test is appropriate.

9.2 Consider the null hypothesis $H_0 : p = 0.5$. A representative sample of size $n = 14$ was selected and contained 6 "successes." First, calculate the value of the sample proportion, $\hat{p}$, and calculate the value of the test statistic, z. Finally, check the necessary conditions to determine if the large-sample z test is appropriate.

SECTION 9.2 Guided Practice—Large-Sample Hypothesis Test for a Population Proportion

In this section, you will consider a typical situation that involves carrying out a large-sample z test for a population proportion. Consider this problem from a statistics textbook:

The article "Streaming Overtakes Live TV Among Consumer Viewing Preferences" (*Variety*, April 22, 2015) states that "U.S. consumers are more inclined to stream entertainment from an Internet service than tune in to live TV." This statement is based on a survey of a representative sample of 2076 U.S. consumers. Of those surveyed, 1100 indicated that they prefer to stream TV shows rather than watch TV programs live. Do the sample data provide convincing evidence that a majority of U.S. consumers prefer to stream TV shows rather than to watch them live? Test the relevant hypotheses using a 0.05 significance level.

If you were to encounter a problem like this, you should start by reading the problem carefully. You can use the three-read strategy introduced in Chapter 7.

First Read: Context

1. **What is the general context?**
   A survey was conducted to learn about the proportion of U.S. consumers who prefer to stream TV shows rather than watch TV shows live. People in a representative sample of U.S. consumers were surveyed.

2. **Are there any terms that you don't understand?**
   This question is about testing a hypothesis, so if you don't recall the steps required to carry out a hypothesis test, you would want to go to the appropriate section in your textbook to review the steps. If you are unsure about what it means to "stream TV shows," you could do an online search to find out that streaming means viewing video or other content from the Internet. If the term "consumer" is unclear in this context, you could also look up the definition of the term to find out that a consumer is someone who purchases goods or services for personal use.

Second Read: Tasks

1. **What question is being asked? What am I being asked to do?**
   This problem asks you to decide if the sample data provide convincing evidence that a majority of U.S. consumers prefer to stream TV shows rather than to watch them live. You are to do this by testing the relevant hypotheses using a 0.05 significance level.

Third Read: Details

1. **What data were collected? Were the data collected in a reasonable way?**
   Data were collected from a sample of 2076 U.S. consumers. The problem states that you can assume that the sample is representative of the population of U.S. consumers.

2. **Are the actual data provided? What was the sample size (or the sample sizes if there are more than one sample)?**
   The sample size was $n = 2076$ (the 2076 consumers surveyed). The actual data (the 2076 individual responses) are not given, but you know that 1100 of the consumers in the sample said that they preferred streaming to watching TV live.

3. **Are any summary statistics or additional information given? If so, which statistics and what are the values of these statistics?**
   The sample proportion is not given, but you know that 1100 of the 2076 responses indicated a preference for streaming, and this is the information needed to calculate the sample proportion $\hat{p}$.

Now you are ready to tackle the tasks required to complete this exercise. The question states: "Do the sample data provide convincing evidence that a majority of U.S. consumers prefer to stream TV shows rather than to watch them live? Test the relevant hypotheses using a 0.05 significance level." This is asking you to test hypotheses about a population proportion, p. The population is U.S. consumers, and you are asked to decide if there is convincing evidence that a majority prefer to stream. You are also told to use a significance level of 0.05 when carrying out the test.

Most textbooks recommend a sequence of steps for carrying out a hypothesis test. When you carry out a hypothesis test, you should follow the recommended sequence of steps. Although the order of the steps may vary somewhat from one text to another, they generally include the following:

Hypotheses: You will need to specify the null and the alternative hypothesis for the hypothesis test and define any symbols that are used in the context of the problem.

Method: You will need to specify what method you plan to use to carry out the test. For example, you might say that you plan to use the large-sample z test for a population proportion if that is a test that would allow you to answer the question posed. You will also need to specify the significance level you will use to reach a conclusion.

Check Conditions: Most hypothesis test procedures are only appropriate when certain conditions are met. Once you have determined the method you plan to use, you will need to verify that any required conditions are met.

Calculate: To reach a conclusion in a hypothesis test, you will need to calculate the value of a test statistic and the associated P-value.

Communicate Results: The final step is to use the calculated values of the test statistic and the P-value to reach a conclusion and to provide an answer to the question posed. You should always provide a conclusion in words.

These steps are illustrated here for the given example.

Hypotheses: You want to determine if there is evidence that a majority of U.S. consumers prefer to stream TV shows rather than to watch them live. A majority means more than half, or a proportion that is greater than 0.5. The population is U.S. consumers, and you can use p to represent the proportion of U.S. consumers who prefer to stream TV shows rather than to watch them live. The question being asked usually translates into the alternative hypothesis, so you would specify an alternative hypothesis of $H_a: p > 0.5$. The null hypothesis would then be obtained by replacing the $>$ in the alternative hypothesis with an $=$. This results in hypotheses of

$$H_0: p = 0.5$$
$$H_a: p > 0.5$$

Method: Because the hypotheses are about a population proportion, you would consider the large-sample test for a population proportion. The test statistic is

$$z = \frac{\hat{p} - 0.5}{\sqrt{\frac{0.5(1 - 0.5)}{n}}}$$

A significance level of $\alpha = 0.05$ was specified.

Check: There are two conditions necessary for the large-sample z test for a population proportion to be appropriate. You are told that the sample is representative of U.S. consumers, and the sample size is $n = 2076$. Using the value of p_0 from the null hypothesis ($p_0 = 0.5$), you can check the sample size condition:

$$np_0 = (2076)(0.5) = 1038$$
$$n(1 - p_0) = (2076)(1 - 0.5) = (2076)(0.5) = 1038$$

Because both np_0 and $n(1 - p_0)$ are greater than or equal to 10, the sample size condition is met.

Calculations: The sample proportion is $\hat{p} = \frac{1100}{2076} = 0.530$ and $n = 2076$, so the value of the test statistic is

$$z = \frac{0.530 - 0.5}{\sqrt{\frac{0.5(1 - 0.5)}{2076}}} = 2.73.$$

Once you know the value of the test statistic, you need to find the associated *P*-value. Your statistics text will have an explanation of how the *P*-value is determined.

This is an upper-tailed test (the inequality in H_0 is $>$), so the *P*-value is the area under the z curve and to the right of 2.73. Therefore, the *P*-value is $P(z \geq 2.73) = 0.003$.

Communicate Results: Because the *P*-value of 0.003 is less than the selected significance level of 0.05, you reject the null hypothesis. There is convincing evidence that a majority of U.S. consumers prefer to stream TV shows rather than to watch them live.

SECTION 9.2 EXERCISES

9.3 Use the three-read strategy to understand the following exercise from a statistics textbook, and then proceed to complete the exercise.

The article "Theaters Losing Out to Living Rooms" appeared in the *San Luis Obispo Tribune* on June 17, 2005, and states that movie attendance declined in 2005. The Associated Press found that 730 of 1000 randomly selected adult Americans prefer to watch movies at home rather than at a movie theater. Is there convincing evidence that a majority (that is, more than 50%) of adult Americans prefer to watch movies at home? Test the relevant hypotheses using a 0.05 significance level.

9.4 Use the three-read strategy to understand the following exercise from a statistics textbook, and then proceed to complete the exercise.

In a survey conducted by CareerBuilders.com, employers were asked if they had ever sent an employee home because they were dressed inappropriately (June 17, 2008, careerbuilders.com). A total of 2765 employers responded to the survey, with 968 saying that they had sent an employee home for inappropriate attire. In a press release, CareerBuilder makes the claim that more than one-third of employers have sent an employee home to change clothes. Do the sample data provide convincing evidence in support of this claim? Test the relevant hypotheses using $\alpha = 0.05$. For purposes of this exercise, you can assume that the sample is representative of employers in the United States.

10 Estimating a Difference in Two Proportions—The Math You Need to Know

SECTION 10.1 Evaluating Expressions

Many statistical studies are carried out to estimate the difference between two population or treatment proportions. If you want to use a confidence interval to provide an estimate, you may need to be able to evaluate the expression for the large-sample confidence interval for a difference in two proportions.

Note: Depending on your access to technology, such as a graphing calculator or a statistics software package, you may not need to evaluate this expression by hand. But just in case your statistics class is not relying on technology to do the calculations, this section considers how to evaluate the expression.

Large-Sample Confidence Interval for a Difference in Proportions

$$(\hat{p}_1 - \hat{p}_2) \pm (z \text{ critical value})\sqrt{\frac{\hat{p}_1(1-\hat{p}_1)}{n_1} + \frac{\hat{p}_2(1-\hat{p}_2)}{n_2}}$$

When you have samples from each of two populations or treatments, subscripts are added to the notation used for the sample statistics to distinguish between them. The notation is summarized in the table below:

Notation	Meaning
p_1	Actual value of the proportion for population or treatment 1
p_2	Actual value of the proportion for population or treatment 2
$\hat{p}_1$	Sample proportion for sample from population or treatment 1
$\hat{p}_2$	Sample proportion for sample from population or treatment 2
n_1	Sample size for sample from population or treatment 1
n_2	Sample size for sample from population or treatment 2

The difference in the actual values of the population or treatment proportions is written as $p_1 - p_2$. To calculate a confidence interval for a difference in two population or treatment proportions, $p_1 - p_2$, you need to know the values of the two sample proportions, $\hat{p}_1$ and $\hat{p}_2$, and the sample sizes, n_1 and n_2. You also need to know the confidence level, which determines the z critical value that is used to calculate the confidence interval.

Before using the given expression to calculate a confidence interval, you need to verify that the sample sizes are large enough for this confidence interval to be appropriate. If you have data from two populations, you also need to confirm that the samples are random samples or that they were selected in a way that makes it reasonable to think that the samples are representative of the populations. If you have data from an experiment, you need to confirm that the experiment used random assignment to the treatment groups.

The sample sizes are large enough to use this confidence interval if $n_1\hat{p}_1$, $n_1(1 - \hat{p}_1)$, $n_2\hat{p}_2$, and $n_2(1 - \hat{p}_2)$ are each greater than or equal to 10. For example, suppose you wanted to use data from a random sample from each of two populations. Suppose that the sample from population 1 had $n_1 = 60$ and $\hat{p}_1 = 0.26$, and the sample from population 2 had $n_2 = 70$ and $\hat{p}_2 = 0.18$.

You would use these values to check the sample size conditions as follows:

$$\begin{aligned}
n_1\hat{p}_1 &= (60)(0.26) = 15.6 \\
n_1(1 - \hat{p}_1) &= (60)(1 - 0.26) = (60)(0.74) = 44.4 \\
n_2\hat{p}_2 &= (70)(0.18) = 12.6 \\
n_2(1 - \hat{p}_2) &= (70)(1 - 0.18) = (70)(0.82) = 57.4
\end{aligned}$$

Because each of these is greater than or equal to 10, the sample sizes of 60 and 70 are large enough to justify use of the large-sample confidence interval for a difference in population proportions.

Recall that an interval is defined by its endpoints—it has a lower endpoint and an upper endpoint. Notice that the expression for the confidence interval includes the symbol $\pm$, which is read as "plus and minus." This means that you will do two calculations, once using $+$ and once using $-$. Each one of these calculations results in one of the two interval endpoints.

To calculate the endpoints of the confidence interval, you can start by determining the appropriate z critical value for the specified confidence level. Your statistics text should have instructions for how this is done. The z critical values for the most widely used confidence levels are given in the following table:

Confidence Level	z Critical Value
90%	1.645
95%	1.96
99%	2.58

Once you have determined the appropriate z critical value, you can substitute the values of the two sample proportions, $\hat{p}_1$ and $\hat{p}_2$, the sample sizes n_1 and n_2, and the z critical value into the expression for the confidence interval. For example, suppose you want to calculate a 90% confidence interval for a difference in population proportions and the sample from population 1 had $n_1 = 60$ and $\hat{p}_1 = 0.26$ and the sample from population 2 had $n_2 = 70$ and $\hat{p}_2 = 0.18$. For a 90% confidence level, the z critical value is 1.645. Substituting these values into the expression for the confidence interval results in the following:

$$\begin{aligned}
&(\hat{p}_1 - \hat{p}_2) \pm (z \text{ critical value})\sqrt{\frac{\hat{p}_1(1 - \hat{p}_1)}{n_1} + \frac{\hat{p}_2(1 - \hat{p}_2)}{n_2}} \\
&= (0.26 - 0.18) \pm (1.645)\sqrt{\frac{(0.26)(1 - 0.26)}{60} + \frac{(0.18)(1 - 0.18)}{70}}
\end{aligned}$$

Begin by evaluating the parts of the expression that appear under the square root. Calculate $(1 - 0.26) = 0.74$. Then multiply the two numbers in the numerator of the first term, and finally divide that number by 60. This is the sequence of steps shown here:

$$\frac{(0.26)(1 - 0.26)}{60} = \frac{(0.26)(0.74)}{60} = \frac{0.192}{60} = 0.003$$

(rounded to three decimal places)

Evaluate the second term under the square root symbol in a similar way:

$$\frac{(0.18)(1-0.18)}{70} = \frac{(0.18)(0.82)}{70} = \frac{0.148}{70} = 0.002$$

(rounded to three decimal places)

Next, add these two values together, and find the square root.

$$\sqrt{\frac{(0.26)(1-0.26)}{60} + \frac{(0.18)(1-0.18)}{70}} = \sqrt{0.003 + 0.002} = \sqrt{0.005} = 0.071$$

This number is multiplied by 1.645, as shown here:

$$(1.645)(0.071) = 0.117$$

This number is the "plus and minus" part of the confidence interval. To calculate the endpoints of the interval, you would evaluate

$$(0.26 - 0.18) \pm 0.117$$
$$0.08 \pm 0.117$$

The lower endpoint of the interval is $0.08 - 0.117 = -0.037$. The upper endpoint of the interval is $0.08 + 0.117 = 0.197$.

The 90% confidence interval estimate of the difference in proportions is $(-0.037, 0.197)$. You would interpret this interval by saying that based on the data from the samples, you can be 90% confident that the actual value of the difference in population proportions is somewhere between -0.037 and 0.197.

Check It Out!

a. Suppose that separate random samples are selected from two populations. The random sample from the first population has a sample size $n_1 = 94$ and a sample proportion $\hat{p}_1 = 0.57$. The random sample from the second population has a sample size $n_2 = 75$ and a sample proportion $\hat{p}_2 = 0.44$. First, use this information to calculate and to interpret a 99% confidence interval for the difference in the two population proportions, which is represented as $p_1 - p_2$. Then check to determine if the necessary conditions are satisfied.

b. Representative samples are selected from two populations. The representative sample from the first population has sample size $n_1 = 58$ and sample proportion $\hat{p}_1 = 0.47$. The representative sample from the second population has sample size $n_2 = 99$ and sample proportion $\hat{p}_2 = 0.16$. First, use this information to calculate and to interpret a 95% confidence interval for the difference in the two population proportions, which is represented as $p_1 - p_2$. Then check to determine if the necessary conditions are satisfied.

Answers

a. For a 99% confidence level, the z critical value is 2.58. Substituting the values for n_1, $\hat{p}_1$, n_2, and $\hat{p}_2$ into the expression for the confidence interval results in the following:

$$(\hat{p}_1 - \hat{p}_2) \pm (z \text{ critical value})\sqrt{\frac{\hat{p}_1(1-\hat{p}_1)}{n_1} + \frac{\hat{p}_2(1-\hat{p}_2)}{n_2}}$$

$$= (0.57 - 0.44) \pm (2.58)\sqrt{\frac{(0.57)(1-0.57)}{94} + \frac{(0.44)(1-0.44)}{75}}$$

Begin by evaluating the parts of the expression that appear under the square root. To evaluate the first term, calculate

$$\frac{(0.57)(1-0.57)}{94} = \frac{(0.57)(0.43)}{94} = \frac{0.2451}{94} = 0.003$$

(rounded to three decimal places)

Evaluate the second term under the square root symbol in a similar way:

$$\frac{(0.44)(1-0.44)}{75} = \frac{(0.44)(0.56)}{75} = \frac{0.2464}{75} = 0.003$$

(rounded to three decimal places)

Then add and take the square root to obtain

$$\sqrt{\frac{(0.57)(1-0.57)}{94} + \frac{(0.44)(1-0.44)}{75}} = \sqrt{0.003 + 0.003} = \sqrt{0.006} = 0.077$$

Multiply by 2.58 to obtain the "plus and minus" part of the expression:

$$(2.58)(0.077) = 0.199$$

Next, calculate the endpoints of the interval:

$$(0.57 - 0.44) \pm 0.199$$
$$0.13 \pm 0.199$$

The lower endpoint of the interval is $0.13 - 0.199 = -0.069$, and the upper endpoint of the interval is $0.13 + 0.199 = 0.329$.

The 99% confidence interval estimate of the difference in proportions is (−0.069, 0.329). You would interpret this interval by saying that based on the data from the samples, you can be 99% confident that the actual value of the difference in population proportions is somewhere between −0.069 and 0.329.

One condition is to have random (or at least justifiably representative) samples from both populations. This condition is satisfied because the problem states that the samples were random samples.

The second condition is the sample size condition, which you can check as follows:

$$n_1\hat{p}_1 = (94)(0.57) = 54$$
$$n_1(1-\hat{p}_1) = (94)(1-0.57) = (94)(0.43) = 40$$
$$n_2\hat{p}_2 = (75)(0.44) = 33$$
$$n_2(1-\hat{p}_2) = (75)(1-0.44) = (75)(0.56) = 42$$

Because all of these are greater than or equal to 10, the sample sizes of 94 and 75 are large enough to allow you to safely use the large-sample confidence interval for a difference in population proportions.

b. For a 95% confidence level, the z critical value is 1.96. Substituting the values for n_1, $\hat{p}_1$, n_2, and $\hat{p}_2$ into the expression for the confidence interval results in the following:

$$(\hat{p}_1 - \hat{p}_2) \pm (z \text{ critical value})\sqrt{\frac{\hat{p}_1(1-\hat{p}_1)}{n_1} + \frac{\hat{p}_2(1-\hat{p}_2)}{n_2}}$$
$$= (0.47 - 0.16) \pm (1.96)\sqrt{\frac{(0.47)(1-0.47)}{58} + \frac{(0.16)(1-0.16)}{99}}$$

Begin by evaluating the parts of the expression that appear under the square root. Evaluate the first term by calculating

$$\frac{(0.47)(1-0.47)}{58} = \frac{(0.47)(0.53)}{58} = \frac{0.2491}{58} = 0.004$$

(rounded to three decimal places)

Evaluate the second term under the square root symbol in a similar way:

$$\frac{(0.16)(1 - 0.16)}{99} = \frac{(0.16)(0.84)}{99} = \frac{0.1344}{99} = 0.001$$

(rounded to three decimal places)

Then add and take the square root to obtain

$$\sqrt{\frac{(0.47)(1 - 0.47)}{58} + \frac{(0.16)(1 - 0.16)}{99}} = \sqrt{0.004 + 0.001} = \sqrt{0.005} = 0.071$$

Multiply by 1.96 to obtain the "plus and minus" part of the expression:

$$(1.96)(0.071) = 0.139$$

Then calculate the endpoints of the interval:

$$(0.47 - 0.16) \pm 0.139$$
$$0.31 \pm 0.139$$

The lower endpoint of the interval is $0.31 - 0.139 = 0.171$, and the upper endpoint of the interval is $0.31 + 0.139 = 0.449$.

The 95% confidence interval estimate of the difference in proportions is (0.171, 0.449). You would interpret this interval by saying that based on the data from the samples, you can be 95% confident that the actual value of the difference in population proportions is somewhere between 0.171 and 0.449.

One condition is to have random (or at least justifiably representative) samples from both populations. This condition is satisfied because the problem states that the samples may be considered representative of their respective populations.

The second condition is the sample size condition, which you can check as follows:

$$n_1\hat{p}_1 = (58)(0.47) = 27$$
$$n_1(1 - \hat{p}_1) = (58)(1 - 0.47) = (58)(0.53) = 31$$
$$n_2\hat{p}_2 = (99)(0.16) = 16$$
$$n_2(1 - \hat{p}_2) = (99)(1 - 0.16) = (99)(0.84) = 83$$

Because all of these are greater than or equal to 10, the sample sizes of 58 and 99 are large enough to allow you to safely use the large-sample confidence interval for a difference in population proportions.

SECTION 10.1 EXERCISES

10.1 An experiment is carried out in order to estimate the difference in treatment proportions. Individuals are randomly assigned to one of two treatment groups. Treatment Group 1 has a sample size of $n_1 = 59$ and a sample proportion of $\hat{p}_1 = 0.20$. Treatment Group 2 has a sample size of $n_2 = 12$ and a sample proportion of $\hat{p}_2 = 0.58$. First, use this information to calculate and to interpret a 90% confidence interval for the difference in the two treatment proportions, which is represented as $p_1 - p_2$. Then check to determine if the necessary conditions are satisfied.

10.2 A random sample is selected from each of two populations in order to estimate the difference in population proportions. The random sample from the first population has sample size $n_1 = 494$ and sample proportion $\hat{p}_1 = 0.188$. The random sample from the second population has sample size $n_2 = 724$ and sample proportion $\hat{p}_2 = 0.113$. First, use this information to calculate and to interpret a 95% confidence interval for the difference in the two population proportions, which is represented as $p_1 - p_2$. Then check to determine if the necessary conditions are satisfied.

SECTION 10.2 Guided Practice—Large-Sample Confidence Interval for a Difference in Two Proportions

Before considering a guided example that involves calculating and interpreting a confidence interval for the difference in two proportions, take a minute to review how intervals that estimate a difference are interpreted.

Interpreting confidence intervals for a difference in proportions is a bit more complicated than interpreting a confidence interval for a single proportion. The following table explains how an interval for a difference in proportions can be interpreted in three different cases:

Case	Interpretation	Example
Both endpoints of the confidence interval for $p_1 - p_2$ are positive.	If $p_1 - p_2$ is positive, it means that you think p_1 is greater than p_2 and the interval gives an estimate of how much greater.	(0.24, 0.36) You think that p_1 is greater than p_2 by somewhere between 0.24 and 0.36.
Both endpoints of the confidence interval for $p_1 - p_2$ are negative.	If $p_1 - p_2$ is negative, it means that you think p_1 is less than p_2 and the interval gives an estimate of how much less.	(−0.14, −0.06) You think that p_1 is less than p_2 by somewhere between 0.06 and 0.14. (Or equivalently, you think that p_2 is greater than p_1 by somewhere between 0.06 and 0.14.)
0 is included in the confidence interval.	If the confidence interval includes 0, a plausible value for $p_1 - p_2$ is 0. This suggests that p_1 and p_2 might be equal.	(−0.04, 0.09) Because 0 is included in the confidence interval, it is plausible that the two population proportions might be equal.

Now consider this problem from a statistics textbook:

A Harris Poll press release dated November 1, 2016, summarized results of a survey of 2463 adults and 510 teens age 13 to 17 ("American Teens No Longer More Likely than Adults to Believe in God, Miracles, Heaven, Jesus, Angels, or the Devil," theharrispoll.com, retrieved December 12, 2016). It was reported that 19% of the teens surveyed and 26% of the adults surveyed indicated that they believe in reincarnation. The samples were selected to be representative of American adults and teens. Use the data from this survey to estimate the difference in the proportion of adults who believe in reincarnation and the proportion of teens who believe in reincarnation. Use a confidence level of 95% and be sure to interpret your interval in context.

If you were to encounter a problem like this, you should start by reading the problem carefully. You can use the three-read strategy introduced in Chapter 7.

First Read: Context

1. **What is the general context?**
A survey was conducted to learn about the difference in the proportions of adults and the proportion of teens who believe in reincarnation.

2. **Are there any terms that you don't understand?**
This question is about constructing a confidence interval, so if you aren't comfortable with the meaning of this term you would want to go to the appropriate section in your textbook to review the definition. If any of the other non-statistical terms are unfamiliar (such as "reincarnation"), you can look up the meaning of those terms in a dictionary to make sure that you fully understand the context.

Second Read: Tasks

1. **What question is being asked? What am I being asked to do?**
This problem asks you to construct and interpret a 95% confidence interval for the difference in the proportion of adults who believe in reincarnation and the proportion of teens who believe in reincarnation.

Third Read: Details

1. **What data were collected? Were the data collected in a reasonable way?**
Data were collected from a sample of 2463 adults and a sample of 510 teens. Each person in the two samples indicated whether they believe in reincarnation. The problem states that you can assume that the samples are representative of American adults and American teens.
2. **Are the actual data provided? What was the sample size (or the sample sizes if there are more than one sample)?**
The sample size was 2463 for the sample of adults and 510 for the sample of teens. The actual data (the individual responses) are not provided, but you are told that 26% of the adults and 19% of the teens indicated belief in reincarnation.
3. **Are any summary statistics or additional information given? If so, which statistics and what are the values of these statistics?**
The sample proportions are not given, but the percentage who believe in reincarnation is given for each sample (26% for adults and 19% for teens) and the sample proportions can be determined from these percentages.

Now you are ready to tackle the tasks required to complete this exercise. This problem asks you to construct and interpret a 95% confidence interval for the difference in the proportion of adults who believe in reincarnation and the proportion of teens who believe in reincarnation.

The expression used to calculate a confidence interval for a difference in proportions is

$$(\hat{p}_1 - \hat{p}_2) \pm (z \text{ critical value})\sqrt{\frac{\hat{p}_1(1-\hat{p}_1)}{n_1} + \frac{\hat{p}_2(1-\hat{p}_2)}{n_2}}$$

To calculate this confidence interval, you need to know the values of the two sample proportions, $\hat{p}_1$ and $\hat{p}_2$, the sample sizes n_1 and n_2, and the confidence level, which determines the z critical value that is used to calculate the confidence interval.

Start by deciding which population you will refer to as population 1 and which you will refer to as population 2. For example, you might choose to think of adults as population 1 and teens as population 2. Then whenever you see a subscript of 1, you would know that it is referring to adults. The subscript 2 will refer to teens. For adults, you know that the sample size was $n_1 = 2463$ and that 26% of those in this sample believe in reincarnation. You can determine the value of the sample proportion for this sample by converting the given percentage (26%) to a proportion. To do this, you drop the % symbol and divide by 100 to obtain $\hat{p}_1 = 0.26$. For teens, you know that the sample size was $n_2 = 510$ and that 19% of those in this sample believe in reincarnation. You can determine the value of the sample proportion for this sample by converting the given percentage (19%) to a proportion to obtain $\hat{p}_2 = 0.19$.

Before using the confidence interval formula, you need to verify that the sample sizes are large enough for the large-sample confidence interval to be appropriate. You also need to confirm that the samples are random samples or that they were selected in a way that makes it reasonable to think that the samples are representative of the populations.

The problem states that you can assume that the samples are representative of the population of adults and the population of teens. The sample sizes are large enough to use the large-sample confidence interval formula if $n_1\hat{p}_1$, $n_1(1-\hat{p}_1)$, $n_2\hat{p}_2$, and $n_2(1-\hat{p}_2)$ are each greater than or equal to 10.

Check the sample size conditions as follows:

$$\begin{aligned} n_1\hat{p}_1 &= (2463)(0.26) = 640 \\ n_1(1-\hat{p}_1) &= (2463)(1-0.26) = (2463)(0.74) = 1823 \\ n_2\hat{p}_2 &= (510)(0.19) = 97 \\ n_2(1-\hat{p}_2) &= (510)(1-0.19) = (70)(0.81) = 413 \end{aligned}$$

Because each of these is greater than or equal to 10, the sample sizes of 2463 and 510 are large enough to use the large-sample confidence interval for a difference in proportions.

To calculate the endpoints of the confidence interval, you can start by determining the appropriate z critical value for the specified confidence level. For a 95% confidence level, the corresponding z critical value is 1.96. (Your statistics text should have instructions for how to find this value.)

Substitute the values of the two sample proportions, $\hat{p}_1$ and $\hat{p}_2$, the sample sizes n_1 and n_2, and the z critical value into the expression for the confidence interval:

$$(\hat{p}_1 - \hat{p}_2) \pm (z \text{ critical value})\sqrt{\frac{\hat{p}_1(1-\hat{p}_1)}{n_1} + \frac{\hat{p}_2(1-\hat{p}_2)}{n_2}}$$

$$= (0.26 - 0.19) \pm (1.96)\sqrt{\frac{(0.26)(1-0.26)}{2463} + \frac{(0.19)(1-0.19)}{510}}$$

To evaluate this expression, start by evaluating the parts of the expression that appears under the square root. To evaluate the first term, calculate $(1 - 0.26) = 0.74$, and then multiply the two numbers in the numerator of the first term. Finally, divide that number by the sample size, 2463. This is the sequence of steps shown here:

$$\frac{(0.26)(1-0.26)}{2463} = \frac{(0.26)(0.74)}{2463} = \frac{0.192}{2463} = 0.00008$$

Then evaluate the second term under the square root symbol in a similar way:

$$\frac{(0.19)(1-0.19)}{510} = \frac{(0.19)(0.81)}{510} = \frac{0.154}{510} = 0.00030$$

Next, you would add these two values together and then find the square root:

$$\sqrt{\frac{(0.26)(1-0.26)}{2463} + \frac{(0.19)(1-0.19)}{510}} = \sqrt{0.00008 + 0.00030} = \sqrt{0.00038} = 0.019$$

This number is multiplied by 1.96, as shown here:

$$(1.96)(0.019) = 0.037$$

This number is the "plus and minus" part of the confidence interval. To calculate the endpoints of the interval, you would evaluate

$$(0.26 - 0.19) \pm 0.037$$
$$0.07 \pm 0.037$$

The lower endpoint of the interval is $0.07 - 0.037 = 0.033$. The upper endpoint of the interval is $0.07 + 0.037 = 0.107$.

The 95% confidence interval estimate of the difference in proportions is (0.033, 0.107).

You also need to interpret the confidence interval in context. You would interpret this interval by saying that based on the data from these samples, you can be 95% confident that the actual value of the difference in the proportion of American adults who believe in reincarnation and the proportion of American teens who believe in reincarnation falls between 0.033 and 0.107. Because both endpoints of the confidence interval are positive, you would say that the proportion of adults who believe in reincarnation is greater than the proportion of teens age 13 to 17 who believe in reincarnation by somewhere between 0.033 and 0.107.

You could also interpret the interval in terms of percentages, saying that you are 95% confident that the difference in the percentage of American adults who believe in reincarnation and the percentage of American teens who believe in reincarnation is somewhere between 3.3 percentage points and 10.7 percentage points. Because both endpoints of the confidence interval are positive, you would say that the percent of adults who believe in reincarnation is greater than the percent of teens age 13 to 17 who believe in reincarnation by somewhere between 3.3 percentage points and 10.7 percentage points.

SECTION 10.2 EXERCISES

10.3 Use the three-read strategy to understand the following exercise from a statistics textbook, and then proceed to complete the exercise.

"Smartest People Often Dumbest About Sunburns" is the headline of an article that appeared in the *San Luis Obispo Tribune* (July 19, 2006). The article states that "those with a college degree reported a higher incidence of sunburn than those without a high school degree—43% versus 25%." Suppose that these percentages were based on independent random samples of size 200 from each of the two groups of interest (college graduates and those without a high school degree). Check the required conditions to determine if they are satisfied. Then estimate the difference in the proportion of people with a college degree who reported sunburn and the corresponding proportion for those without a high school degree, using a 90% confidence interval.

10.4 Use the three-read strategy to understand the following exercise from a statistics textbook and then proceed to complete the exercise.

Public Agenda conducted a survey of 1379 parents and 1342 students in grades 6 to 12 regarding the importance of science and mathematics in the school curriculum (Associated Press, February 15, 2006). It was reported that 50% of students thought that understanding science and having strong math skills are essential for them to succeed in life after school, whereas 62% of the parents thought these were essential. The two samples—parents and students—were independently selected random samples. Using a confidence level of 95%, estimate the difference between the proportion of students who think that understanding science and having math skills are essential and the corresponding proportion for parents. Check the necessary conditions to determine if they are satisfied.

11 Testing Hypotheses About a Difference in Two Proportions—The Math You Need to Know

SECTION 11.1 Evaluating Expressions

Many statistical studies are carried out to test hypotheses about the difference between two population or treatment proportions. To determine if there is convincing evidence that two population or treatment proportions are not equal, you may need to calculate the value of the test statistic as part of the process for carrying out the test.

Note: Depending on your access to technology, such as a graphing calculator or a statistics software package, you may not need to evaluate this expression by hand. But just in case your statistics class is not relying on technology to do the calculations, this section considers how to evaluate the expression.

Test Statistic for Testing Hypotheses About a Difference in Population Proportions

$$z = \frac{\hat{p}_1 - \hat{p}_2}{\sqrt{\frac{\hat{p}_c(1-\hat{p}_c)}{n_1} + \frac{\hat{p}_c(1-\hat{p}_c)}{n_2}}}$$

$$\hat{p}_c = \frac{n_1\hat{p}_1 + n_2\hat{p}_2}{n_1 + n_2}$$

When you have samples from each of two populations or treatments, subscripts are added to the notation used for the sample statistics to distinguish between them. The notation is summarized in the table below:

Notation	Meaning
p_1	Actual proportion for population or treatment 1
p_2	Actual proportion for population or treatment 2
$\hat{p}_1$	Sample proportion for sample from population or treatment 1
$\hat{p}_2$	Sample proportion for sample from population or treatment 2
n_1	Sample size for sample from population or treatment 1
n_2	Sample size for sample from population or treatment 2

The difference in the actual population or treatment proportions is written as $p_1 - p_2$.

A hypothesis test about a difference in two proportions uses data from a sample from each of two populations or treatments to decide between two competing hypotheses—a null hypothesis (denoted by H_0) and an alternative hypothesis (denoted by H_a). In this test, the null hypothesis specifies that the two population or treatment proportions are equal. This null hypothesis can be written in two different ways: $H_0: p_1 = p_2$ or $H_0: p_1 - p_2 = 0$, because if $p_1 = p_2$, the difference between them is equal to 0.

To calculate the value of the test statistic, you need to know the values of the two sample proportions, $\hat{p}_1$ and $\hat{p}_2$, the sample sizes n_1 and n_2, and the value of the combined estimate of the common proportion, $\hat{p}_c$. The null hypothesis states that the two population or treatment proportions are equal, and $\hat{p}_c$ is an estimate of this common value. The value of $\hat{p}_c$ is calculated by evaluating the following expression:

$$\hat{p}_c = \frac{n_1\hat{p}_1 + n_2\hat{p}_2}{n_1 + n_2}$$

Before using the large-sample z test statistic to carry out a hypothesis test, you need to verify that the sample sizes are large enough for this test to be appropriate. If you have data from two populations, you also need to confirm that the samples are random samples or that they were selected in a way that makes it reasonable to think that the samples are representative of the populations. If you have data from an experiment, you need to confirm that the experiment used random assignment to the treatment groups.

The sample sizes are large enough to use the large-sample z test statistic if $n_1\hat{p}_1$, $n_1(1 - \hat{p}_1)$, $n_2\hat{p}_2$, and $n_2(1 - \hat{p}_2)$ are all greater than or equal to 10.

For example, suppose you wanted to use data from random samples from two populations. Suppose that the sample from population 1 had $n_1 = 40$ and $\hat{p}_1 = 0.525$, and the sample from population 2 had $n_2 = 55$ and $\hat{p}_2 = 0.400$.

Use these values to check the sample size conditions as follows:

$$\begin{aligned} n_1\hat{p}_1 &= (40)(0.525) = 21 \\ n_1(1 - \hat{p}_1) &= (40)(1 - 0.525) = (40)(0.475) = 19 \\ n_2\hat{p}_2 &= (55)(0.400) = 22 \\ n_2(1 - \hat{p}_2) &= (55)(1 - 0.400) = (55)(0.600) = 33 \end{aligned}$$

Because all of these are greater than or equal to 10, the sample sizes of 40 and 55 are large enough to use the large-sample z test statistic to test a hypothesis about a difference in two proportions.

To calculate the value of the test statistic, start by finding the value of $\hat{p}_c$. Substitute the values of the two sample proportions, $\hat{p}_1$ and $\hat{p}_2$, and the sample sizes n_1 and n_2 into the expression for the estimate of the common proportion, $\hat{p}_c$. This results in

$$\hat{p}_c = \frac{n_1\hat{p}_1 + n_2\hat{p}_2}{n_1 + n_2} = \frac{(40)(0.525) + (55)(0.400)}{40 + 55}$$

Complete the calculation by determining the value of the numerator and the value of the denominator and then dividing, as shown here:

$$\hat{p}_c = \frac{n_1\hat{p}_1 + n_2\hat{p}_2}{n_1 + n_2} = \frac{(40)(0.525) + (55)(0.400)}{40 + 55} = \frac{21 + 22}{95} = \frac{43}{95} = 0.453$$

(rounded to three decimal places)

Now you can calculate the value of the z test statistic by substituting the values of the two sample proportions, $\hat{p}_1$ and $\hat{p}_2$, the sample sizes n_1 and n_2, and the value of $\hat{p}_c$ into the expression for the test statistic. This results in

$$z = \frac{\hat{p}_1 - \hat{p}_2}{\sqrt{\frac{\hat{p}_c(1 - \hat{p}_c)}{n_1} + \frac{\hat{p}_c(1 - \hat{p}_c)}{n_2}}} = \frac{0.525 - 0.400}{\sqrt{\frac{(0.453)(1 - 0.453)}{40} + \frac{(0.453)(1 - 0.453)}{55}}}$$

To evaluate this expression, start by evaluating the parts of the expression that appear under the square root. To evaluate the first term, first calculate $(1 - 0.453) = 0.547$. Then multiply the two numbers in the numerator of the first term, and finally divide that number by 40. This is the sequence of steps shown here:

$$\frac{(0.453)(1 - 0.453)}{40} = \frac{(0.453)(0.547)}{40} = \frac{0.248}{40} = 0.006$$

(rounded to three decimal places)

Then evaluate the second term under the square root symbol in a similar way:

$$\frac{(0.453)(1 - 0.453)}{55} = \frac{(0.453)(0.547)}{55} = \frac{0.248}{55} = 0.005$$

(rounded to three decimal places)

Next, you would add these two values together and then find the square root:

$$\sqrt{\frac{(0.453)(1 - 0.453)}{40} + \frac{(0.453)(1 - 0.453)}{55}} = \sqrt{0.006 + 0.005} = \sqrt{0.011} = 0.105$$

This number is the value of the denominator of the z test statistic. You can now complete the calculation as shown here:

$$z = \frac{\hat{p}_1 - \hat{p}_2}{\sqrt{\frac{\hat{p}_c(1 - \hat{p}_c)}{n_1} + \frac{\hat{p}_c(1 - \hat{p}_c)}{n_2}}} = \frac{0.525 - 0.400}{\sqrt{\frac{(0.453)(1 - 0.453)}{40} + \frac{(0.453)(1 - 0.453)}{55}}}$$

$$= \frac{0.125}{0.105} = 1.19 \text{ (rounded to two decimal places)}$$

The value of the test statistic for this example is $z = 1.19$.

Check It Out!

a. Suppose that you have collected data from representative samples taken from two separate populations. The sample from population 1 has sample size $n_1 = 86$ and sample proportion of successes $\hat{p}_1 = 0.86$. The sample from population 2 has sample size $n_2 = 88$ and sample proportion of successes $\hat{p}_2 = 0.76$. Calculate the value of the test statistic that could be used to decide if there is convincing evidence that the actual proportion of successes in population 1 is greater than actual proportion of successes in population 2. Check the necessary conditions to determine if they are satisfied.

b. In an experiment, subjects were randomly assigned into two treatment groups. The sample size for treatment 1 is $n_1 = 78$ with sample proportion of successes $\hat{p}_1 = 0.77$. The sample size for treatment 2 is $n_2 = 46$ with sample proportion of successes $\hat{p}_2 = 0.33$. Calculate the value of the test statistic to be used to decide if there is convincing evidence that the actual proportion of successes for treatment 1 is greater than the actual proportion of successes for treatment 2. Check the necessary conditions to determine if they are satisfied.

Answers

a. Use the values provided to check the sample size conditions as follows:

$$n_1\hat{p}_1 = (86)(0.86) = 74$$

$$n_1(1 - \hat{p}_1) = (86)(1 - 0.86) = (86)(0.14) = 12$$

$$n_2\hat{p}_2 = (88)(0.76) = 67$$

$$n_2(1 - \hat{p}_2) = (88)(1 - 0.76) = (88)(0.24) = 21$$

Because both of these are greater than or equal to 10, the sample sizes of 86 and 88 are large enough to use the large-sample z test statistic to test a hypothesis about a difference in two population proportions.

Calculate the value of the test statistic by first finding the value of $\hat{p}_c$. Substitute the values of the two sample proportions, $\hat{p}_1$ and $\hat{p}_2$, and the sample sizes n_1 and n_2 into the expression for the estimate of the common proportion, $\hat{p}_c$. This results in

$$\hat{p}_c = \frac{n_1\hat{p}_1 + n_2\hat{p}_2}{n_1 + n_2} = \frac{(86)(0.86) + (88)(0.76)}{86 + 88} = \frac{74 + 67}{174} = \frac{141}{174} = 0.810$$

(rounded to three decimal places)

Now you can calculate the value of the z test statistic by substituting the values of the two sample proportions, $\hat{p}_1$ and $\hat{p}_2$, the sample sizes n_1 and n_2, and the value of $\hat{p}_c$ into the expression for the test statistic. This results in

$$z = \frac{\hat{p}_1 - \hat{p}_2}{\sqrt{\frac{\hat{p}_c(1-\hat{p}_c)}{n_1} + \frac{\hat{p}_c(1-\hat{p}_c)}{n_2}}} = \frac{0.86 - 0.76}{\sqrt{\frac{(0.81)(1-0.81)}{86} + \frac{(0.81)(1-0.81)}{88}}}$$

To evaluate this expression, start by evaluating the parts of the expression that appears under the square root. Evaluate the first term to obtain

$$\frac{(0.81)(1-0.81)}{86} = \frac{(0.81)(0.19)}{86} = \frac{0.1539}{86} = 0.002$$

(rounded to three decimal places)

Then evaluate the second term under the square root symbol in a similar way:

$$\frac{(0.81)(1-0.81)}{88} = \frac{(0.81)(0.19)}{88} = \frac{0.1539}{88} = 0.002$$

(rounded to three decimal places)

Next, add and then take the square root to obtain the value of the denominator of the test statistic:

$$\sqrt{\frac{(0.81)(1-0.81)}{86} + \frac{(0.81)(1-0.81)}{88}} = \sqrt{0.002 + 0.002} = \sqrt{0.004} = 0.063$$

Complete the calculation as shown here:

$$z = \frac{\hat{p}_1 - \hat{p}_2}{\sqrt{\frac{\hat{p}_c(1-\hat{p}_c)}{n_1} + \frac{\hat{p}_c(1-\hat{p}_c)}{n_2}}} = \frac{0.86 - 0.76}{\sqrt{\frac{(0.81)(1-0.81)}{86} + \frac{(0.81)(1-0.81)}{88}}} = \frac{0.10}{0.063} = 1.59$$

The value of the test statistic for this example is $z = 1.59$.

b. Use the values provided to check the sample size conditions as follows:

$$n_1\hat{p}_1 = (78)(0.77) = 60$$
$$n_1(1-\hat{p}_1) = (78)(1-0.77) = (78)(0.23) = 18$$
$$n_2\hat{p}_2 = (46)(0.33) = 15$$
$$n_2(1-\hat{p}_2) = (46)(1-0.33) = (46)(0.67) = 31$$

Because both of these are greater than or equal to 10, the sample sizes of 78 and 46 are large enough to use the large-sample z test statistic to test a hypothesis about a difference in two population proportions.

Calculate the value of the test statistic by first finding the value of $\hat{p}_c$. Substitute the values of the two sample proportions, $\hat{p}_1$ and $\hat{p}_2$, and the sample sizes n_1 and n_2 into the expression for the estimate of the common proportion, $\hat{p}_c$. This results in

$$\hat{p}_c = \frac{n_1\hat{p}_1 + n_2\hat{p}_2}{n_1 + n_2} = \frac{(78)(0.77) + (46)(0.33)}{78 + 46} = \frac{60 + 15}{124} = \frac{75}{124} = 0.605$$

(rounded to three decimal places)

Now you can calculate the value of the z test statistic by substituting the values of the two sample proportions, $\hat{p}_1$ and $\hat{p}_2$, the sample sizes n_1 and n_2, and the value of $\hat{p}_c$ into the expression for the test statistic. This results in

$$z = \frac{\hat{p}_1 - \hat{p}_2}{\sqrt{\frac{\hat{p}_c(1-\hat{p}_c)}{n_1} + \frac{\hat{p}_c(1-\hat{p}_c)}{n_2}}} = \frac{0.77 - 0.33}{\sqrt{\frac{(0.605)(1-0.605)}{78} + \frac{(0.605)(1-0.605)}{46}}}$$

Begin by evaluating the parts of the expression that appears under the square root. Evaluate the first term under the square root symbol as follows:

$$\frac{(0.605)(1-0.605)}{78} = \frac{(0.605)(0.395)}{78} = \frac{0.2390}{78} = 0.003$$

(rounded to three decimal places)

Evaluate the second term under the square root symbol in a similar way:

$$\frac{(0.605)(1-0.605)}{46} = \frac{(0.605)(0.395)}{46} = \frac{0.2390}{46} = 0.005$$

(rounded to three decimal places)

Next, add these two values and take the square root to obtain the value of the denominator of the test statistic:

$$\sqrt{\frac{(0.605)(1-0.605)}{78} + \frac{(0.605)(1-0.605)}{46}} = \sqrt{0.003 + 0.005} = \sqrt{0.008} = 0.089$$

Complete the calculation as shown here:

$$z = \frac{\hat{p}_1 - \hat{p}_2}{\sqrt{\frac{\hat{p}_c(1-\hat{p}_c)}{n_1} + \frac{\hat{p}_c(1-\hat{p}_c)}{n_2}}} = \frac{0.77 - 0.33}{\sqrt{\frac{(0.605)(1-0.605)}{78} + \frac{(0.605)(1-0.605)}{46}}}$$

$$= \frac{0.44}{0.089} = 4.94$$

The value of the test statistic for this example is $z = 4.94$.

SECTION 11.1 EXERCISES

11.1 Random samples are selected from two populations. The sample from population 1 has sample size $n_1 = 85$ and sample proportion of successes $\hat{p}_1 = 0.44$. The sample from population 2 has sample size $n_2 = 81$ and sample proportion of successes $\hat{p}_2 = 0.95$. Calculate the value of the test statistic that could be used to decide if there is convincing evidence that the actual proportion of successes in population 1 is less than the actual proportion of successes in population 2. Check the necessary conditions to determine if they are satisfied.

11.2 In an experiment, subjects were randomly assigned into two treatment groups. Treatment group 1 had sample size $n_1 = 79$ with sample proportion of successes $\hat{p}_1 = 0.77$. Treatment group 2 has sample size $n_2 = 38$ with sample proportion of successes

$\hat{p}_2 = 0.11$. Calculate the value of the test statistic that would be used to decide if there is convincing evidence that the actual proportion of successes for individuals undergoing treatment 1 is greater than the actual proportion of successes for individuals undergoing treatment 2. Check the necessary conditions to determine if they are satisfied.

SECTION 11.2 Guided Practice—Large Sample Hypothesis Test for a Difference in Two Proportions

In this section, you will consider a typical situation that involves carrying out a large sample z test for a difference in population proportions. Consider this problem from a statistics textbook:

> Gallup surveyed adult Americans about their consumer debt ("Americans' Big Debt Burden Growing, Not Evenly Distributed," February 4, 2016, gallup.com, retrieved December 15, 2016). They reported that 61% of Gen Xers (those born between 1965 and 1980) and 46% of Millennials (those born between 1981 and 1996) did not pay off their credit cards each month, and therefore carried a balance from month to month. Suppose that these percentages were based on representative samples of 300 Gen Xers and 450 Millennials. Is there convincing evidence that the proportion of Gen Xers who do not pay off their credit cards each month is greater than this proportion for Millennials? Test the appropriate hypotheses using a significance level of 0.05.

If you were to encounter a problem like this, you should start by reading the problem carefully. You can use the three-read strategy introduced in Chapter 7.

First Read: Context

1. **What is the general context?**
A survey was conducted to learn about the difference in the proportions of people who do not pay off their credit cards each month for two different age groups. The two groups compared are Gen Xers (those born between 1965 and 1980) and Millennials (those born between 1981 and 1996).

2. **Are there any terms that you don't understand?**
This question is about testing a hypothesis, so if you don't recall the steps required to carry out a hypothesis test, you would want to go to the appropriate section in your textbook to review the steps. If you don't recall the meaning of the term significance level, you should also go to your statistics text to review that definition. The terms Gen Xers and Millennials are defined in the problem.

Second Read: Tasks

1. **What question is being asked? What am I being asked to do?**
This problem asks you to decide if the sample data provide convincing evidence that the proportion of Gen Xers who do not pay off their credit cards each month is greater than this proportion for Millennials. You are to test the appropriate hypotheses using a significance level of 0.05.

Third Read: Details

1. **What data were collected? Were the data collected in a reasonable way?**
Data were collected from a sample of 300 Gen Xers and a sample of 450 Millennials. The problem states that the samples are representative of the populations of Gen Xers and Millennials.

2. **Are the actual data provided? What was the sample size (or the sample sizes if there are more than one sample)?**
The sample size was 300 for the sample of Gen Xers and the sample size was 450 for the sample of Millennials. The actual data (the individual responses) are not provided, but you are told that 61% of the Gen Xers and 46% of the Millennials do not pay off their credit cards each month.

3. Are any summary statistics or additional information given? If so, which statistics and what are the values of these statistics?

The sample proportions are not given, but the percent who do not pay off their credit cards is given for each sample (61% for Gen Xers and 46% for Millennials) and the sample proportions can be determined from these percentages.

Now you are ready to tackle the tasks required to complete this exercise. The problem states "Is there convincing evidence that the proportion of Gen Xers who do not pay off their credit cards each month is greater than this proportion for Millennials? Test the appropriate hypotheses using a significance level of 0.05." This is asking you to test hypotheses about a difference in population proportions. The populations of interest are Millennials and Gen Xers, and you are asked to decide if there is convincing evidence that the proportion who do not pay off their credit cards is greater for Gen Xers. You are also told to use a significance level of 0.05 when carrying out the test.

Many textbooks recommend a sequence of steps for carrying out a hypothesis test. When you carry out a hypothesis test, you should follow the recommended sequence of steps. Although the order of the steps may vary somewhat from one text to another, they generally include the following:

Hypotheses: You will need to specify the null and the alternative hypothesis for the hypothesis test and define any symbols that are used in the context of the problem.

Method: You will need to specify what method you plan to use to carry out the test. For example, you might say that you plan to use the large-sample z test for a difference in two population proportions if that is a test that would allow you to answer the question posed. You will also need to specify the significance level you will use to reach a conclusion.

Check Conditions: Most hypothesis test procedures are only appropriate when certain conditions are met. One you have determined the method you plan to use, you will need to verify that any required conditions are met.

Calculate: To reach a conclusion in a hypothesis test, you will need to calculate the value of a test statistic and the associated P-value.

Communicate Results: The final step is to use the calculated values of the test statistic and the P-value to reach a conclusion and to provide an answer to the question posed. You should always provide a conclusion in words.

These steps are illustrated here for the given example.

Hypotheses: You want to determine if there is evidence that the proportion of Gen Xers who do not pay off their credit cards each month is greater than the proportion for Millennials. Start by deciding which population you will refer to as population 1 and which population you will refer to as population 2. For example, you might choose to think of Gen Xers as population 1 and Millennials as population 2. Then whenever you see a subscript of 1, you would know that it is referring to Gen Xers. The subscript 2 will refer to Millennials. For Gen Xers, you know that the sample size was $n_1 = 300$ and that 61% of those in this sample do not pay off their credit cards each month. You can determine the value of the sample proportion for this sample by converting the given percentage (61%) to a proportion. To do this, you drop the % symbol and divide by 100 to obtain $\hat{p}_1 = 0.61$. For Millennials, you know that the sample size was $n_2 = 450$ and that 46% of those in this sample do not pay off their credit cards each month. You can determine the value of the sample proportion for this sample by converting the given percentage (46%) to a proportion and obtain $\hat{p}_2 = 0.46$.

The question asks if the proportion is greater for Gen Xers than for Millennials, so you are asked if there is evidence that p_1 is greater than p_2. If this is the case, the value of $p_1 - p_2$ would be greater than 0. The question being asked usually translates into the alternative hypothesis, so you would specify an alternative hypothesis of

$H_a: p_1 - p_2 > 0$. The null hypothesis would then be obtained by replacing the > in the alternative hypothesis with an =. This results in hypotheses of

$$H_0: p_1 - p_2 = 0$$
$$H_a: p_1 - p_2 > 0$$

Method: Because the hypotheses are about a difference in population proportions, you would consider the large-sample test for a difference in population proportions. The test statistic is

$$z = \frac{\hat{p}_1 - \hat{p}_2}{\sqrt{\frac{\hat{p}_c(1-\hat{p}_c)}{n_1} + \frac{\hat{p}_c(1-\hat{p}_c)}{n_2}}}$$

A significance level of $\alpha = 0.05$ was specified.

Check: There are two conditions necessary for the large-sample z test for a difference in population proportions to be appropriate. You are told that the samples are representative of the population of Gen Xers and the population of Millennials.

For the sample from population 1 (Gen Xers), $n_1 = 300$ and $\hat{p}_1 = 0.61$. For the sample from population 2 (Millennials), $n_2 = 450$ and $\hat{p}_2 = 0.46$.

Use these values to check the sample size conditions as follows:

$$n_1\hat{p}_1 = (300)(0.61) = 183$$
$$n_1(1 - \hat{p}_1) = (300)(1 - 0.61) = (300)(0.39) = 117$$
$$n_2\hat{p}_2 = (450)(0.46) = 207$$
$$n_2(1 - \hat{p}_2) = (450)(1 - 0.46) = (450)(0.54) = 243$$

Because both of these are greater than or equal to 10, the sample sizes of 300 and 450 are large enough to use the large-sample z test statistic to test a hypothesis about a difference in two proportions.

Calculations: To calculate the value of the test statistic, you need to know the values of the two sample proportions, $\hat{p}_1$ and $\hat{p}_2$, the sample sizes n_1 and n_2, and the value of the combined estimate of the common proportion, $\hat{p}_c$. Start by finding the value of $\hat{p}_c$. Substitute the values of the two sample proportions, $\hat{p}_1$ and $\hat{p}_2$, and the sample sizes n_1 and n_2 into the expression for the estimate of the common proportion, $\hat{p}_c$. This results in

$$\hat{p}_c = \frac{n_1\hat{p}_1 + n_2\hat{p}_2}{n_1 + n_2} = \frac{(300)(0.61) + (450)(0.46)}{300 + 450}$$

Complete the calculation by determining the value of the numerator and the value of the denominator and then dividing, as shown here:

$$\hat{p}_c = \frac{n_1\hat{p}_1 + n_2\hat{p}_2}{n_1 + n_2} = \frac{(300)(0.61) + (450)(0.46)}{300 + 450} = \frac{183 + 207}{750} = \frac{390}{750} = 0.52$$

Now you can calculate the value of the z test statistic by substituting the values of the two sample proportions, $\hat{p}_1$ and $\hat{p}_2$, the sample sizes n_1 and n_2, and the value of $\hat{p}_c$ into the expression for the test statistic. This results in

$$z = \frac{\hat{p}_1 - \hat{p}_2}{\sqrt{\frac{\hat{p}_c(1-\hat{p}_c)}{n_1} + \frac{\hat{p}_c(1-\hat{p}_c)}{n_2}}} = \frac{0.61 - 0.46}{\sqrt{\frac{(0.52)(1-0.52)}{300} + \frac{(0.52)(1-0.52)}{450}}}$$

To evaluate this expression, start by evaluating the parts of the expression that appear under the square root. To evaluate the first term, first calculate $(1 - 0.52) = 0.48$. Then multiply the two numbers in the numerator of the first term, and finally divide that number by 300. This is the sequence of steps shown here:

$$\frac{(0.52)(1-0.52)}{300} = \frac{(0.52)(0.48)}{300} = \frac{0.250}{300} = 0.0008$$

Then evaluate the second term under the square root symbol in a similar way:

$$\frac{(0.52)(1-0.52)}{450} = \frac{(0.52)(0.48)}{450} = \frac{0.250}{450} = 0.0006$$

Next, you would add these two values together and then find the square root:

$$\sqrt{\frac{(0.52)(1-0.52)}{300} + \frac{(0.52)(1-0.52)}{450}} = \sqrt{0.0008 + 0.0006} = \sqrt{0.0014} = 0.037$$

This number is the value of the denominator of the z test statistic. You can now complete the calculation as shown here:

$$z = \frac{\hat{p}_1 - \hat{p}_2}{\sqrt{\frac{\hat{p}_c(1-\hat{p}_c)}{n_1} + \frac{\hat{p}_c(1-\hat{p}_c)}{n_2}}} = \frac{0.61 - 0.46}{\sqrt{\frac{(0.52)(1-0.52)}{300} + \frac{(0.52)(1-0.52)}{450}}} = \frac{0.15}{0.037}$$

$$= 4.05 \text{ (rounded to two decimal places)}$$

The value of the test statistic for this example is $z = 4.05$.

Once you know the value of the test statistic, you need to find the associated P-value. Your statistics text will have an explanation of how the P-value is determined.

This is an upper-tailed test (the inequality in H_a is $>$), so the P-value is the area under the z curve and to the right of 4.05. Therefore, the P-value is $P(z \geq 4.05) \approx 0$.

Communicate Results: Because the P-value is approximately 0 and this value is less than the selected significance level of 0.05, you reject the null hypothesis. There is convincing evidence that the proportion of Gen Xers who do not pay off their credit cards each month is greater than the proportion of Millennials who do not pay off their credit cards each month.

SECTION 11.2 EXERCISES

11.3 Use the three-read strategy to understand the following exercise from a statistics textbook, and then proceed to complete the exercise.

The authors of the article titled "Adolescents and MP3 Players: Too Many Risks, Too Few Precautions," which was published in the journal *Pediatrics* ([2009]: pages e953–e958), concluded that more boys than girls listen to music at high volumes. This conclusion was based on data from independent random samples of 764 Dutch boys and 748 Dutch girls age 12 to 19. Of the boys, 397 reported that they almost always listen to music at a high volume setting. Of the girls, 331 reported listening to music at a high volume setting. Do the sample data support the authors' conclusion that the proportion of Dutch boys who listen to music at high volume is greater than this proportion for Dutch girls? Test the relevant hypotheses using a significance level of $\alpha = 0.05$. Remember to check the necessary conditions to determine if they are satisfied.

11.4 Use the three-read strategy to understand the following exercise from a statistics textbook, and then proceed to complete the exercise.

Some commercial airplanes recirculate approximately 50% of the cabin air in order to increase fuel efficiency. The authors of the paper "Aircraft Cabin Air Recirculation and Symptoms of the Common Cold" (*Journal of the American Medical Association* [2002]: 483–486) studied 1100 airline passengers who flew from San Francisco to Denver. Some passengers traveled on airplanes that recirculated air, and others traveled on planes that did not recirculate air. Of the 517 passengers who flew on planes that did not recirculate air, 108 reported post-flight respiratory symptoms, while 110 of the 583 passengers on planes that did recirculate air reported such symptoms. The question of interest is whether the proportions of passengers with post-flight respiratory symptoms differ for planes that do and do not recirculate air. Test the relevant hypotheses using a significance level of $\alpha = 0.01$. You may assume that it is reasonable to regard these two samples as being independent representative samples of the two populations of interest, but remember to check the sample size condition.

12 Estimating a Population Mean—The Math You Need to Know

SECTION 12.1 Evaluating Expressions

There are several expressions that you will encounter as you learn about methods for estimating a population mean.

Note: Depending on your access to technology, such as a graphing calculator or a statistics software package, you may not need to evaluate these expressions by hand. But just in case your statistics class is not relying on technology to do these calculations, this section looks at how to evaluate the expressions.

Margin of Error When Estimating a Population Mean (using a 95% confidence level)

$$\text{margin of error} = 1.96\frac{s}{\sqrt{n}}$$

Before you use the given formula to calculate the margin of error associated with an estimate of a population mean, you need to verify that the sample size is large enough or that it is reasonable to assume that the population distribution is approximately normal. You also need to confirm that the sample is a random sample or that it was selected in a way that makes it reasonable to think that the sample is representative of the population.

The sample size is large enough to use this formula if n is greater than or equal to 30 (note that some texts are more conservative and use n at least 40). If n is less than 30, this margin of error formula is only appropriate if it is reasonable to assume that the population distribution is approximately normal. A graphical display of the data distribution (such as a dotplot, a boxplot, or a normal probability plot) can be used to help you decide if this assumption is reasonable.

To calculate the margin of error, you need to know the sample size n and the value of the sample standard deviation s. Sometimes the value of s will be given and sometimes you will have to use data to calculate its value. For a review of how to calculate the standard deviation, see Section 3.5.

For example, suppose that you have a random sample of $n = 34$ students at a two-year college and that each student was asked how many miles they travel to get from home to campus. The sample mean distance was $\bar{x} = 23.4$ miles and the sample standard deviation was 18.0 miles. Because the sample was a random sample and the sample size is greater than or equal to 30, it is appropriate to use the margin of error formula. Substitute the values of n and s into the formula for margin of error to obtain

$$\text{margin of error} = 1.96\frac{s}{\sqrt{n}} = (1.96)\left(\frac{18.0}{\sqrt{34}}\right)$$

Begin by evaluating $\sqrt{34} = 5.381$, and then complete the calculation as shown here:

$$\text{margin of error} = 1.96\frac{s}{\sqrt{n}} = (1.96)\left(\frac{18.0}{\sqrt{34}}\right) = (1.96)\left(\frac{18.0}{5.381}\right)$$

$$= (1.96)(3.345) = 6.556$$

(rounded to three decimal places)

The margin of error can be interpreted as the maximum likely estimation error. You would estimate that the mean distance traveled to campus for students at this college is 23.4 miles, and it is unlikely that this estimate differs from the actual value of the population mean by more than 6.556 miles.

Sometimes the margin of error is calculated using the value 2 in place of 1.96 in the expression for margin of error. This simplifies the calculations and doesn't usually change the calculated value by much (it results in a slightly larger value for margin of error).

Check It Out!

a. The Cardiac Care Network in Ontario, Canada, collected information on the time between when a cardiac patient is recommended for heart surgery and when the surgery is performed ("Wait Times Data Guide," Ministry of Health and Long-Term Care, Ontario, Canada, 2006). The reported mean wait time (in days) for a random sample of 539 patients waiting for bypass surgery was $\bar{x} = 19$ days, and the sample standard deviation was $s = 10$ days. (This standard deviation was estimated from information in the report.) The Cardiac Care Network was interested in estimating the actual mean wait time for the population of patients recommended for bypass surgery.

   Use the information provided to calculate the margin of error (for a confidence level of 95%), and interpret the margin of error in context.

b. In June 2009, Harris Interactive conducted its Great Schools Survey. In this survey, the sample consisted of 1086 adults who were parents of school-age children. The sample was selected to be representative of the population of parents of school-age children. One question on the survey asked respondents how much time per month (in hours) they spent volunteering at their children's school during the previous school year. The following summary statistics for time volunteered per month were given: $n = 1086$, $\bar{x} = 5.6$, and $s = 5.2$. (The standard deviation was estimated from information in the survey report.) Researchers were interested in estimating the actual population mean time per month that parents of school-age children spend volunteering at their children's school.

   Use the information provided to calculate the margin of error (for a 95% confidence level), and interpret the margin of error in context.

Answers

a. Because the sample was a random sample and the sample size is greater than or equal to 30, it is appropriate to use the margin of error formula. Substitute the values of n and s into the formula for margin of error to obtain

$$\text{margin of error} = 1.96\frac{s}{\sqrt{n}} = (1.96)\left(\frac{10}{\sqrt{539}}\right)$$

Complete the calculation as shown here:

$$\text{margin of error} = 1.96\frac{s}{\sqrt{n}} = (1.96)\left(\frac{10}{\sqrt{539}}\right) = (1.96)\left(\frac{10}{23.22}\right)$$

$$= (1.96)(0.431) = 0.84$$

(rounded to two decimal places)

The margin of error can be interpreted as the maximum likely estimation error. You would estimate that the mean wait time for the population of patients recommended for bypass surgery is 19 days, and it is unlikely that this estimate differs from the actual value of the population mean by more than 0.84 days.

b. Because the sample was a random sample and the sample size is greater than or equal to 30, it is appropriate to use the margin of error formula. Substitute the values of n and s into the formula for margin of error to obtain

$$\text{margin of error} = 1.96\frac{s}{\sqrt{n}} = (1.96)\left(\frac{5.2}{\sqrt{1086}}\right)$$

Complete the calculation as shown here:

$$\text{margin of error} = 1.96\frac{s}{\sqrt{n}} = (1.96)\left(\frac{5.2}{\sqrt{1086}}\right) = (1.96)\left(\frac{5.2}{32.95}\right)$$

$$= (1.96)(0.158) = 0.310$$

(rounded to three decimal places)

The margin of error can be interpreted as the maximum likely estimation error. You would estimate that the population mean time per month that parents of school-age children spend volunteering at their children's school is 5.6 hours, and it is unlikely that this estimate differs from the actual value of the population mean by more than 0.310 hours.

Confidence Interval for a Population Mean

$$\bar{x} \pm (t \text{ critical value})\frac{s}{\sqrt{n}}$$

To calculate a confidence interval for a population mean, you need to know the value of the sample mean, $\bar{x}$, the sample size n, and the confidence level, which determines the t critical value that is used to calculate the confidence interval. Before using the given expression to calculate a confidence interval, you need to verify that the sample size is large enough to calculate a confidence interval using the given formula or that it is reasonable to think that the population distribution is approximately normal. You also need to confirm that the sample is a random sample or that it was selected in a way that makes it reasonable to think that the sample is representative of the population.

The sample size is large enough to use this formula if n is greater than or equal to 30 (note that some texts are more conservative and use n greater than 40). If n is less than 30, this confidence interval formula is only appropriate if it is reasonable to assume that the population distribution is approximately normal. A plot of the data (dotplot, boxplot, or normal probability plot) can be used to help you decide if this assumption is reasonable.

Recall that an interval is defined by its endpoints—it has a lower endpoint and an upper endpoint. Notice that the expression for the confidence interval includes the symbol $\pm$, which is read as "plus and minus." This means that you will do two calculations, once using $+$ and once using $-$. Each one of these calculations results in one of the two interval endpoints.

To calculate the endpoints of the confidence interval, you can start by determining the appropriate t critical value for the specified confidence level. Your statistics textbook should have instructions for how this is done.

Once you have determined the appropriate t critical value, you can substitute the values of n and s into the expression for the confidence interval.

For example, suppose you want to calculate a 95% confidence interval for the mean distance traveled to campus for the random sample of students described in the discussion of margin of error ($n = 34$, $\bar{x} = 23.4$, and $s = 18.0$). To determine the t critical value, you first must calculate the degrees of freedom (df) associated with this interval. For a confidence interval for a population mean, $df = n - 1$, which for this example would be $df = 34 - 1 = 33$. A table of t critical values or technology could then be used to find the t critical value. For a 95% confidence level and $df = 33$, the t critical value is 2.04.

Substituting values into the expression for the confidence interval results in the following:

$$\bar{x} \pm (t \text{ critical value})\frac{s}{\sqrt{n}} = 23.4 \pm (2.04)\left(\frac{18.0}{\sqrt{34}}\right)$$

Start by evaluating $\sqrt{34} = 5.831$, and then complete the calculation following the sequence of steps shown here:

$$23.4 \pm (2.04)\left(\frac{18.0}{\sqrt{34}}\right)$$

$$23.4 \pm (2.04)\left(\frac{18.0}{5.831}\right)$$

$$23.4 \pm (2.04)(3.087)$$

$$23.4 \pm 6.297$$

To calculate the endpoints of the interval, you would evaluate

$$23.4 \pm 6.297$$

The lower endpoint of the interval is $23.4 - 6.297 = 17.103$. The upper endpoint of the interval is $23.4 + 6.297 = 29.697$.

The 95% confidence interval estimate of the population mean travel distance is (17.103, 29.697). You would interpret this interval by saying that based on the data from the sample, you can be 95% confident that the actual value of the population mean travel distance is somewhere between 17.103 and 29.697 miles.

Check It Out!

a. Recall Check It Out! Part (a), above. The Cardiac Care Network was interested in estimating the actual mean wait time for the population of patients recommended for bypass surgery.

Calculate a 95% confidence interval for the mean wait time for the population of patients recommended for bypass surgery, and interpret the interval in context. You can use the summary statistics and from the previous Check It Out! Part (a) to get started ($n = 539$, $\bar{x} = 19$ days, and $s = 10$ days).

b. Recall Check It Out! Part (b), above. Researchers were interested in estimating the actual population mean time per month that parents of school-age children spend volunteering at their children's school.

Calculate a 95% confidence interval for the population mean time per month that parents of school-age children spend volunteering at their children's school, and interpret the interval in context. You can use the summary statistics from the previous Check It Out! Part (b) to get started ($n = 1086$, $\bar{x} = 5.6$ hours, and $s = 5.2$ hours).

Answers

a. Calculate the degrees of freedom (df) associated with this interval. For a confidence interval for a population mean, $df = n - 1$, which is $df = 539 - 1 = 538$. A table of t critical values or technology could then be used to find the t critical value. For a 95% confidence level and $df = 538$, the t critical value is 1.964. This t critical value is very close to the z critical value of 1.96, because the sample size and degrees of freedom are so large.

Substituting values into the expression for the confidence interval results in the following:

$$\bar{x} \pm (t \text{ critical value})\frac{s}{\sqrt{n}} = 19 \pm (1.964)\left(\frac{10}{\sqrt{539}}\right)$$

Complete the calculation following the sequence of steps shown here:

$$19 \pm (1.964)\left(\frac{10}{\sqrt{539}}\right)$$

$$19 \pm (1.964)\left(\frac{10}{23.22}\right)$$

$$19 \pm (1.964)(0.431)$$

$$19 \pm 0.85$$

Then calculate the endpoints of the interval:

$$19 \pm 0.85$$

The lower endpoint of the interval is $19 - 0.85 = 18.15$. The upper endpoint of the interval is $19 + 0.85 = 19.85$.

The 95% confidence interval estimate of the population mean wait time is (18.15, 19.85). You would interpret this interval by saying that based on the data from the sample, you can be 95% confident that the actual value of the population mean wait time for patients recommended for bypass surgery is somewhere between 18.15 and 19.85 days.

b. Calculate the degrees of freedom (df) associated with this interval. For a confidence interval for a population mean, $df = n - 1$, which is $df = 1086 - 1 = 1085$. A table of t critical values or technology could then be used to find the t critical value. For a 95% confidence level and $df = 1085$, the t critical value is 1.962. This t critical value is very close to the z critical value of 1.96, because the sample size and degrees of freedom are so large.

Substituting values into the expression for the confidence interval results in the following:

$$\bar{x} \pm (t \text{ critical value})\frac{s}{\sqrt{n}} = 5.6 \pm (1.962)\left(\frac{5.2}{\sqrt{1086}}\right)$$

Complete the calculation as shown here:

$$5.6 \pm (1.962)\left(\frac{5.2}{\sqrt{1086}}\right)$$

$$5.6 \pm (1.962)\left(\frac{5.2}{32.95}\right)$$

$$5.6 \pm (1.962)(0.158)$$

$$5.6 \pm 0.310$$

Then calculate the endpoints of the interval:

$$5.6 \pm 0.310$$

The lower endpoint of the interval is $5.6 - 0.310 = 5.290$. The upper endpoint of the interval is $5.6 + 0.310 = 5.910$.

The 95% confidence interval estimate of the population mean time per month is (5.290, 5.910). You would interpret this interval by saying that based on the data from the sample, you can be 95% confident that the actual value of the population mean time per month that parents of school-age children spend volunteering at their children's school is somewhere between 5.290 and 5.910 hours.

Determining Sample Size

$$n = \left(\frac{1.96\sigma}{M}\right)^2$$

When estimating a population mean, the larger the sample size, the smaller the margin of error will be. This makes sense because you would expect that an estimate based on a larger sample size would tend to be closer to the actual value of the population mean than an estimate based on a smaller sample.

Before collecting any data, you might want to determine how large the sample size should be to achieve a specified margin of error. When this is the case, the expression given above is used to determine sample size.

Notice that σ, which is the symbol used to represent the population standard deviation, appears in the formula for sample size. Because the value of the population standard deviation will not be known, to use this expression to calculate a sample size you need to have an estimate of the population standard deviation. This might be the sample standard deviation from a pilot study, or if you think that the population distribution is not too skewed, you might estimate the population standard deviation by estimating the range of values in the population and then dividing that estimate by 4. You also need to know the desired value for the margin of error, M.

For example, you saw that the margin of error for the estimate of the population mean travel distance based on the data from the sample of $n = 34$ students previously described was 6.556 miles. Suppose that you wanted to carry out a second study that would provide you with a more precise estimate, and you want to determine the sample size necessary to estimate a population mean with a margin of error of $M = 2$ miles. You can use the sample standard deviation $s = 18.0$ as the estimate for σ. Begin by substituting the values for M and your estimate for σ into the expression. Substituting $M = 2$ and $\sigma = 18.0$ into the expression results in

$$n = \left(\frac{1.96\sigma}{M}\right)^2 = \left(\frac{(1.96)(18.0)}{2}\right)^2$$

To evaluate this expression, you would first calculate the value of $\left(\frac{(1.96)(18.0)}{2}\right)$. You would then square this number. This sequence of steps illustrated here:

$$n = \left(\frac{(1.96)(18)}{2}\right)^2 = \left(\frac{35.280}{2}\right)^2 = (17.640)^2 = 311.170$$

In sample size calculations, you always round up to the next whole number to obtain the necessary sample size. For this example, a sample size of $n = 312$ would be needed to achieve a margin of error of $M = 2$.

If you did not have an estimate of the population standard deviation from a pilot study, you might have estimated the population range and divided that number by 4. For example, you might think that the smallest travel distance is about 1 mile and that the longest travel distance is about 50 miles. This would result in an estimate of the population range of $50 - 1 = 49$. Dividing this number by 4 results in $\frac{49}{4} = 12.25$. You could then use this number as the estimate for σ in the sample size calculation. This would result in

$$n = \left(\frac{(1.96)(12.25)}{2}\right)^2 = \left(\frac{24.01}{2}\right)^2 = (12.005)^2 = 144.120$$

and you would use a sample size of 145.

Check It Out!

a. Recall Check It Out! Part (a), above. The Cardiac Care Network was interested in estimating the actual mean wait time for the population of patients recommended for bypass surgery.

   Determine the sample size necessary to estimate the population mean wait time for the population of patients recommended for bypass surgery with a margin of error of $M = 0.5$ days and with a 95% confidence level. Use the summary statistics from the previous Check It Out! Part (a) to get started ($n = 539$, $\bar{x} = 19$, and $s = 10$).

b. Recall Check It Out! Part (b), above. Researchers were interested in estimating the actual population mean time per month that parents of school-age children spend volunteering at their children's school.

   Determine the sample size necessary to estimate the population mean time per month that parents of school-age children spend volunteering at their children's school for the population represented by the sample with a margin of error of $M = 0.2$ hours and with a 95% confidence level. Use the summary statistics from the previous Check It Out! Part (b) to get started ($n = 1086$, $\bar{x} = 5.6$ hours, and $s = 5.2$ hours).

Answers

a. Begin by substituting the values for M and your estimate for σ into the expression. Substituting $M = 0.5$ and $\sigma = 10$ into the expression results in

$$n = \left(\frac{1.96\sigma}{M}\right)^2 = \left(\frac{(1.96)(10)}{0.5}\right)^2$$

   Then evaluate the expression as shown here:

$$n = \left(\frac{(1.96)(10)}{0.5}\right)^2 = \left(\frac{19.6}{0.5}\right)^2 = (39.2)^2 = 1536.64$$

   In sample size calculations, you always round up to the next whole number to obtain the necessary sample size. For this example, a sample size of at least $n = 1537$ would be needed to achieve a margin of error of $M = 0.5$.

b. Begin by substituting the values for M and your estimate for σ into the expression. Substituting $M = 0.2$ and $\sigma = 5.2$ into the expression results in:

$$n = \left(\frac{1.96\sigma}{M}\right)^2 = \left(\frac{(1.96)(5.2)}{0.2}\right)^2$$

   Then evaluate the expression as shown here:

$$n = \left(\frac{(1.96)(5.2)}{0.2}\right)^2 = \left(\frac{10.192}{0.2}\right)^2 = (50.96)^2 = 2596.92$$

   In sample size calculations, you always round up to the next whole number to obtain the necessary sample size. For this example, a sample size of at least $n = 2597$ would be needed to achieve a margin of error of $M = 0.2$.

SECTION 12.1 EXERCISES

12.1 In a study of academic procrastination, the authors of the paper "Correlates and Consequences of Behavioral Procrastination" (*Procrastination, Current Issues and New Directions* [2000]) reported that for a sample of 411 undergraduate students at a midsize public university, the sample mean time spent studying for the final exam in an introductory psychology course was 7.74 hours and the sample standard deviation of study times was 3.40 hours. Assume that this sample is representative of students taking introductory psychology at this university.

a. Use the information provided to calculate the margin of error (for a 95% confidence level), and interpret the margin of error in context.

b. Calculate a 95% confidence interval for the mean time spent studying for the final exam in an introductory psychology course at this university, and interpret the interval in context.

c. Determine the sample size necessary to estimate the population mean time spent studying for the final exam in an introductory psychology course at this university with a margin of error of $M = 0.25$ hours and with 95% confidence.

12.2 The authors of the paper "Deception and Design: The Impact of Communication Technology on Lying Behavior" (*Proceedings of Computer Human Interaction* [2004]) asked 30 students in an upper-division communications course at a large university to keep a journal for 7 days, recording each social interaction and whether or not they told any lies during that interaction. A lie was defined as "any time you intentionally try to mislead someone." The paper reported that the sample mean number of lies per day for the 30 students was 1.58, and the sample standard deviation of the number of lies per day was 1.02.

a. Use the information provided to calculate the margin of error (for a 95% confidence level), and interpret the margin of error in context.

b. Calculate a 95% confidence interval for the actual mean number of lies per day, for the population represented by the sample of students in this study, and interpret the interval in context.

c. Determine the sample size necessary to estimate the actual population mean number of lies per day with a margin of error half as large as the margin of error that you calculated in part (a), using a 95% confidence level.

SECTION 12.2 Guided Practice—Margin of Error

In this section, you will consider a typical situation that involves finding a margin of error. Consider this problem from a statistics textbook:

> The paper "Patterns and Composition of Weight Change in College Freshmen" (*College Student Journal* [2015]: 553–564) reported that the freshman year weight gain for the students in a representative sample of 103 freshmen at a midwestern university was 5.7 pounds and that the standard deviation of the weight gain was 6.8 pounds. An estimate of the mean weight gain for freshmen at this university is 5.7 pounds. What is the value of the margin of error associated with this estimate, and how would you interpret this value? Use a 95% confidence level.

If you were to encounter a problem like this, you should start by reading the problem carefully. You can use the three-read strategy introduced in Chapter 7.

First Read: Context

1. **What is the general context?**
   A study was conducted to learn about mean weight gain during the freshman year for students at a particular midwestern university.
2. **Are there any terms that you don't understand?**
   This question is about margin of error, so if you don't recall the definition of margin of error, you would want to go to the appropriate section in your textbook to review the definition.

Second Read: Tasks

1. **What question is being asked? What am I being asked to do?**
   This problem asks you to calculate a margin of error and to interpret the value of the margin of error.

Third Read: Details

1. **What data were collected? Were the data collected in a reasonable way?**
   Data were collected from a sample of 103 college students. The problem states that the sample is representative of college students at the university.
2. **Are the actual data provided? What was the sample size (or the sample sizes if there are more than one sample)?**
   The sample size was $n = 103$ (the 103 students in the sample). The actual data (the 103 individual weight gain values) are not given.
3. **Are any summary statistics or additional information given? If so, which statistics and what are the values of these statistics?**
   The sample mean and the sample standard deviation are both given. The sample mean was $\bar{x} = 5.7$ pounds and the sample standard deviation was $s = 6.8$ pounds.

Now you are ready to tackle the tasks required to complete this exercise.

Calculate the Margin of Error

Because the sample was a representative sample and the sample size is greater than or equal to 30, it is appropriate to use the margin of error formula. Substitute the values of n and s into the formula for margin of error to obtain

$$\text{margin of error} = 1.96\frac{s}{\sqrt{n}} = (1.96)\left(\frac{6.8}{\sqrt{103}}\right)$$

Start by evaluating $\sqrt{103} = 10.149$, and then complete the calculation as shown here:

$$\text{margin of error} = 1.96\frac{s}{\sqrt{n}} = (1.96)\left(\frac{6.8}{\sqrt{103}}\right) = (1.96)\left(\frac{6.8}{10.149}\right) = (1.96)(0.670) = 1.313$$

(rounded to three decimal places)

Interpret the Margin of Error in Context

The margin of error can be interpreted as the maximum likely estimation error. You would estimate that the mean weight gain for freshmen at this university is 5.7 pounds, and it is unlikely that this estimate differs from the actual value of the population mean by more than 1.313 pounds.

SECTION 12.2 EXERCISES

12.3 Use the three-read strategy to understand the following exercise from a statistics textbook, and then proceed to complete the exercise.

The article "How Business Students Spend Their Time—Do They Really Know?" (*Research in Higher Education Journal* [May 2009]: 1–10) describes a study of 212 business students at a large public university. Each student kept a log of time spent on various activities during a 1-week period. For these 212 students, the sample mean time spent studying was 9.66 hours and the sample standard deviation was 6.62 hours. Suppose that this sample was a random sample from the population of all business majors at this university and that you are interested in learning about the value of μ, the mean time spent studying for this population. Calculate the value of the margin of error, and interpret the margin of error in context. Use a 95% confidence level.

12.4 Use the three-read strategy to understand the following exercise from a statistics textbook, and then proceed to complete the exercise, with a 95% confidence level.

The authors of the paper "Short-Term Health and Economic Benefits of Smoking Cessation: Low Birth Weight" (*Pediatrics* [1999]:1312–1320) investigated the medical cost associated with babies born to mothers who smoke. The paper included estimates of mean medical cost for low-birth-weight babies for different ethnic groups. For a sample of 654 Hispanic low-birth-weight babies, the mean medical cost was $55,007 and the standard error $\left(\frac{s}{\sqrt{n}}\right)$ was $3011. For a sample of 13 Native American low-birth-weight babies, the mean and standard error were $73,418 and $29,577, respectively. Calculate the margin of error for estimating the actual population mean using the Hispanic sample, and then calculate the margin of error for estimating the actual population mean using the Native American sample. Explain why the two margins of error are so different.

SECTION 12.3 Guided Practice—Confidence Interval for a Population Mean

Consider this problem from a statistics textbook:

> Students in a representative sample of 65 first-year students selected from a large university in England participated in a study of academic procrastination ("Study Goals and Procrastination Tendencies at Different Stages of the Undergraduate Degree," *Studies in Higher Education* [2016}: 2028–2043). Each student in the sample completed the Tuckman Procrastination Scale, which measures procrastination tendencies. Scores on this scale can range from 16 to 64, with scores over 40 indicating high levels of procrastination. For the 65 first-year students in this study, the mean score on the procrastination scale was 37.02 and the standard deviation was 6.44.

a. Construct a 95% confidence interval estimate of μ, the mean procrastination scale for first-year students at this college.

b. Based on your interval, is 40 a plausible value for the population mean score? What does this imply about the population of first-year students?

If you were to encounter a problem like this, you should start by reading the problem carefully. You can use the three-read strategy introduced in Chapter 7.

First Read: Context

1. **What is the general context?**
   A survey was conducted to learn about the academic procrastination for first-year students at a university in England. Each student in the sample completed the Tuckman Procrastination Scale, which measures procrastination tendencies. High scores on this scale mean a greater tendency to procrastinate.

2. **Are there any terms that you don't understand?**
   This question is about constructing a confidence interval, so if you aren't comfortable with the meaning of this term you would want to go to the appropriate section in your textbook to review the definition. If any of the other non-statistical terms are unfamiliar, such as procrastination, you can look up the meaning of those terms in a dictionary to make sure that you fully understand the context. "Procrastination" means putting off or delaying.

Second Read: Tasks

1. **What question is being asked? What am I being asked to do?**
   This problem asks you to do two things:
   1. Construct a 95% confidence interval estimate of μ, the mean procrastination scale score for first-year students at this college.
   2. Decide if 40 is a plausible value for the population mean score and indicate what this implies about the population of first-year students.

Third Read: Details

1. **What data were collected? Were the data collected in a reasonable way?**
   Data were collected from 65 first-year students at a university in England. Each student in the sample completed the Tuckman Procrastination Scale, which measures procrastination tendencies. The problem states that that the sample is representative of students at this university.

2. **Are the actual data provided? What was the sample size (or the sample sizes if there are more than one sample)?**
   The sample size was $n = 65$. The actual data (the 65 individual procrastination scale scores) are not provided.

3. **Are any summary statistics or additional information given? If so, which statistics and what are the values of these statistics?**
   The sample mean and the sample standard deviation are both given. The sample mean was $\bar{x} = 37.02$ and the sample standard deviation was $s = 6.44$.

Now you are ready to tackle the tasks required to complete this exercise. This problem asks you to construct a 95% confidence interval estimate of μ, the mean procrastination scale score for first-year students at this college. You are also asked to decide if 40 is a plausible value for the population mean score and indicate what this implies about the population of first-year students.

Because the sample was a representative sample and the sample size is greater than or equal to 30, it is appropriate to use the confidence interval formula. To determine the t critical value, you first must calculate the degrees of freedom (df) associated with this interval. For a confidence interval for a population mean, $df = n - 1$, which for this example would be $df = 65 - 1 = 64$. A table of t critical values or technology could then be used to find the t critical value. For a 95% confidence level and $df = 64$, the t critical value is 2.00. Substituting values into the expression for the confidence interval results in the following:

$$\bar{x} \pm (t \text{ critical value})\frac{s}{\sqrt{n}} = 37.02 \pm (2.00)\left(\frac{6.44}{\sqrt{65}}\right)$$

Begin by evaluating $\sqrt{65} = 8.062$, and then complete the calculation following the sequence of steps shown here:

$$37.02 \pm (2.00)\left(\frac{6.44}{\sqrt{65}}\right)$$

$$37.02 \pm (2.00)\left(\frac{6.44}{8.062}\right)$$

$$37.02 \pm (2.00)(0.799)$$

$$37.02 \pm 1.598$$

To calculate the endpoints of the interval, you would evaluate

$$37.02 \pm 1.598$$

The lower endpoint of the interval is $37.02 - 1.598 = 35.422$. The upper endpoint of the interval is $37.02 + 1.598 = 38.618$.

The 95% confidence interval estimate of the population mean procrastination scale score is (35.422, 38.618). You would interpret this interval by saying that based on the data from the sample, you can be 95% confident that the actual value of the population procrastination score is somewhere between 35.422 and 38.618.

The value of 40 is not included in the confidence interval. This means that based on the sample data, you don't think 40 is a plausible value for the population mean.

SECTION 12.3 EXERCISES

12.5 Use the three-read strategy to understand the following exercise from a statistics textbook, and then proceed to complete the exercise.

The authors of the paper "Driven to Distraction" (*Psychological Science* [2001]:462–466) describe a study to evaluate the effect of using a cell phone on reaction time. Subjects were asked to perform a simulated driving task while talking on a cell phone. While performing this task, occasional red and green lights flashed on the computer screen. If a green light flashed, subjects were to continue driving, but if a red light flashed, subjects were to brake as quickly as possible and the reaction time (in msec) was recorded. The following summary statistics were read from a graph that appeared in the paper:

$$n = 548, \bar{x} = 530 \text{ msec, and } s = 70 \text{ msec.}$$

Construct and interpret a 95% confidence interval for the mean time to react to a red light while talking on a cell phone. What assumption must be made in order to generalize this confidence interval to the population of all drivers?

12.6 Use the three-read strategy to understand the following exercise from a statistics textbook, and then proceed to complete the exercise.

Seventy-seven students at the University of Virginia were asked to keep a diary of a conversation with their mothers, recording any lies they told during this conversation (*San Luis Obispo Telegram-Tribune*, August 16, 1995). The mean number of lies per conversation was 0.5. Suppose that the standard deviation (which was not reported in the article) was 0.4, and that this group of 77 is representative of the population of students at this university. Construct a 95% confidence interval for the mean number of lies per conversation for this population. Interpret the interval in context.

SECTION 12.4 Guided Practice—Determining Sample Size

Consider this problem from a statistics textbook:

> The paper "Alcohol Consumption, Sleep, and Academic Performance Among College Students" (*Journal of Studies on Alcohol and Drugs* [2009]: 355–363) describes a study of 236 students who were randomly selected from a list of students enrolled at a liberal arts college in the northeastern region of the U.S. Each student in the sample responded to a number of questions about their sleep patterns. For these 236 students, the sample mean time spent sleeping per night was reported to be 7.71 hours and the sample standard deviation of the sleeping times was 1.03 hours. Suppose that you are interested in learning about the value of μ, the population mean time spent sleeping per night for students at your college. You would like your estimate to have a margin of error of 0.25 hours. How many students should you include in a random sample of students from your college? You can assume that the standard deviation of sleep times at your college is similar to the standard deviation for the students in the sample at the liberal arts college. Use a 95% confidence level.

If you were to encounter a problem like this, you should start by reading the problem carefully. You can use the three-read strategy introduced in Chapter 7.

First Read: Context

1. **What is the general context?**
   You are planning to collect data to estimate the mean time spent sleeping each night for students at your college and want to know how large a sample should be selected.
2. **Are there any terms that you don't understand?**
   The term margin of error is used. If you don't recall the definition of this term, you should go to the appropriate section of your statistics textbook to review the meaning of this term.

Second Read: Tasks

1. **What question is being asked? What am I being asked to do?**
   You are asked to determine how many students should be included in the sample to achieve a given margin of error.

Third Read: Details

1. **What data were collected? Were the data collected in a reasonable way?**
   No data have been collected yet—you are just in the process of planning for data collection.
2. **Are the actual data provided? What was the sample size (or the sample sizes if there are more than one sample)?**
   A sample has not been selected yet—you want to determine the size of the sample that should be selected.
3. **Are any summary statistics or additional information given? If so, which statistics and what are the values of these statistics?**
   You know that you want to choose a sample size that would result in a margin of error of 0.25. Summary statistics are given from a similar study carried out at a different college. You are told that you can assume that the standard deviation of sleep times at your college is similar to the standard deviation at the other college, which was 1.03 hours.

Now you are ready to tackle the tasks required to complete this exercise. This problem asks you to determine how many students should be included in a random sample of students from your college if you want to estimate the mean number of hours slept each night with a margin of error of 0.25 hours. To answer this question, you can use the formula for sample size:

$$n = \left(\frac{1.96\sigma}{M}\right)^2$$

You can use the sample standard deviation $s = 1.03$ from the similar study at the liberal arts college as the estimate for σ. Begin by substituting the values for M and your estimate for σ into the expression. Substituting $M = 0.25$ and $\sigma = 1.03$ into the expression results in

$$n = \left(\frac{1.96\sigma}{M}\right)^2 = \left(\frac{(1.96)(1.03)}{0.25}\right)^2$$

To evaluate this expression, you would first calculate the value of $\left(\frac{(1.96)(1.03)}{0.25}\right)$. You would then square this number. This sequence of steps illustrated here:

$$n = \left(\frac{(1.96)(1.03)}{0.25}\right)^2 = \left(\frac{2.018}{0.25}\right)^2 = (8.072)^2 = 65.157$$

In sample size calculations, you always round up to the next whole number to obtain the necessary sample size. For this example, a sample size of 66 would be needed to achieve a margin of error of $M = 0.25$ hours.

SECTION 12.4 EXERCISES

12.7 Use the three-read strategy to understand the following exercise from a statistics textbook, and then proceed to complete the exercise.

The Bureau of Alcohol, Tobacco, and Firearms (BATF) has been concerned about lead levels in California wines. In a previous testing of wine specimens, lead levels ranging from 50 to 700 parts per billion were recorded. How many wine specimens should be tested if the BATF wishes to estimate the mean lead level for California wines with a margin of error of 10 parts per billion? Use a 95% confidence level. (Hint: Refer to Section 12.1 for information about estimating a standard deviation by dividing the range by four.)

12.8 Use the three-read strategy to understand the following exercise from a statistics textbook, and then proceed to complete the exercise.

Samples of two different models of cars were selected, and the actual speed for each car was determined when the speedometer registered 50 mph. The resulting 95% confidence intervals for mean actual speed were (51.3, 52.7) for model 1 and (49.4, 50.6) for model 2. Assuming that the two sample standard deviations were equal, which confidence interval is based on the larger sample size? Explain your reasoning.

13 Testing Hypotheses About a Population Mean—The Math You Need to Know

SECTION 13.1 Evaluating Expressions

In order to determine if there is convincing evidence against a claim about a population mean, you may choose to calculate the value of a test statistic as part of the process for carrying out a test.

Note: Depending on your access to technology, such as a graphing calculator or a statistics software package, you may not need to evaluate the expression for the test statistic by hand. But just in case your statistics class is not relying on technology to do the calculation, this section illustrates how to evaluate this expression.

Test Statistic for Testing Hypotheses About a Population Mean

$$t = \frac{\bar{x} - \mu_0}{\frac{s}{\sqrt{n}}}$$

A hypothesis test about a population mean uses data from a sample to decide between two competing hypotheses—a null hypothesis (denoted by H_0) and an alternative hypothesis (denoted by H_a). The null hypothesis specifies a particular hypothesized value of the population mean, denoted by μ_0. The null hypothesis will be of the form $H_0 : \mu = \mu_0$. For example, you might want to carry out a hypothesis test where the null hypothesis states that population mean is 100. The null hypothesis would then be $H_0 : \mu = 100$. For purposes of calculating the value of the test statistic, the value of μ_0 for this null hypothesis is $\mu_0 = 100$.

Before calculating the value of the test statistic, you need to verify that the sample size is large enough for the one-sample t test to be appropriate. The sample size is large enough to use the one-sample t test for a population mean if $n \geq 30$. If the sample size is less than 30, you must be willing to assume that the population distribution is at least approximately normal. You also need to confirm that the sample is a random sample or that it was selected in a way that makes it reasonable to think that the sample is representative of the population.

To calculate the value of the test statistic, you need to know the values of the sample size, n, the sample mean, $\bar{x}$, and the sample standard deviation, s. Sometimes these values will be given, but other times you might need to calculate them from given data. For a review of how to evaluate the expressions for $\bar{x}$ and s, see Sections 3.4 and 3.5 in Chapter 3.

Suppose that you want to test the null hypothesis $H_0 : \mu = 75$ and that a sample of size $n = 50$ resulted in a sample mean of $\bar{x} = 82$ and a sample standard deviation of $s = 16$. To calculate the value of the test statistic, substitute the values of $\bar{x}$ (the value of the sample mean), s (the value of the sample standard deviation), μ_0 (the hypothesized value from the null hypothesis), and n into the formula for the test statistic to obtain

$$t = \frac{\bar{x} - \mu_0}{\frac{s}{\sqrt{n}}} = \frac{82 - 75}{\frac{16}{\sqrt{50}}}$$

Begin by evaluating the part of the expression in the numerator of the test statistic: $82 - 75 = 7$. The next step is to evaluate the part of the expression that appears in the denominator. First calculate $\sqrt{50} = 7.071$. Then calculate the value of the denominator. Here is the sequence of steps:

$$\frac{16}{\sqrt{50}} = \frac{16}{7.071} = 2.263 \text{ (rounded to three decimal places)}$$

Complete the calculation of the value of the test statistic by dividing the 7 in the numerator by the value in the denominator (2.263) to obtain

$$t = \frac{\bar{x} - \mu_0}{\frac{s}{\sqrt{n}}} = \frac{82 - 75}{\frac{16}{\sqrt{50}}} = \frac{7}{2.263} = 3.09 \text{ (rounded to two decimal places)}$$

The value of the test statistic for this example is $t = 3.09$.

Check It Out!

a. A random sample of size $n = 14$ is selected from a population that is known to be normally distributed. The sample mean is $\bar{x} = 167$, and the sample standard deviation is $s = 51$. Suppose that you want to test the null hypothesis $H_0 : \mu = 150$. Calculate the value of the test statistic.

b. A sample of size $n = 78$ is selected to be representative of a population of measurements. Calculate the value of the test statistic for testing the null hypothesis $H_0 : \mu = 90$ when the value of the sample mean is $\bar{x} = 89.1$ and the sample standard deviation is $s = 11.7$.

Answers

a. Substitute the values of $\bar{x}$ (the value of the sample mean), s (the value of the sample standard deviation), μ_0 (the hypothesized value from the null hypothesis), and n into the formula for the test statistic to obtain

$$t = \frac{\bar{x} - \mu_0}{\frac{s}{\sqrt{n}}} = \frac{167 - 150}{\frac{51}{\sqrt{14}}}$$

Calculate the value of the test statistic as shown here:

$$t = \frac{\bar{x} - \mu_0}{\frac{s}{\sqrt{n}}} = \frac{167 - 150}{\frac{51}{\sqrt{14}}} = \frac{17}{13.629} = 1.25 \text{ (rounded to two decimal places)}$$

The value of the test statistic for this example is $t = 1.25$

b. Substitute the values of $\bar{x}$ (the value of the sample mean), s (the value of the sample standard deviation), μ_0 (the hypothesized value from the null hypothesis), and n into the formula for the test statistic to obtain

$$t = \frac{\bar{x} - \mu_0}{\frac{s}{\sqrt{n}}} = \frac{89.1 - 90}{\frac{11.7}{\sqrt{78}}}$$

Calculate the value of the test statistic as shown here:

$$t = \frac{\bar{x} - \mu_0}{\frac{s}{\sqrt{n}}} = \frac{89.1 - 90}{\frac{11.7}{\sqrt{78}}} = \frac{-0.9}{1.325} = -0.68 \text{ (rounded to two decimal places)}$$

The value of the test statistic for this example is $t = -0.68$.

SECTION 13.1 EXERCISES

13.1 A random sample of size $n = 80$ is selected from a population of measurements. Calculate the value of the test statistic for testing the null hypothesis $H_0 : \mu = 100$ when the value of the sample mean is $\bar{x} = 95$ and the sample standard deviation is $s = 4.8$.

13.2 A representative sample of size $n = 23$ is selected from a population that is known to be normally distributed. The sample mean is $\bar{x} = 7.7$, and the sample standard deviation is $s = 2.4$. Suppose that you want to test the null hypothesis $H_0 : \mu = 5$. Calculate the value of the test statistic.

SECTION 13.2 Guided Practice—One-Sample Hypothesis Test for a Population Mean

In this section, you will consider a typical situation that involves carrying out a one-sample t test for a population mean. Consider this problem from a statistics textbook:

> The authors of the paper "Changes in Quantity, Spending, and Nutritional Characteristics of Adult, Adolescent and Child Urban Corner Store Purchases After an Environmental Intervention" (*Preventative Medicine* [2015]: 81–85) wondered if increasing the availability of healthy food options would also increase the amount people spend at the corner store. They collected data from a representative sample of 5949 purchases at corner stores in Philadelphia after the stores increased their healthy food options. The sample mean amount spent for this sample of purchases was $2.86 and the sample standard deviation was $5.40.
>
> Before the stores increased the availability of healthy foods, the population mean total amount spent per purchase was thought to be about $2.80. Do the data from this study provided convincing evidence that the population mean amount spent per purchase is greater after the change to increase healthy food options? Carry out a hypothesis test with a significance level of 0.05.

If you were to encounter a problem like this, you should start by reading the problem carefully. You can use the three-read strategy introduced in Chapter 7.

First Read: Context

1. **What is the general context?**
   Researchers wondered if increasing healthy food options at corner stores would increase the amount of money that people spent at the store. The looked at a representative sample of purchases at corner stores in Philadelphia that had increased healthy food options.

2. **Are there any terms that you don't understand?**
   This question is about testing a hypothesis, so if you don't recall the steps required to carry out a hypothesis test, you would want to go to the appropriate section in your textbook to review these steps. If you are unsure about what a corner store is, you could do an online search to find out that a corner store is a small store selling groceries and household goods in a mainly residential area.

Second Read: Tasks

1. **What question is being asked? What am I being asked to do?**
   This problem asks you to decide if the sample data provide convincing evidence that the mean amount of money spent on a corner store purchase after the introduction of healthier food options is greater than it was before the healthier food options were introduced. You are to do this by carrying out a hypothesis test using a significance level of 0.05.

Third Read: Details

1. **What data were collected? Were the data collected in a reasonable way?**
   The data are from a sample of purchases made at corner stores in Philadelphia. You are told that the sample can be regarded as representative of the corner store purchases in Philadelphia.

2. **Are the actual data provided? What was the sample size (or the sample sizes if there are more than one sample)?**
   The sample size was $n = 5949$ (the 5949 individual purchases). The actual data (the amounts of the 5949 purchases) are not given.

3. **Are any summary statistics or additional information given? If so, which statistics and what are the values of these statistics?**
   The value of the sample mean and the value of the sample standard deviation are given. They are $\bar{x} = \$2.86$ and $s = \$5.40$. You also know that the sample size was $n = 5949$.

Now you are ready to tackle the tasks required to complete this exercise. The problem states "Do the data from this study provided convincing evidence that the population mean amount spent per purchase is greater after the change to increase healthy food options? Carry out a hypothesis test with a significance level of 0.05." This is asking you to test hypotheses about a population mean, μ. The population is all purchases made at Philadelphia corner stores that have introduced healthier food options. You are asked to decide if there is convincing evidence that the mean amount spent per purchase is greater than it was before the introduction of healthier food options. The mean amount per purchase before the introduction of healthier food options was \$2.80. You are also told to use a significance level of 0.05 when carrying out the test.

Many textbooks recommend a sequence of steps for carrying out a hypothesis test. When you carry out a hypothesis test, you should follow the recommended sequence of steps. Although the order of the steps may vary somewhat from one text to another, they generally include the following:

Hypotheses: You will need to specify the null and the alternative hypothesis for the hypothesis test and define any symbols that are used in the context of the problem.

Method: You will need to specify what method you plan to use to carry out the test. For example, you might say that you plan to use the one-sample *t* test for a population mean if that is a test that would allow you to answer the question posed. You will also need to specify the significance level you will use to reach a conclusion.

Check Conditions: Many hypothesis test procedures are only appropriate when certain conditions are met. Once you have determined the method you plan to use, you will need to verify that any required conditions are met.

Calculate: To reach a conclusion in a hypothesis test, you will need to calculate the value of a test statistic and the associated *P*-value.

Communicate Results: The final step is to use the computed values of the test statistic and the *P*-value to reach a conclusion and to provide an answer to the question posed. You should always provide a conclusion in words.

These steps are illustrated here for the given example.

Hypotheses: You want to determine if there is evidence that the mean amount spent on a corner store purchase in Philadelphia is greater after the introduction of healthier food options. The mean amount per purchase before the introduction of healthier food options was \$2.80. You can use μ to represent the mean amount spent on a corner store purchase in Philadelphia after the introduction of healthier food options. The question being asked usually translates into the alternative hypothesis, so the alternative hypothesis would be $H_a : \mu > 2.80$. The null hypothesis would then be obtained by replacing the $>$ in the alternative hypothesis with an $=$. This results in hypotheses of

$$H_0 : \mu = 2.80$$
$$H_a : \mu > 2.80$$

Method: Because the hypotheses are about a population mean, you would consider the one-sample t test for a population mean. The test statistic is

$$t = \frac{\bar{x} - 2.80}{\frac{s}{\sqrt{n}}}$$

A significance level of $\alpha = 0.05$ was specified.

Check: There are two conditions necessary for the one-sample t test for a population mean to be appropriate. You are told that the sample is representative of corner store purchases in Philadelphia after the introduction of healthier food options. The sample size is $n = 5949$. Because the sample size is greater than or equal to 30, the sample size condition is met.

Calculations: The value of the sample mean and the value of the sample standard deviation are $\bar{x} = \$2.86$ and $s = \$5.40$. You also know that the sample size was $n = 5949$. The value of the test statistic is

$$t = \frac{\bar{x} - 2.80}{\frac{s}{\sqrt{n}}} = \frac{2.86 - 2.80}{\frac{5.40}{\sqrt{5949}}} = \frac{0.06}{\frac{5.40}{77.130}}$$

$$= \frac{0.06}{0.070} = 0.86 \text{ (rounded to two decimal places)}$$

Once you know the value of the test statistic, you need to find the associated P-value. Your statistics text will have an explanation of how the P-value is determined.

This is an upper-tailed test (the inequality in H_a is $>$), so the P-value an area under a t curve. The degrees of freedom (df) associated with a one-sample t test is $n - 1$, so $df = 5949 - 1 = 5948$. The P-value is the area under the t curve with $df = 5948$ and to the right of 0.86. Therefore, the P-value is $P(t \geq 0.86) = 0.198$.

Communicate Results: Because the P-value of 0.198 is greater than the selected significance level of 0.05, you fail to reject the null hypothesis. There is not convincing evidence that the mean amount spent on a corner store purchase in Philadelphia is greater after the introduction of healthier food options.

SECTION 13.2 EXERCISES

13.3 A study was conducted by researchers at Penn State University who investigated whether time perception, an indication of a person's ability to concentrate, is impaired during nicotine withdrawal. The study results were summarized in the paper "Smoking Abstinence Impairs Time Estimation Accuracy in Cigarette Smokers" (*Psychopharmacology Bulletin* [2003]: 90–95). After a 24-hour smoking abstinence, $n = 20$ smokers were asked to estimate how much time had passed during a 45-second period. Suppose the resulting data on perceived elapsed time (in seconds) yielded a sample mean of $\bar{x} = 59.3$ seconds, and sample standard deviation $s = 9.84$ seconds.

Do the data from this study provided convincing evidence that the population mean estimated time is greater than 45 seconds? Use the three-read strategy to understand the problem, and then carry out a hypothesis test with a significance level of $\alpha = 0.05$.

13.4 A credit bureau analysis of undergraduate students, credit records found that the average number of credit cards in an undergraduate's wallet was 4.1 ("Undergraduate Students and Credit Cards in 2004," Nellie Mae, May 2005). It was also reported that in a random sample of 132 undergraduates, the sample mean number of credit cards that the students said they carried was 2.6. The sample standard deviation was not reported, but for purposes of this exercise, suppose that it was 1.2. Is there convincing evidence at the 0.01 level that the mean number of credit cards that undergraduates report carrying is less than the credit bureau's figure of 4.1? Use the three-read strategy to understand the problem, and then carry out an appropriate hypothesis test. Use a significance level of $\alpha = 0.05$.

14 Estimating a Difference in Means Using Paired Samples—The Math You Need to Know

SECTION 14.1 Evaluating Expressions

When samples are paired, each observation in one sample can be paired in a meaningful way with a particular observation in the other sample. When you have data from paired samples, you will work with the differences obtained by subtracting the value from one sample from the corresponding value in the other sample for each pair. The single set of differences can be used to both estimate the difference in population or treatment means and to test hypotheses about the difference in population or treatment means. In this chapter, you will look at the expressions you will encounter when calculating a confidence interval for a difference in means using data from paired samples.

Note: Depending on your access to technology, such as a graphing calculator or a statistics software package, you may not need to evaluate this expression by hand. But just in case your statistics class is not relying on technology to do these calculations, this section illustrates how to evaluate the expression.

Confidence Interval for a Difference in Means Using Paired Samples

$$\bar{x}_d \pm (t \text{ critical value})\frac{s_d}{\sqrt{n_d}}$$

Recall that when you have data from paired samples, you will work with the differences obtained by subtracting each value from one sample from the corresponding value in the other sample to form a set of differences. In the given expression for the confidence interval, $\bar{x}_d$ represents the mean value of the sample of differences, s_d represents the standard deviation of the sample of differences, and n_d is the number of differences (which will be equal to the sample size of each of the two paired samples). Sometimes the values of $\bar{x}_d$ and s_d will be given, but in other cases you may have to calculate the differences and the values of $\bar{x}_d$ and s_d. For a review of how to calculate the values of the sample mean and the sample standard deviation, see Chapter 3.

To calculate a confidence interval for a difference in means using paired samples, you need to know the values of the mean and the standard deviation for the sample of differences, $\bar{x}_d$ and s_d, the sample size n_d, and the confidence level, which determines the t critical value that is used to calculate the confidence interval.

Before using the given expression to calculate a confidence interval, you should also verify that the sample size is large enough to calculate a confidence interval using the given formula or that it is reasonable to think that the population distribution of differences is approximately normal. You also need to confirm that the sample is a random sample or that it was selected in a way that makes it reasonable to think that the sample is representative of the population of differences.

The sample size is large enough to use the paired-samples confidence interval formula if n_d is greater than or equal to 30 (note that some texts are more conservative and use n_d greater than or equal to 40). If n_d is less than 30, this confidence interval formula is only appropriate if it is reasonable to assume that the population distribution of differences is approximately normal. A plot of the differences (a dotplot, a boxplot, or a normal probability plot) can be used to help you decide if this assumption is reasonable.

Recall that an interval is defined by its endpoints—it has a lower endpoint and an upper endpoint. Notice that the expression for the confidence interval includes the symbol $\pm$, which is read as "plus and minus." This means that you will do two calculations, once using $+$ and once using $-$. Each one of these calculations results in one of the two interval endpoints.

To calculate the endpoints of the confidence interval, start by determining the appropriate t critical value for the specified confidence level. Your statistics text should have instructions for how this is done.

Once you have determined the appropriate t critical value, substitute the values of n_d, $\bar{x}_d$, and s_d into the expression for the confidence interval.

For example, the authors of the paper "Driving Performance While Using a Mobile Phone: A Simulation Study of Greek Professional Drivers" (*Transportation Research Part F* [2016]: 164–170) used data from a study in which 50 Greek male taxi drivers drove in a driving simulator. In the simulator, they were asked to drive following a lead car. On one drive, they had no distractions and the average distance between the driver's car and the lead car was recorded. In a second drive, the drivers talked on a mobile phone while driving. This resulted in paired samples—one sample of the distances with no distractions and one sample of distances when using a mobile phone. The 50 sample differences (no distraction $-$ talking on mobile phone) were calculated and used to calculate the sample mean and the sample standard deviation of the differences. The mean of the 50 sample differences was 0.47 meters and the standard deviation of the sample differences was 1.22 meters. The authors assumed that this sample of 50 drivers is representative of Greek taxi drivers.

In this setting, $\bar{x}_d = 0.47$, $s_d = 1.22$, and $n_d = 50$.

Suppose that you want to use this information to construct and interpret a 95% confidence interval for the difference in the means for following distance for Greek taxi drivers while driving with no distractions and while driving and texting. To determine the t critical value, you first must calculate the degrees of freedom (df) associated with this interval. For a paired-samples confidence interval for a difference in means, $df = n_d - 1$, which for this example would be $df = 50 - 1 = 49$. A table of t critical values or technology could then be used to find the t critical value. For a 95% confidence level and $df = 49$, the t critical value is 2.010.

Substituting values into the expression for the paired-samples confidence interval results in the following:

$$\bar{x}_d \pm (t \text{ critical value})\frac{s_d}{\sqrt{n_d}} = 0.47 \pm (2.010)\left(\frac{1.22}{\sqrt{50}}\right)$$

Start by evaluating $\sqrt{50} = 7.071$, and then complete the calculation following the sequence of steps shown here:

$$0.47 \pm (2.010)\left(\frac{1.22}{\sqrt{50}}\right)$$

$$0.47 \pm (2.010)\left(\frac{1.22}{7.071}\right)$$

$$0.47 \pm (2.010)(0.173)$$

$$0.47 \pm 0.348$$

To calculate the endpoints of the interval, you would evaluate

$$0.47 \pm 0.348$$

The lower endpoint of the interval is $0.47 - 0.348 = 0.122$. The upper endpoint of the interval is $0.47 + 0.348 = 0.818$.

The 95% confidence interval estimate of the difference in means for following distance (no distraction − talking on mobile phone) is (0.122, 0.818). You would interpret this interval by saying that based on the data from the sample, you can be 95% confident that the actual difference in means for following distance for Greek taxi drivers is somewhere between 0.122 and 0.818 meters. Because the differences were calculated as (no distraction – talking on mobile phone) and both endpoints of the confidence interval are positive, you could also say that the mean following distance with no distractions is greater than the mean following distance while talking on a mobile phone by somewhere between 0.122 and 0.818 meters.

Check It Out!

a. Researchers collected measurements on the same 107 individuals at time 1 and again at time 2. The difference for each individual was calculated, time 2 − time 1. You may assume that the differences are representative of their population. The sample mean of the paired differences is $\bar{x}_d = 57$, and the sample standard deviation is $s_d = 47$. Use this information to construct and then interpret a 90% confidence interval for the population mean difference, μ_d, for time 2 − time 1.

b. A study involved measuring the weights of twins, with the two siblings labeled thing 1 (first born) and thing 2 (second born). There were 131 pairs of twins in the study, and the paired differences were calculated as thing 1 − thing 2. The sample mean of the paired differences was $\bar{x}_d = 8.4$ pounds, and the sample standard deviation of the paired differences was $s_d = 1.3$ pounds.

   Use this information to construct and then interpret a 95% confidence interval for the population mean weight difference, μ_d, for thing 1 − thing 2.

Answers

a. The t critical value depends on the degrees of freedom (df), which is $df = 107 - 1 = 106$ for this example. A table of t critical values or technology could be used to find the t critical value. For a 90% confidence level and $df = 106$, the t critical value is 1.659.

   Substituting values into the expression for the paired-samples confidence interval results in the following:

$$\bar{x}_d \pm (t \text{ critical value})\frac{s_d}{\sqrt{n_d}} = 57 \pm (1.659)\left(\frac{47}{\sqrt{107}}\right)$$

Complete the calculation of the confidence interval:

$$57 \pm (1.659)\left(\frac{47}{\sqrt{107}}\right)$$

$$57 \pm (1.659)\left(\frac{47}{10.34}\right)$$

$$57 \pm (1.659)(4.545)$$

$$57 \pm 7.54$$

The lower endpoint of the interval is $57 - 7.54 = 49.46$. The upper endpoint of the interval is $57 + 7.54 = 64.54$.

The 90% confidence interval estimate of the population mean difference (time 2 − time 1) is (49.46, 64.54). You can be 90% confident that the population mean difference falls between 49.46 and 64.54.

b. The degrees of freedom (df) for a paired-samples confidence interval is $df = n_d - 1$, which for this example is $df = 131 - 1 = 130$. A table of t critical values or technology could then be used to find the t critical value. For a 95% confidence level and $df = 130$, the t critical value is 1.978.

Substituting values into the expression for the paired-samples confidence interval results in the following:

$$\bar{x}_d \pm (t \text{ critical value})\frac{s_d}{\sqrt{n_d}} = 8.4 \pm (1.978)\left(\frac{1.3}{\sqrt{131}}\right)$$

Complete the calculation of the confidence interval:

$$8.4 \pm (1.978)\left(\frac{1.3}{\sqrt{131}}\right)$$

$$8.4 \pm (1.978)\left(\frac{1.3}{11.45}\right)$$

$$8.4 \pm (1.978)(0.114)$$

$$8.4 \pm 0.23$$

The lower endpoint of the interval is $8.4 - 0.23 = 8.17$. The upper endpoint of the interval is $8.4 + 0.23 = 8.63$.

The 95% confidence interval estimate of the population mean weight difference (thing 1 − thing 2) is (8.17, 8.63). You can be 95% confident that the population mean weight difference falls between 8.17 pounds and 8.63 pounds.

SECTION 14.1 EXERCISES

14.1 Reaction time (in milliseconds) was recorded for each of $n_d = 34$ individuals both before and after drinking a beverage containing caffeine. The paired difference for each individual was calculated, Before – After. You may assume that the differences are representative of their population. The sample mean of the paired differences was $\bar{x}_d = 6.7$, and the sample standard deviation was $s_d = 0.8$. Use this information to construct and interpret a 90% confidence interval for the population mean difference in reaction time, μ_d, for Before – After.

14.2 In a research study, $n_d = 64$ participants compared two brands of soft drink and rated each on a scale from 1 to 10. One brand was labeled "Soda" and the other was labeled "Pop." The paired differences, "Pop" – "Soda" were calculated. You may assume that the differences represent the population of all paired differences for these two brands, with ratings subtracted in the same order. The sample mean of the paired differences was $\bar{x}_d = 0.64$, and the sample standard deviation was $s_d = 0.71$. Use this information to construct and interpret a 95% confidence interval for the population mean difference, μ_d, for "Pop" – "Soda."

SECTION 14.2 Guided Practice—Confidence Interval for a Difference in Means—Paired Samples

Consider this problem from a statistics textbook:

> The article "Puppy Love? It's Real, Study Says" (*USA TODAY*, April 17, 2015) describes a study into how people communicate with their pets. The conclusion stated in the title of the article was based on research published in *Science* ("Oxytocin-Gaze Positive Loop and the Coevolution of Human-Dog Bonds," April 17, 2015). Researchers measured the oxytocin levels (in picograms per milligram, pg/mg) of 22 dog owners before and again after a 30-minute interaction with their dogs. (Oxytocin is a hormone known to play a role in parent–child bonding.) The difference in oxytocin level (Before − After) was calculated for each of the 22 dog owners. Suppose that the mean and standard deviation of the differences (approximate values based on a graph in the paper) were $\bar{x}_d = 27$ pg/mg and $s_d = 30$ pg/mg.
>
> Assume that it is reasonable to regard the 22 dog owners who participated in this study as representative of dog owners in general. Estimate the difference in the means for oxytocin level of dog owners before and after 30 minutes of interaction with their dogs using a 90% confidence interval.

If you were to encounter a problem like this, you should start by reading the problem carefully. You can use the three-read strategy introduced in Chapter 7.

First Read: Context

1. **What is the general context?**
   A study was conducted to see how oxytocin levels change when people interact with their dogs. Oxytocin level was measured before the people in the study interacted with their dogs and again after they had interacted with their dog for 30 minutes.

2. **Are there any terms that you don't understand?**
   This question is about constructing a confidence interval, so if you aren't comfortable with the meaning of this term you would want to go to the appropriate section in your textbook to review the definition. You might not have been familiar with the word "oxytocin," but this word is defined in the problem as a hormone known to play a role in parent–child bonding. If you are unsure what a hormone is, you can look up the meaning of that term in a dictionary to make sure that you fully understand the context.

Second Read: Tasks

1. **What question is being asked? What am I being asked to do?**
   This problem asks you to estimate the difference in the means for oxytocin level of dog owners before and after 30 minutes of interaction with their dogs using a 90% confidence interval.

Third Read: Details

1. **What data were collected? Were the data collected in a reasonable way?**
   Data were collected from 22 dog owners. Each dog owner participating in the study had oxytocin level measured before and after a 30-minute interaction with his or her dog. The problem states that that the sample can be regarded as representative of dog owners. Even though there was a single group of dog owners participating in the study, there are actually two paired samples in this problem because oxytocin level was measured before and after the interaction for each person. This results in two samples of measurements (one of Before measurements and one of After measurements).

2. **Are the actual data provided? What was the sample size (or the sample sizes if there are more than one sample)?**
   The actual data (the Before and After oxytocin measurements) are not provided. There are two paired samples (one of Before measurements and one of After measurements). Each sample consists of 22 oxytocin measurements. This would result in a set of 22 differences, so the sample size for the sample of differences is $n_d = 22$.

3. **Are any summary statistics or additional information given? If so, which statistics and what are the values of these statistics?**
   The sample mean and the sample standard deviation for the sample of $n_d = 22$ differences are both given. The sample mean difference was $\bar{x}_d = 27$ and the sample standard deviation of the differences was $s_d = 30$.

Now you are ready to tackle the tasks required to complete this exercise. This problem asks you to construct a 90% confidence interval estimate of the difference in mean oxytocin level before and after a 30-minute interaction with a dog.

You are told that you can regard the sample as a representative sample of dog owners. The sample size is only 22, so you would also need to assume that the difference distribution is normal. If the actual data had been provided, you could decide if this assumption was reasonable by making a dotplot (or a boxplot or a normal probability plot) of the sample differences and looking to see if the distribution was reasonably symmetric with no outliers.

To determine the t critical value, you first must calculate the degrees of freedom (df) associated with this interval. For a paired-samples confidence interval for a difference in means, $df = n_d - 1$, which for this example would be $df = 22 - 1 = 21$. A table of t critical values or technology could then be used to find the t critical value. For a 90% confidence level and $df = 21$, the t critical value is 1.72. Substituting values into the expression for the confidence interval results in the following:

$$\bar{x}_d \pm (t \text{ critical value})\frac{s_d}{\sqrt{n_d}} = 27 \pm (1.72)\left(\frac{30}{\sqrt{22}}\right)$$

Begin by evaluating $\sqrt{22} = 4.690$, and then complete the calculation following the sequence of steps shown here:

$$27 \pm (1.72)\left(\frac{30}{\sqrt{22}}\right)$$

$$25 \pm (1.72)\left(\frac{30}{4.690}\right)$$

$$25 \pm (1.72)(6.397)$$

$$25 \pm 11.003$$

To calculate the endpoints of the interval, you would evaluate

$$27 \pm 11.003$$

The lower endpoint of the interval is $27 - 11.003 = 15.997$. The upper endpoint of the interval is $27 + 11.003 = 38.003$.

The 90% confidence interval estimate of the difference in means for oxytocin level before interacting with a dog and after a 30 minute interaction with a dog is (15.997, 38.003). You would interpret this interval by saying that based on the data from the sample, you can be 90% confident that the actual difference in means for oxytocin level before and after interacting with a dog for 30 minutes is somewhere between 15.997 pg/mg and 38.003 pg/mg. Because the differences were calculated as (Before − After) and both endpoints of the confidence interval are positive, you could also say that the mean oxytocin level before the 30-minute interaction with the dog is greater than the mean oxytocin level after the 30-minute interaction by somewhere between 15.997 pg/mg and 38.003 pg/mg.

SECTION 14.2 EXERCISES

14.3 Use the three-read strategy to understand the following exercise from a statistics textbook, and then proceed to complete the exercise.

The paper "The Truth About Lying in Online Dating Profiles" (*Proceedings, Computer-Human Interactions* [2007]: 1–4) describes an investigation including 40 men with online dating profiles. Each participating man's height (in inches) was measured and the actual height was compared to the height given in the man's online profile. The differences between the online profile height and the actual height (Profile − Actual) were used to calculate the values in the accompanying table:

Sample mean	$\bar{x}_d = 0.57$
Sample standard deviation	$s_d = 0.81$
Sample size	$n_d = 40$

Construct a 99% confidence interval for the population mean difference in profile height and actual height, and interpret the interval in context.

14.4 Use the three-read strategy to understand the following exercise from a statistics textbook, and then proceed to complete the exercise.

Do girls think they don't need to take as many science classes as boys? The article "Intentions of Young Students to Enroll in Science Courses in the Future: An Examination of Gender Differences" (*Science Education* [1999]: 55–76) describes a survey of randomly selected children in grades 4, 5, and 6. The 224 girls participating in the survey each indicated the number of science courses they intended to take in the future, and they also indicated the number of science courses they thought boys their age should take in the future. For each girl, the authors calculated the difference between the number of science classes she intends to take and the number she thinks boys should take.

The sample mean of the differences was −0.83 (indicating girls intended, on average, to take fewer science classes than they thought boys should take), and the sample standard deviation of the differences was 1.51. Construct and interpret a 95% confidence interval for the population mean difference.

15 Testing Hypotheses About a Difference in Means Using Paired Samples—The Math You Need to Know

SECTION 15.1 Evaluating Expressions

When samples are paired, each observation in one sample can be paired in a meaningful way with a particular observation in the other sample. When you have data from paired samples, you will work with the differences obtained by subtracting the value from one sample from the corresponding value in the other sample for each pair. The single set of differences can be used to both estimate the difference in population or treatment means and to test hypotheses about the difference in population or treatment means. In this chapter, you will look at the expression for the test statistic that can be used to test hypotheses about a difference in means using data from paired samples.

Note: Depending on your access to technology, such as a graphing calculator or a statistics software package, you may not need to evaluate the expression for the test statistic by hand. But just in case your statistics class is not relying on technology to do the calculation, this section illustrates how to evaluate the expression.

Test Statistic for a Paired-Samples *t* Test for a Difference in Means

$$t = \frac{\bar{x}_d - \mu_0}{\frac{s_d}{\sqrt{n_d}}}$$

A paired-samples hypothesis test for a difference in population or treatment means uses data from paired samples to decide between two competing hypotheses—a null hypothesis (denoted by H_0) and an alternative hypothesis (denoted by H_a). The null hypothesis specifies a particular hypothesized value for the difference in means, denoted by μ_0. The null hypothesis will be of the form $H_0 : \mu_d = \mu_0$, where μ_d represents the actual value of the difference in means. For example, you might want to carry out a hypothesis test where the null hypothesis states that actual difference in means is 0. The null hypothesis would then be $H_0 : \mu_d = 0$, and this hypothesis is equivalent to saying that the two population or treatment means are equal. For purposes of calculating the value of the test statistic, the value of μ_0 for this null hypothesis is $\mu_0 = 0$.

Recall that when you have data from paired samples, you will work with the differences obtained by subtracting each value from one sample from the corresponding value in the other sample to form a set of differences. In the given expression for the paired-samples *t* test statistic, $\bar{x}_d$ represents the mean value of the sample of differences, s_d represents the standard deviation of the sample of differences, and n_d is the number of differences (which will be equal to the size of each of the two paired samples). Sometimes the values of $\bar{x}_d$ and s_d will be given, but other times you may have to calculate the differences and the values of $\bar{x}_d$ and s_d. For a review of how to calculate the values of the sample mean and the sample standard deviation, see Chapter 3.

Before using the given expression to test hypotheses about a difference in means, you should also verify that the sample size is large enough for the paired-samples *t* test to be appropriate. The sample size is large enough to use the paired-samples *t* test statistic if n_d is greater than or equal to 30 (note that some texts are more conservative and use n_d greater than or equal to 40). If n_d is less than 30, the paired-samples *t* test is only appropriate if it is reasonable to assume that the population distribution of differences is approximately normal. A plot of the sample differences (dotplot, boxplot, or normal probability plot) can be used to help you decide if this assumption is reasonable. You also need to confirm that the sample of differences is a random sample of the population of differences or that it is reasonable to think that the sample is representative of the population of differences.

Suppose that you want to use data from paired samples to test the null hypothesis $H_0 : \mu_d = 0$ and that paired samples of size $n_d = 40$ were used to obtain differences and to calculate a sample mean of $\bar{x}_d = 8$ and a sample standard deviation of $s_d = 6$. To calculate the value of the test statistic, substitute the values of $\bar{x}_d$ (the value of the mean of the sample of differences), s_d (the value of the standard deviation of the sample of differences), μ_0 (the hypothesized value from the null hypothesis), and n_d into the formula for the test statistic to obtain

$$t = \frac{\bar{x}_d - \mu_0}{\frac{s_d}{\sqrt{n_d}}} = \frac{8 - 0}{\frac{6}{\sqrt{40}}}$$

Begin by evaluating the part of the expression in the numerator of the test statistic: $8 - 0 = 8$. The next step is to evaluate the part of the expression that appears in the denominator. First calculate $\sqrt{40} = 6.325$. Then calculate the value of the denominator. This is the sequence of steps shown here:

$$\frac{6}{\sqrt{40}} = \frac{6}{6.325} = 0.949 \text{ (rounded to three decimal places)}$$

Complete the calculation of the value of the test statistic by dividing the 8 in the numerator by the value in the denominator (0.949) to obtain

$$t = \frac{\bar{x}_d - \mu_0}{\frac{s_d}{\sqrt{n_d}}} = \frac{8 - 0}{\frac{6}{\sqrt{40}}} = \frac{8}{0.949} = 8.43 \text{ (rounded to two decimal places)}$$

The value of the test statistic for this example is $t = 8.43$.

Check It Out!

a. Suppose that researchers collected $n_d = 106$ pairs of measurements from individuals in a Before/After study. The sample mean of the differences, After − Before, was $\bar{x}_d = 56$, and the sample standard deviation was $s_d = 65$. Calculate the value of the test statistic for testing the null hypothesis $H_0 : \mu_d = 0$, where μ_d is the population mean difference, After − Before.

b. Suppose that each person in a random sample of $n_d = 38$ high school students compared how far they can throw a ball (in feet) with their dominant hand (D), to how far they can throw the same ball with their non-dominant hand (N). The sample mean difference (D − N) was $\bar{x}_d = 9.5$ feet, and the sample standard deviation of the differences was $s_d = 3.2$. Calculate the value of the test statistic for testing the null hypothesis $H_0 : \mu_d = 0$, where μ_d is the population mean difference, D − N.

Answers

a. Substitute the values of $\bar{x}_d$, s_d, and n_d into the formula for the test statistic:

$$t = \frac{\bar{x}_d - \mu_0}{\frac{s_d}{\sqrt{n_d}}} = \frac{56 - 0}{\frac{65}{\sqrt{106}}}$$

Complete the calculation of the value of the test statistic as shown here:

$$t = \frac{\bar{x}_d - \mu_0}{\frac{s_d}{\sqrt{n_d}}} = \frac{56 - 0}{\frac{65}{\sqrt{106}}} = \frac{56}{6.313} = 8.87 \text{ (rounded to two decimal places)}$$

The value of the test statistic for this example is $t = 8.87$.

b. Substitute the values of $\bar{x}_d$, s_d, and n_d into the formula for the test statistic:

$$t = \frac{\bar{x}_d - \mu_0}{\frac{s_d}{\sqrt{n_d}}} = \frac{9.5 - 0}{\frac{3.2}{\sqrt{38}}}$$

Complete the calculation of the value of the test statistic as shown here:

$$t = \frac{\bar{x}_d - \mu_0}{\frac{s_d}{\sqrt{n_d}}} = \frac{9.5 - 0}{\frac{3.2}{\sqrt{38}}} = \frac{9.5}{0.519} = 18.30 \text{ (rounded to two decimal places)}$$

The value of the test statistic for this example is $t = 18.30$.

SECTION 15.1 EXERCISES

15.1 Use the three-read strategy to understand the following exercise from a statistics textbook, and then proceed to complete the exercise.

A representative sample of 89 overweight adults entered a 6-month weight loss program, and had their before and after weights measured. The sample mean difference in weight, Before − After, was 8.4 pounds, and the sample standard deviation of the differences was 1.77 pounds. Calculate the value of the test statistic for testing the null hypothesis $H_0 : \mu_d = 0$, where μ_d is the population mean weight difference, Before − After.

15.2 Use the three-read strategy to understand the following exercise from a statistics textbook, and then proceed to complete the exercise.

Blood sugar level was measured for each person in a sample of size $n = 64$. Measurements were made before and after eating a sugary snack. The difference in blood sugar reading (After − Before) was calculated for each person. The sample mean of the differences was $\bar{x}_d = 74$ minutes, and the sample standard deviation of the differences was $s_d = 30$ minutes. Calculate the value of the test statistic for testing the null hypothesis $H_0 : \mu_d = 0$, where μ_d is the population mean difference in blood sugar level, After − Before.

SECTION 15.2 Guided Practice—Paired-Samples Hypothesis Test for a Difference in Means

In this section, you will consider a typical situation that involves carrying out a paired-samples t test for a difference in means. Consider this problem from a statistics textbook:

> The authors of the paper "Concordance of Self-Report and Measured Height and Weight of College Students" (*Journal of Nutrition, Education and Behavior* [2015]: 94–98) carried out a study to see if male college students tend to over-report height. Each man in a sample of 634 male college students selected from eight different universities was asked how tall he was. The actual height was then measured. The sample mean difference between the reported height and actual measured height (Reported – Actual) was 0.6 inches and the standard deviation of the differences was 0.8 inches. For purposes of this exercise, you can assume that the sample was representative of male college students. Carry out a hypothesis test to determine if there is a significant difference in the mean reported height and the mean actual height for male college students. Use a significance level of $\alpha = 0.01$.

If you were to encounter a problem like this, you should start by reading the problem carefully. You can use the three-read strategy introduced in Chapter 7.

First Read: Context

1. **What is the general context?**
   Researchers wondered if male college students tend to over-report their height. They recorded reported height and measured actual height for a sample of 634 male college students.
2. **Are there any terms that you don't understand?**
   This question is about testing a hypothesis, so if you don't recall the steps required to carry out a hypothesis test, you would want to go to the appropriate section in your textbook to review the steps.

Second Read: Tasks

1. **What question is being asked? What am I being asked to do?**
   This problem asks you carry out a hypothesis test to determine if there is a significant difference in the mean reported height and the mean actual height for male college students. You are also told you should use a significance level of $\alpha = 0.01$.

Third Read: Details

1. **What data were collected? Were the data collected in a reasonable way?**
   The data are from a sample of 634 male college students. The problem states that the sample can be regarded as representative of male college students. Even though there was a single group of male college students participating in the study, there are actually two paired samples in this problem because both reported height and actual height were recorded for each person. This results in two samples of measurements (one of reported heights and one of actual heights).
2. **Are the actual data provided? What was the sample size (or the sample sizes if there are more than one sample)?**
   The actual data (the reported and actual heights) are not provided. There are two paired samples (one of reported heights and one of actual heights). Each sample consists of 634 observations. This would result in a set of 634 differences, so the sample size for the sample of differences is $n_d = 634$.
3. **Are any summary statistics or additional information given? If so, which statistics and what are the values of these statistics?**
   The sample mean and the sample standard deviation for the sample of $n_d = 634$ differences are both given. The sample mean difference was $\bar{x}_d = 0.6$ inches and the sample standard deviation of the differences was $s_d = 0.8$ inches.

Now you are ready to tackle the tasks required to complete this exercise. The problem states "Carry out a hypothesis test to determine if there is a significant difference in the mean reported height and the mean actual height for male college students. Use a significance level of $\alpha = 0.01$." This is asking you to test hypotheses about a difference in population means using data from paired samples. You are asked to determine if there is a significant difference in the mean reported height and the mean actual height for male college students. The null hypothesis will be of the form $H_0 : \mu_d = \mu_0$, where μ_d represents the actual value of the difference in means and μ_0 is the hypothesized difference.

Recall that when you have data from paired samples, you will work with the differences obtained by subtracting each value from one sample from the corresponding value in the other sample to form a set of differences. In the given expression for the paired-samples t test statistic, $\bar{x}_d$ represents the mean value of the sample of differences, s_d represents the standard deviation of the sample of differences, and n_d is the number of differences (which will be equal to the size of each of the two paired samples).

Many textbooks recommend a sequence of steps for carrying out a hypothesis test. When you carry out a hypothesis test, you should follow the recommended sequence of steps. Although the order of the steps may vary somewhat from one text to another, they generally include the following:

Hypotheses: You will need to specify the null and the alternative hypothesis for the hypothesis test and define any symbols that are used in the context of the problem.

Method: You will need to specify what method you plan to use to carry out the test. For example, you might say that you plan to use the paired-samples t test for a difference in means if that is a test that would allow you to answer the question posed. You will also need to specify the significance level you will use to reach a conclusion.

Check Conditions: Most hypothesis test procedures are only appropriate when certain conditions are met. Once you have determined the method you plan to use, you will need to verify that any required conditions are met.

Calculate: To reach a conclusion in a hypothesis test, you will need to calculate the value of a test statistic and the associated P-value.

Communicate Results: The final step is to use the calculated values of the test statistic and the P-value to reach a conclusion and to provide an answer to the question posed. You should always provide a conclusion in words.

These steps are illustrated here for the given example.

Hypotheses: You want to determine if there is a significance difference in the mean reported height and the mean actual height for male college students. You can use μ_d to represent the difference in means for height (Reported height − Actual height). The question being asked usually translates into the alternative hypothesis. Here the question asks if the mean reported height is different from the mean actual height. This would be the case when $\mu_d \neq 0$ (because $\mu_d = 0$ says the means are the equal). The alternative hypothesis would be $H_a : \mu_d \neq 0$. The null hypothesis would then be obtained by replacing the $\neq$ in the alternative hypothesis with an $=$. This results in hypotheses of

$$H_0 : \mu_d = 0$$
$$H_a : \mu_d \neq 0$$

Method: Because the hypotheses are about a difference in means and the samples are paired, you would consider the paired-samples t test for a difference in means. The test statistic is

$$t = \frac{\bar{x}_d - 0}{\frac{s_d}{\sqrt{n_d}}}$$

A significance level of $\alpha = 0.01$ was specified.

Check: There are two conditions necessary for the paired-samples t test for a difference in means to be appropriate. You are told that the sample is representative of male college students. There are 634 observations in each sample, so the number of differences is $n_d = 634$. Because the sample size is greater than or equal to 30, the sample size condition is met.

Calculations: The value of the mean of the sample of differences and the value of the standard deviation of the sample of differences are $\bar{x}_d = 0.6$ inches and $s_d = 0.8$ inches. You also know that the sample size was $n_d = 634$. The value of the test statistic is

$$t = \frac{\bar{x}_d - 0}{\frac{s_d}{\sqrt{n_d}}} = \frac{0.6 - 0}{\frac{0.8}{\sqrt{634}}} = \frac{0.6}{\frac{0.8}{25.179}} = \frac{0.6}{0.032} = 18.75 \text{ (rounded to two decimal places)}$$

Once you know the value of the test statistic, you need to find the associated P-value. Your statistics text will have an explanation of how the P-value is determined.

This is a two-tailed test (the inequality in H_a is $\neq$), so the P-value is an area under a t curve. The degrees of freedom (df) associated with a paired-samples t test is $n_d - 1$, so $df = 634 - 1 = 633$. The P-value is the two times area under the t curve with $df = 633$ and to the right of 18.75. Therefore, the P-value is $2(P(t \geq 18.75)) \approx 0$.

Communicate Results: Because the P-value of 0 is less than the selected significance level of 0.01, you reject the null hypothesis. There is a significant difference in the mean reported height and the mean actual height for male college students. In fact, male college students over-report their height, on average.

SECTION 15.2 EXERCISES

15.3 In the study described in the paper "Exposure to Diesel Exhaust Induces Changes in EEG in Human Volunteers" (*Particle and Fibre Toxicology* [2007]), 10 healthy men were exposed to diesel exhaust for 1 hour. A measure of brain activity (called median power frequency, or MPF) was recorded at one location in the brain both before and after the diesel exhaust exposure. The resulting data are given in the accompanying table. For purposes of this exercise, assume that it is reasonable to regard the sample of 10 men as representative of healthy adult males.

Subject	Before	After
1	6.4	8.0
2	8.7	12.6
3	7.4	8.4
4	8.7	9.0
5	9.8	8.4
6	8.9	11.0
7	9.3	14.4
8	7.4	11.3
9	6.6	7.1
10	8.9	11.2

Use the three-read strategy to understand the problem, and then carry out a hypothesis test to determine if there is a significant difference in the means for MPF for healthy adult males before and after diesel exhaust exposure. Use a significance level of $\alpha = 0.05$.

15.4 Ultrasound is often used in the treatment of soft tissue injuries. In a study to investigate the effect of ultrasound therapy on knee extension, range of motion was measured for people in a representative sample of physical therapy patients both before and after ultrasound therapy. A subset of the data appearing in the paper "Location of Ultrasound Does Not Enhance Range of Motion Benefits of Ultrasound and Stretch Treatment" (University of Virginia Thesis, Trae Tashiro, 2003) is given in the accompanying table.

	Range of Motion						
Patient	1	2	3	4	5	6	7
Before Ultrasound	31	53	45	57	50	43	32
After Ultrasound	32	59	46	64	49	45	40

Use the three-read strategy to understand the problem, and then carry out a hypothesis test to determine if the population mean difference in Range of Motion (Before − After) is significantly different from zero. Use a significance level of $\alpha = 0.01$.

16 Estimating a Difference in Two Means Using Independent Samples—The Math You Need to Know

SECTION 16.1 Evaluating Expressions

Many statistical studies are carried out to estimate the difference between two population or treatment means. If you want to use a confidence interval to provide an estimate, you may need to be able to evaluate the expressions used to calculate the two-sample t confidence interval for a difference in two means and the associated degrees of freedom.

Note: Depending on your access to technology, such as a graphing calculator or a statistics software package, you may not need to evaluate these expressions by hand. But just in case your statistics class is not relying on technology to do the calculations, this section considers how to evaluate the expressions.

Degrees of Freedom (df) for the Two-Sample t Confidence Interval for a Difference in Means Using Independent Samples

$$df = \frac{(V_1 + V_2)^2}{\frac{V_1^2}{n_1 - 1} + \frac{V_2^2}{n_2 - 1}} \quad \text{where } V_1 = \frac{s_1^2}{n_1} \quad \text{and} \quad V_2 = \frac{s_2^2}{n_2}$$

To evaluate the expression for degrees of freedom (df) for the two-sample t confidence interval, you need to know the two sample sizes, n_1 and n_2, and the values of the two sample standard deviations, s_1 and s_2. The first step is to calculate the values of V_1 and V_2, since these values are needed in the calculation of df. Once you have calculated these values, you can substitute them into the expression for df.

For example, suppose that you want to determine the df that would be used for a two-sample t confidence interval for a difference in two population means. You have two independent random samples, one from population 1 and one from population 2. The sample from population 1 is size $n_1 = 40$ and the sample standard deviation for this sample is $s_1 = 10$. The sample from population 2 is size $n_2 = 35$ and the sample standard deviation for this sample is $s_2 = 12$.

To evaluate the expression for *df*, first calculate the values of V_1 and V_2 by substituting the corresponding values for sample size and standard deviation into the expressions for V_1 and V_2 and evaluate these expressions to obtain

$$V_1 = \frac{s_1^2}{n_1} = \frac{(10)^2}{40} = \frac{100}{40} = 2.500$$

$$V_2 = \frac{s_2^2}{n_2} = \frac{(12)^2}{35} = \frac{144}{35} = 4.114$$

Now substitute these values and the two sample sizes into the expression for *df*:

$$df = \frac{(V_1 + V_2)^2}{\frac{V_1^2}{n_1 - 1} + \frac{V_2^2}{n_2 - 1}} = \frac{(2.500 + 4.114)^2}{\frac{(2.500)^2}{40 - 1} + \frac{(4.114)^2}{35 - 1}}$$

To evaluate this expression, first calculate the value of the numerator by adding the two terms inside the parentheses and then squaring that number: $(2.500 + 4.114)^2 = (6.614)^2 = 43.745$. There are two terms in the denominator, so evaluate each of them and then add to determine the value of the denominator. These steps are shown here:

$$\frac{V_1^2}{n_1 - 1} = \frac{(2.500)^2}{40 - 1} = \frac{6.250}{39} = 0.160$$

$$\frac{V_2^2}{n_2 - 1} = \frac{(4.114)^2}{35 - 1} = \frac{16.925}{34} = 0.498$$

$$\frac{V_1^2}{n_1 - 1} + \frac{V_2^2}{n_2 - 1} = 0.160 + 0.498 = 0.658$$

Finally, divide the numerator value by the denominator value:

$$df = \frac{(V_1 + V_2)^2}{\frac{V_1^2}{n_1 - 1} + \frac{V_2^2}{n_2 - 1}} = \frac{(2.500 + 4.114)^2}{\frac{(2.500)^2}{40 - 1} + \frac{(4.114)^2}{35 - 1}} = \frac{43.745}{0.658} = 66.482$$

The final step in determining *df* is to round the *df* down to an integer, so in this example you would round down to $df = 66$. Note that for *df*, you always round down.

Check It Out!

a. Two independent random samples are selected. The first, from population 1, contains $n_1 = 96$ measurements and has sample standard deviation $s_1 = 52$. The second sample is from population 2, and it contains $n_2 = 90$ measurements with a sample standard deviation $s_2 = 64$. Calculate the value of the degrees of freedom, *df*, for the two-sample *t* confidence interval.

b. Measurements taken on a random sample of $n_1 = 45$ individuals from population A had sample standard deviation $s_1 = 5.7$. The $n_2 = 26$ measurements in an independent random sample of individuals from population B had sample standard deviation $s_2 = 6.3$. Use these values to calculate the value of the degrees of freedom, *df*, for the two-sample *t* confidence interval.

Answers

a. Calculate the values of V_1 and V_2. For V_1, use the values of $n_1 = 96$ and $s_1 = 52$, and for V_2 use $n_2 = 90$ and $s_2 = 64$ in the formulas, as follows.

$$V_1 = \frac{s_1^2}{n_1} = \frac{(52)^2}{96} = \frac{2704}{96} = 28.167$$

$$V_2 = \frac{s_2^2}{n_2} = \frac{(64)^2}{90} = \frac{4096}{90} = 45.511$$

Substitute these values and the two sample sizes into the expression for *df*:

$$df = \frac{(V_1 + V_2)^2}{\dfrac{V_1^2}{n_1 - 1} + \dfrac{V_2^2}{n_2 - 1}} = \frac{(28.167 + 45.511)^2}{\dfrac{(28.167)^2}{96 - 1} + \dfrac{(45.511)^2}{90 - 1}}$$

Evaluate this expression as shown here:

$$df = \frac{(V_1 + V_2)^2}{\dfrac{V_1^2}{n_1 - 1} + \dfrac{V_2^2}{n_2 - 1}} = \frac{(28.167 + 45.511)^2}{\dfrac{(28.167)^2}{96 - 1} + \dfrac{(45.511)^2}{90 - 1}} = \frac{5428.448}{31.623} = 171.661$$

The final step in determining *df* is to round the *df* down to an integer, so in this example you would round down to $df = 171$.

b. Calculate the values of V_1 and V_2. For V_1, use the values of $n_1 = 45$ and $s_1 = 5.7$, and for V_2 use $n_2 = 26$ and $s_2 = 6.3$ in the formulas:

$$V_1 = \frac{s_1^2}{n_1} = \frac{(5.7)^2}{45} = \frac{32.49}{45} = 0.722$$

$$V_2 = \frac{s_2^2}{n_2} = \frac{(6.3)^2}{26} = \frac{39.69}{26} = 1.527$$

Substitute these values and the two sample sizes into the expression for *df*:

$$df = \frac{(V_1 + V_2)^2}{\dfrac{V_1^2}{n_1 - 1} + \dfrac{V_2^2}{n_2 - 1}} = \frac{(0.722 + 1.527)^2}{\dfrac{(0.722)^2}{45 - 1} + \dfrac{(1.527)^2}{26 - 1}}$$

Evaluate this expression as shown here:

$$df = \frac{(V_1 + V_2)^2}{\dfrac{V_1^2}{n_1 - 1} + \dfrac{V_2^2}{n_2 - 1}} = \frac{(0.722 + 1.527)^2}{\dfrac{(0.722)^2}{45 - 1} + \dfrac{(1.527)^2}{26 - 1}} = \frac{5.058}{0.105} = 48.171$$

The final step in determining *df* is to round the *df* down to an integer, so in this example you would round down to $df = 48$.

Two-Sample *t* Confidence Interval for a Difference in Means Using Independent Samples

$$(\bar{x}_1 - \bar{x}_2) \pm (t \text{ critical value})\sqrt{\frac{s_1^2}{n_1} + \frac{s_2^2}{n_2}}$$

When you have independent samples from each of two populations or treatments, subscripts are added to the notation used for the sample statistics to distinguish between them. The notation is summarized in the table below:

Notation	Meaning
μ_1	Actual value of the mean for population or treatment 1
μ_2	Actual value of the mean for population or treatment 2
$\bar{x}_1$	Sample mean for sample from population or treatment 1
$\bar{x}_2$	Sample mean for sample from population or treatment 2
n_1	Sample size for sample from population or treatment 1
n_2	Sample size for sample from population or treatment 2
s_1	Sample standard deviation for population or treatment 1
s_2	Sample standard deviation for population or treatment 2

The difference in the actual values of the population or treatment means is written as $\mu_1 - \mu_2$. To calculate a confidence interval for a difference in two population or treatment means, $\mu_1 - \mu_2$, you need to know the values of the two sample means, $\bar{x}_1$ and $\bar{x}_2$, the sample sizes n_1 and n_2, the two sample standard deviations, s_1 and s_2, and the confidence level, which determines the t critical value that is used to calculate the confidence interval.

Before using the given expression to calculate a confidence interval, you need to verify that the sample sizes are large enough for this confidence interval to be appropriate. The sample sizes are large enough to use this confidence interval if n_1 and n_2 are each greater than or equal to 30. If either sample size is less than 30, this confidence interval formula is only appropriate if it is reasonable to assume that the population distributions are approximately normal. A plot of the data for each sample (dotplot, boxplot, or normal probability plot) can be used to help you decide if this assumption is reasonable.

If you have data from two populations, you also need to confirm that the samples are random samples or that they were selected in a way that makes it reasonable to think that the samples are representative of the populations. If you have data from an experiment, you need to confirm that the experiment used random assignment to the treatment groups.

You will also need to calculate the number of degrees of freedom (df) that will be used to determine the appropriate t critical value, as described earlier in this section.

Recall that an interval is defined by its endpoints—it has a lower endpoint and an upper endpoint. Notice that the expression for the confidence interval includes the symbol $\pm$, which is read as "plus and minus." This means that you will do two calculations, once using $+$ and once using $-$. Each one of these calculations results in one of the two interval endpoints.

To calculate the endpoints of the confidence interval, you can start by determining the appropriate t critical value for the specified confidence level. Your statistics text should have instructions for how this is done. Once you have determined the appropriate t critical value, you can substitute the values of the two sample means, $\bar{x}_1$ and $\bar{x}_2$, the two sample standard deviations, s_1 and s_2, the sample sizes n_1 and n_2, and the t critical value into the expression for the confidence interval.

For example, suppose you want to calculate a 95% confidence interval for a difference in population means using independent samples. You have two independent random samples, one from population 1 and one from population 2. The sample from population 1 is size $n_1 = 40$, the sample mean is $\bar{x}_1 = 80$, and the sample standard deviation for this sample is $s_1 = 10$. The sample from population 2 is size $n_2 = 35$, the sample mean is $\bar{x}_2 = 63$, and the sample standard deviation for this sample is $s_2 = 12$. For these sample sizes and standard deviations, the value of df was determined to be $df = 66$ (see calculation earlier in this section). For a 95% confidence level, the t critical value for $df = 66$ is 2.00.

Substituting these values into the expression for the confidence interval results in the following:

$$(\bar{x}_1 - \bar{x}_2) \pm (t \text{ critical value})\sqrt{\frac{s_1^2}{n_1} + \frac{s_2^2}{n_2}}$$

$$= (80 - 63) \pm (2.00)\sqrt{\frac{(10)^2}{40} + \frac{(12)^2}{35}}$$

Begin by evaluating the parts of the expression that appear under the square root. Calculate

$$\frac{s_1^2}{n_1} = \frac{(10)^2}{40} = \frac{100}{40} = 2.500 \quad \text{and} \quad \frac{s_2^2}{n_2} = \frac{(12)^2}{35} = \frac{144}{35} = 4.114$$

Next, add these two values together, and find the square root:

$$\sqrt{2.500 + 4.114} = \sqrt{6.614} = 2.572$$

This number is multiplied by 2.00, as shown here:

$$(2.00)(2.572) = 5.144$$

This number is the "plus and minus" part of the confidence interval. To calculate the endpoints of the interval, you would evaluate

$$(80 - 63) \pm 5.144$$
$$17 \pm 5.144$$

The lower endpoint of the interval is $17 - 5.144 = 11.856$. The upper endpoint of the interval is $17 + 5.144 = 22.144$.

The 95% confidence interval estimate of the difference in means is (11.856, 22.144). You would interpret this interval by saying that based on the data from the samples, you can be 95% confident that the actual value of the difference in population means is somewhere between 11.856 and 22.144.

Check It Out!

a. Suppose you want to calculate a 90% confidence interval for a difference in population means using independent samples. Measurements taken from a random sample of $n_1 = 96$ individuals from population 1 had a sample mean of $\bar{x}_1 = 40$ a sample standard deviation of $s_1 = 52$. An independent random sample of $n_2 = 90$ measurements from population 2 had a sample mean of $\bar{x}_2 = 61$ and a sample standard deviation of $s_2 = 64$. For these sample sizes and standard deviations, the value of *df* was determined to be $df = 171$ (see Check It Out! Part (a), earlier in this section). For a 90% confidence level, the *t* critical value for $df = 171$ is 1.654. Use these values to calculate the endpoints of a 90% confidence interval for the difference in the two population means.

b. Recall that in Check It Out! Part (b) (earlier in this section), a random sample of size $n_1 = 45$ from population A had a sample standard deviation of $s_1 = 5.7$. Suppose that this sample from population A had a sample mean equal to $\bar{x}_1 = 3.8$. An independent random sample of size $n_2 = 26$ from population B had a sample standard deviation of $s_2 = 6.3$. Suppose that the sample from population B had sample mean $\bar{x}_2 = 3.1$. Recall that for these sample sizes and standard deviations, the value of *df* was determined to be $df = 48$. (See Check It Out! Example 2 earlier in this section.) For a 95% confidence level, the *t* critical value for $df = 48$ is 2.011. Use these values to calculate the endpoints of a 95% confidence interval for the difference in the two population means.

Answers

a. Substituting the given values into the expression for the confidence interval results in the following:

$$(\bar{x}_1 - \bar{x}_2) \pm (t \text{ critical value})\sqrt{\frac{s_1^2}{n_1} + \frac{s_2^2}{n_2}}$$
$$= (40 - 61) \pm (1.654)\sqrt{\frac{(52)^2}{96} + \frac{(64)^2}{90}}$$

Evaluate the parts of the expression that appear under the square root. Calculate

$$\frac{s_1^2}{n_1} = \frac{(52)^2}{96} = \frac{2704}{96} = 28.167 \quad \text{and} \quad \frac{s_2^2}{n_2} = \frac{(64)^2}{90} = \frac{4096}{90} = 45.511$$

Next, add these two values together, and find the square root:

$$\sqrt{28.167 + 45.511} = \sqrt{73.688} = 8.583$$

This number is multiplied by 1.654, as shown here:

$$(1.654)(8.583) = 14.196$$

This number is the "plus and minus" part of the confidence interval. To calculate the endpoints of the interval, you would evaluate

$$(40 - 61) \pm 14.196$$
$$-21 \pm 14.196$$

The lower endpoint of the interval is $-21 - 14.196 = -35.196$. The upper endpoint of the interval is $-21 + 14.196 = -6.804$.

The 90% confidence interval estimate of the difference in means is $(-35.196, -6.804)$. You would interpret this interval by saying that based on the data from the samples, you can be 90% confident that the actual value of the difference in population means, subtracted as population 1 − population 2, is somewhere between -35.196 and -6.804.

b. Substituting the given values into the expression for the confidence interval results in the following:

$$(\bar{x}_1 - \bar{x}_2) \pm (t \text{ critical value})\sqrt{\frac{s_1^2}{n_1} + \frac{s_2^2}{n_2}}$$
$$= (3.8 - 3.1) \pm (2.011)\sqrt{\frac{(5.7)^2}{45} + \frac{(6.3)^2}{26}}$$

Evaluate the parts of the expression that appear under the square root. Calculate

$$\frac{s_1^2}{n_1} = \frac{(5.7)^2}{45} = \frac{32.49}{45} = 0.722 \quad \text{and} \quad \frac{s_2^2}{n_2} = \frac{(6.3)^2}{26} = \frac{39.69}{26} = 1.527$$

Next, add these two values together, and find the square root:

$$\sqrt{0.722 + 1.527} = \sqrt{2.249} = 1.500$$

This number is multiplied by 2.011, as shown here:

$$(2.011)(1.500) = 3.016$$

This number is the "plus and minus" part of the confidence interval. To calculate the endpoints of the interval, you would evaluate

$$(3.8 - 3.1) \pm 3.016$$
$$0.7 \pm 3.016$$

The lower endpoint of the interval is $0.7 - 3.016 = -2.316$. The upper endpoint of the interval is $0.7 + 3.016 = 3.716$.

The 95% confidence interval estimate of the difference in means is $(-2.316, 3.716)$. You can be 95% confident that the difference in the population means, for population A − population B, falls between -2.316 and 3.716.

SECTION 16.1 EXERCISES

16.1 Measurements taken from a random sample of $n_1 = 27$ individuals from population Fred had a sample mean of $\bar{x}_1 = 1.8$ and a sample standard deviation of $s_1 = 1.5$. An independent random sample of $n_2 = 53$ measurements from population Barney had a sample mean of $\bar{x}_2 = 4.8$ and a sample standard deviation of $s_2 = 6.9$. First calculate the value of the degrees of freedom (*df*), and then calculate a 99% confidence interval for the difference in population means, Fred – Barney.

16.2 Measurements were taken on a representative sample of $n_1 = 93$ individuals from population A, and the sample mean of the measurements was $\bar{x}_1 = 75$ with sample standard deviation was $s_1 = 29$. An independent representative sample of $n_2 = 38$ measurements from population B had a sample mean of $\bar{x}_2 = 52$ with a sample standard deviation of $s_2 = 17$. First calculate the value of the degrees of freedom (*df*), and then calculate a 95% confidence interval for the difference in population means, A − B.

SECTION 16.2 Guided Practice—Two-Sample *t* Confidence Interval for a Difference in Means Using Independent Samples

Before considering a guided example that involves calculating and interpreting a confidence interval for the difference in two means, take a minute to review how intervals that estimate a difference are interpreted.

Interpreting confidence intervals for a difference in means is a bit more complicated that interpreting a confidence interval for a single mean. The following table explains how an interval for a difference in means can be interpreted in three different cases:

Case	Interpretation	Example
Both endpoints of the confidence interval for $\mu_1 - \mu_2$ are positive.	If $\mu_1 - \mu_2$ is positive, it means that you think that μ_1 is greater than μ_2 and the interval gives an estimate of how much greater.	(8.6, 12.4) You think that μ_1 is greater than μ_2 by somewhere between 8.6 and 12.4.
Both endpoints of the confidence interval for $\mu_1 - \mu_2$ are negative.	If $\mu_1 - \mu_2$ is negative, it means that you think that μ_1 is less than μ_2 and the interval gives an estimate of how much less.	(−12.2, −7.3) You think that μ_1 is less than μ_2 by somewhere between 7.3 and 12.4.
0 is included in the confidence interval.	If the confidence interval includes 0, a plausible value for $\mu_1 - \mu_2$ is 0. This suggests that μ_1 and μ_2 could be equal.	(−3.5, 4.8) Because 0 is included in the confidence interval, it is plausible that the two means could be equal.

Now consider this problem from a statistics textbook:

> The authors of the paper "Influence of Biofeedback Weight Bearing Training in Sit to Stand to Sit and the Limits of Stability on Stroke Patients" (*The Journal of Physical Therapy Science* [2016]: 3011–2014) randomly selected two samples of patients admitted to the hospital after suffering a stroke. One sample was selected from patients who received biofeedback weight training for 8 weeks and the other sample was selected from patients who did not receive this training. At the end of 8 weeks, the time it took (in seconds) to stand from a sitting position and then to sit down again (called sit-stand-sit time) was measured for the people in each sample. Data consistent with summary quantities given in the paper are given below. For purposes of this exercise, you can assume that the samples are representative of the population of stroke patients who receive the biofeedback training and the population of stroke patients who do not receive this training. Use the given data to construct and interpret a 95% confidence interval for the difference in mean sit-stand-sit time for these two populations.
>
> **Biofeedback Group**
>
> 1.9 2.6 4.3 2.1 2.7 4.1 3.2 4.0 3.2 3.5 2.8 3.5 3.5 2.3 3.1
>
> **No Biofeedback Group**
>
> 5.1 4.7 3.9 4.2 4.7 4.3 4.2 5.1 3.4 4.2 5.1 4.4 4.0 3.4 3.9

If you were to encounter a problem like this, you should start by reading the problem carefully. You can use the three-read strategy introduced in Chapter 7.

First Read: Context

1. **What is the general context?**
   A study was carried out to learn about the time it takes for stroke patients to stand from a sitting position and then sit back down again. This is called the sit-stand-sit time. The study compared the sit-stand-sit times for a group of stroke patients who received biofeedback weight training for 8 weeks after having a stroke with a group of stroke patients who did not receive this training.
2. **Are there any terms that you don't understand?**
   This question is about constructing a confidence interval, so if you aren't comfortable with the meaning of this term you would want to go to the appropriate section in your statistics textbook to review the definition. If any of the other non-statistical terms are unfamiliar (such as "biofeedback"), you can look up the meaning of those terms in a dictionary to make sure that you fully understand the context.

Second Read: Tasks

1. **What question is being asked? What am I being asked to do?**
   This problem asks you to construct and interpret a 95% confidence interval for the difference in mean sit-stand-sit time for the population of stroke patients who receive 8 weeks of biofeedback weight training and the population of stroke patients who do not receive this training.

Third Read: Details

1. **What data were collected? Were the data collected in a reasonable way?**
   Data were collected from a sample of 15 stroke patients who received biofeedback weight training for 8 weeks and a sample of 15 stroke patients who did not receive this training. Sit-stand-sit time was recorded for each person in each of the two samples. The problem states that you can assume that the samples are representative of the population of stroke patients who receive the biofeedback training and the population of stroke patients who do not receive this training.
2. **Are the actual data provided? What was the sample size (or the sample sizes if there are more than one sample)?**
   The sample size was 15 for the sample of stroke patients who received biofeedback weight training and 15 for the sample of stroke patients who did not receive biofeedback weight training. The actual data (the sit-stand-sit times for the 15 stroke patients in each sample) are provided.
3. **Are any summary statistics or additional information given? If so, which statistics and what are the values of these statistics?**
   The sample means and standard deviations are not given, so you will need to calculate these values using the given data.

Now you are ready to tackle the tasks required to complete this exercise. This problem asks you to construct and interpret a 95% confidence interval for the difference in the mean sit-stand-sit time for the population of stroke patients who receive 8 weeks of biofeedback weight training and the population of stroke patients who do not receive this training.

The expression used to calculate a confidence interval for a difference in means for independent samples is

$$(x_1 - x_2) \pm (t \text{ critical value})\sqrt{\frac{s_1^2}{n_1} + \frac{s_2^2}{n_2}}$$

To calculate this confidence interval, you need to know the values of the two sample means, $\bar{x}_1$ and $\bar{x}_2$, the two sample standard deviations, s_1 and s_2, the sample sizes n_1 and n_2, and the confidence level and degrees of freedom (df), which determine the t critical value that is used to calculate the confidence interval.

Start by deciding which population you will refer to as population 1 and which will be population 2. For example, you might choose to think of stroke patients who receive 8 weeks of biofeedback weight training as population 1 and stroke patients who do not receive biofeedback weight training as population 2. Then whenever you see a subscript of 1, you would know that it is referring to the biofeedback population. The subscript 2 will refer to the no biofeedback population.

For the sample of stroke patients who received 8 weeks of biofeedback weight training, you know that the sample size was $n_1 = 15$, but you will have to use the given data to calculate the mean and standard deviation for sample 1. These calculations are shown below. For a review of how to calculate the sample mean and sample standard deviation, see Chapter 3.

$$\bar{x}_1 = \frac{\sum x}{n_1} = \frac{46.8}{15} = 3.120$$

$$s_1 = \sqrt{\frac{(x - \bar{x}_1)^2}{n_1 - 1}} = \sqrt{\frac{7.324}{15 - 1}} = \sqrt{\frac{7.324}{14}} = \sqrt{0.523} = 0.723$$

For the sample of stroke patients who did not receive biofeedback weight training, the sample mean and standard deviation are

$$\bar{x}_2 = \frac{\sum x}{n_2} = \frac{64.605}{15} = 4.307$$

$$s_2 = \sqrt{\frac{(x - \bar{x}_2)^2}{n_2 - 1}} = \sqrt{\frac{4.309}{15 - 1}} = \sqrt{\frac{4.309}{14}} = \sqrt{0.308} = 0.555$$

Because the sample sizes are both less than 30, for the two-sample *t* confidence interval to be appropriate you need to be willing to assume that the population distributions of sit-stand-sit times are at least approximately normal. Because the actual data are given, you should construct a graphical display of the data for each sample to help you decide if this assumption is reasonable. The figure below shows boxplots constructed using the data for the two samples. Because the boxplots are roughly symmetric and have no outliers, the assumption that the population distributions are approximately normal seems reasonable.

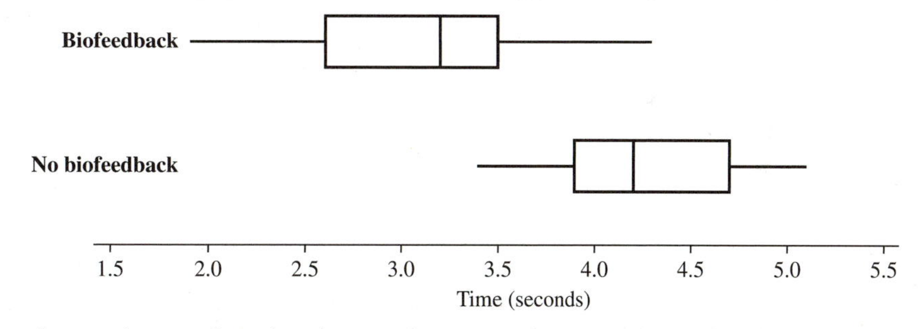

You also need to confirm that the samples are random samples or that they were selected in a way that makes it reasonable to think that the samples are representative of the populations. The problem states that you can assume that the samples are representative of the population of stroke patients who receive the biofeedback training and the population of stroke patients who do not receive this training.

To calculate the endpoints of the confidence interval, you can start by determining the appropriate *t* critical value for the specified confidence level. To do this, you must determine the value of *df*. This calculation is shown below. (For more detail on the calculation of *df* for the two-sample *t* confidence interval, see Section 16.1.)

$$df = \frac{(V_1 + V_2)^2}{\frac{V_1^2}{n_1 - 1} + \frac{V_2^2}{n_2 - 1}} \quad \text{where } V_1 = \frac{s_1^2}{n_1} \text{ and } V_2 = \frac{s_2^2}{n_2}$$

$$V_1 = \frac{s_1^2}{n_1} = \frac{(0.723)^2}{15} = \frac{0.523}{15} = 0.035$$

$$V_2 = \frac{s_2^2}{n_2} = \frac{(0.555)^2}{15} = \frac{0.308}{15} = 0.021$$

$$df = \frac{(V_1 + V_2)^2}{\frac{V_1^2}{n_1 - 1} + \frac{V_2^2}{n_2 - 1}} = \frac{(0.035 + 0.021)^2}{\frac{(0.035)^2}{15 - 1} + \frac{(0.021)^2}{15 - 1}} = \frac{(0.056)^2}{\frac{0.0012}{14} + \frac{0.0004}{14}}$$

$$= \frac{0.00314}{0.00009 + 0.00003} = \frac{0.00314}{0.00012} = 26.167$$

Rounding this value down, you would use $df = 26$.

For a 95% confidence level and $df = 26$, the corresponding t critical value is 2.06. (Your statistics text should have instructions for how to find this value.)

Substituting the values of the two sample means, $\bar{x}_1$ and $\bar{x}_2$, the two sample standard deviations, s_1 and s_2, the sample sizes n_1 and n_2, and the t critical value into the expression for the confidence interval results in the following:

$$(\bar{x}_1 - \bar{x}_2) \pm (t \text{ critical value})\sqrt{\frac{s_1^2}{n_1} + \frac{s_2^2}{n_2}}$$

$$= (3.120 - 4.307) \pm (2.06)\sqrt{\frac{(0.723)^2}{15} + \frac{(0.555)^2}{15}}$$

Begin by evaluating the parts of the expression that appear under the square root. Calculate

$$\frac{s_1^2}{n_1} = \frac{(0.723)^2}{15} = \frac{0.523}{15} = 0.035 \quad \text{and} \quad \frac{s_2^2}{n_2} = \frac{(0.555)^2}{15} = \frac{0.308}{15} = 0.021$$

Now add these two values together, and find the square root:

$$\sqrt{0.035 + 0.021} = \sqrt{0.056} = 0.237$$

This number is multiplied by 2.06, as shown here:

$$(2.06)(0.237) = 0.488$$

This number is the "plus and minus" part of the confidence interval. To calculate the endpoints of the interval, you would evaluate

$$(3.120 - 4.307) \pm 0.488$$
$$-1.187 \pm 0.488$$

The lower endpoint of the interval is $-1.187 - 0.488 = -1.675$. The upper endpoint of the interval is $-1.187 + 0.488 = -0.699$.

The 95% confidence interval estimate of the difference in means is $(-1.675, -0.699)$. You also need to interpret the confidence interval in context. You would interpret this interval by saying that based on the data from these samples, you can be 95% confident that the actual value of the difference in the mean sit-stand-sit time for stroke patients who receive 8 weeks of biofeedback weight training and the mean sit-stand-sit time for stroke patients who do not receive biofeedback weight training falls between -1.675 and -0.699. Because both endpoints of the confidence interval are negative, you would say that the mean sit-stand-sit time for stroke patients who receive 8 weeks of biofeedback weight training is less than the mean sit-stand-sit time for stroke patients who do not receive biofeedback weight training by somewhere between 0.699 seconds and 1.675 seconds.

SECTION 16.2 EXERCISES

16.3 The paper "Ladies First?" A Field Study of Discrimination in Coffee Shops" (*Applied Economics* [2008]: 1–19) describes a study in which researchers observed wait times in coffee shops in Boston. Both wait time and perceived sex of the customer were recorded. The mean wait time for a sample of 145 male customers was 85.2 seconds. The mean wait time for a sample of 141 female customers was 113.7 seconds.

The sample standard deviations (estimated from graphs in the paper) were 50 seconds for the sample of males and 75 seconds for the sample of females. Suppose that these two samples are representative of the populations of wait times for female coffee shop customers and for male coffee shop customers. Use the three-read strategy to understand the problem, and then use the given information to construct a 99% confidence interval for the difference in mean wait times, Male – Female. Interpret the confidence interval in the context of the study.

16.4 Some people believe that talking on a cell phone while driving slows reaction time, increasing the risk of accidents. The study described in the paper "A Comparison of the Cell Phone Driver and the Drunk Driver" (*Human Factors* [2006]: 381–391) investigated the braking reaction time of people driving in a driving simulator. Drivers followed a pace car in the simulator, and when the pace car's brake lights came on, the drivers were supposed to step on the brake. The time between the pace car brake lights coming on and the driver stepping on the brake was measured. Two samples of 40 drivers participated in the study. The 40 people in one sample used a cell phone while driving. The 40 people in the second sample drank a mixture of orange juice and alcohol in an amount calculated to achieve a blood alcohol level of 0.08% (a value considered legally drunk in most states). For the cell phone sample, the mean braking reaction time was 779 milliseconds and the standard deviation was 209 milliseconds. For the alcohol sample, the mean breaking reaction time was 849 milliseconds and the standard deviation was 228 milliseconds. For purposes of this exercise, you can assume that the two samples are representative of the two populations of interest. Use the three-read strategy to understand the problem, and then use the given information to construct and interpret a 95% confidence interval for the difference in mean braking reaction time for the two populations, Cell phone – Drunk. Interpret the confidence interval in the context of the research.

17 Testing Hypotheses About a Difference in Two Means Using Independent Samples—The Math You Need to Know

SECTION 17.1 Evaluating Expressions

Many statistical studies are carried out to test hypotheses about the difference between two population or treatment means. If you want to carry out such a test, you may need to be able to evaluate the expressions used in the calculation of the two-sample t test statistic and the associated degrees of freedom.

Note: Depending on your access to technology, such as a graphing calculator or a statistics software package, you may not need to evaluate these expressions by hand. But just in case your statistics class is not relying on technology to do the calculations, this section considers how to evaluate the expressions.

Degrees of Freedom (df) for the Two-Sample t Test for a Difference in Means Using Independent Samples

$$df = \frac{(V_1 + V_2)^2}{\dfrac{V_1^2}{n_1 - 1} + \dfrac{V_2^2}{n_2 - 1}} \quad \text{where } V_1 = \frac{s_1^2}{n_1} \quad \text{and} \quad V_2 = \frac{s_2^2}{n_2}$$

To evaluate the expression for degrees of freedom (df) for the two-sample t test for a difference in means using independent samples, you need to know the two sample sizes, n_1 and n_2, and the values of the two sample standard deviations, s_1 and s_2. The first step is to calculate the values of V_1 and V_2, since these values are needed in the calculation of df. Once you have calculated these values, you can substitute them into the expression for df.

For example, suppose that you want to determine the df that would be used for a two-sample t test for a difference in two population means. You have two independent random samples, one from population 1 and one from population 2. The sample from population 1 is size $n_1 = 36$, and the sample standard deviation for this sample is $s_1 = 15$.

The sample from population 2 is size $n_2 = 42$, and the sample standard deviation for this sample is $s_2 = 12$.

To evaluate the expression for *df*, first calculate the values of V_1 and V_2 by substituting the corresponding values for sample size and standard deviation into the expressions for V_1 and V_2, and then evaluate these expressions to obtain

$$V_1 = \frac{s_1^2}{n_1} = \frac{(15)^2}{36} = \frac{225}{36} = 6.250$$

$$V_2 = \frac{s_2^2}{n_2} = \frac{(12)^2}{42} = \frac{144}{42} = 3.429$$

Now substitute these values and the two sample sizes into the expression for *df*:

$$df = \frac{(V_1 + V_2)^2}{\frac{V_1^2}{n_1 - 1} + \frac{V_2^2}{n_2 - 1}} = \frac{(6.250 + 3.429)^2}{\frac{(6.250)^2}{36 - 1} + \frac{(3.429)^2}{42 - 1}}$$

To evaluate this expression, first calculate the value of the numerator by adding the two terms inside the parentheses and then squaring that number: $(6.250 + 3.429)^2 = (9.679)^2 = 93.683$. There are two terms in the denominator, so evaluate each of them and then add to determine the value of the denominator. These steps are shown here:

$$\frac{V_1^2}{n_1 - 1} = \frac{(6.250)^2}{36 - 1} = \frac{39.063}{35} = 1.116$$

$$\frac{V_2^2}{n_2 - 1} = \frac{(3.429)^2}{42 - 1} = \frac{11.758}{41} = 0.287$$

$$\frac{V_1^2}{n_1 - 1} + \frac{V_2^2}{n_2 - 1} = 1.116 + 0.287 = 1.403$$

Finally, divide the numerator value by the denominator value:

$$df = \frac{(V_1 + V_2)^2}{\frac{V_1^2}{n_1 - 1} + \frac{V_2^2}{n_2 - 1}} = \frac{(6.250 + 3.429)^2}{\frac{(6.250)^2}{36 - 1} + \frac{(3.429)^2}{42 - 1}} = \frac{93.683}{1.403} = 66.773$$

The final step in determining *df* is to round the *df* down to an integer, so in this example you would round down to $df = 66$. Note that for *df*, you should always round down.

Check It Out!

a. Two independent random samples are selected. The first, from population A, contains $n_1 = 63$ measurements and has sample standard deviation $s_1 = 1.6$. The second sample is from population B, and it contains $n_2 = 48$ measurements with sample standard deviation $s_2 = 1.3$. Calculate the value of the degrees of freedom, *df*, for the two-sample *t* test for a difference in means using independent samples.

b. Measurements taken on a random sample of $n_1 = 33$ individuals from population I had sample standard deviation $s_1 = 78$. The $n_2 = 44$ measurements taken in an independent random sample of individuals from population II had sample standard deviation $s_2 = 50$. Use these values to calculate the value of the degrees of freedom, *df*, for the two-sample *t* test for a difference in means using independent samples.

Answers

a. Calculate the values of V_1 and V_2. For V_1, using the values of $n_1 = 63$ and $s_1 = 1.6$, and for V_2 use $n_2 = 48$ and $s_2 = 1.3$ as follows:

$$V_1 = \frac{s_1^2}{n_1} = \frac{(1.6)^2}{63} = \frac{2.56}{63} = 0.041$$

$$V_2 = \frac{s_2^2}{n_2} = \frac{(1.3)^2}{48} = \frac{1.69}{48} = 0.035$$

Substitute these values and the two sample sizes into the expression for df:

$$df = \frac{(V_1 + V_2)^2}{\dfrac{V_1^2}{n_1 - 1} + \dfrac{V_2^2}{n_2 - 1}} = \frac{(0.041 + 0.035)^2}{\dfrac{(0.041)^2}{63 - 1} + \dfrac{(0.035)^2}{48 - 1}}$$

Evaluate this expression as shown here:

$$df = \frac{(V_1 + V_2)^2}{\dfrac{V_1^2}{n_1 - 1} + \dfrac{V_2^2}{n_2 - 1}} = \frac{(0.041 + 0.035)^2}{\dfrac{(0.041)^2}{63 - 1} + \dfrac{(0.035)^2}{48 - 1}} = \frac{0.0058}{0.000053} = 109.43$$

The final step in determining df is to round the df down to an integer, so in this example you would round down to $df = 109$.

b. Calculate the values of V_1 and V_2. For V_1, use the values of $n_1 = 33$ and $s_1 = 78$, and for V_2 use $n_2 = 44$ and $s_2 = 50$ in the formulas:

$$V_1 = \frac{s_1^2}{n_1} = \frac{(78)^2}{33} = \frac{6084}{33} = 184.36$$

$$V_2 = \frac{s_2^2}{n_2} = \frac{(50)^2}{44} = \frac{2500}{44} = 56.82$$

Substitute these values and the two sample sizes into the expression for df:

$$df = \frac{(V_1 + V_2)^2}{\dfrac{V_1^2}{n_1 - 1} + \dfrac{V_2^2}{n_2 - 1}} = \frac{(184.36 + 56.82)^2}{\dfrac{(184.36)^2}{33 - 1} + \dfrac{(56.82)^2}{44 - 1}}$$

Evaluate this expression as shown here:

$$df = \frac{(V_1 + V_2)^2}{\dfrac{V_1^2}{n_1 - 1} + \dfrac{V_2^2}{n_2 - 1}} = \frac{(184.36 + 56.82)^2}{\dfrac{(184.36)^2}{33 - 1} + \dfrac{(56.82)^2}{44 - 1}} = \frac{58{,}167.8}{1137.2} = 51.15$$

The final step in determining df is to round the df down to an integer, so in this example you would round down to $df = 51$.

Two-Sample t Test Statistic for Testing Hypotheses About a Difference in Means Using Independent Samples

$$t = \frac{(\bar{x}_1 - \bar{x}_2) - (\mu_1 - \mu_2)}{\sqrt{\dfrac{s_1^2}{n_1} + \dfrac{s_2^2}{n_2}}}$$

When you have independent samples from each of two populations or treatments, subscripts are added to the notation used for the sample statistics to distinguish between them. The notation is summarized in the table below:

Notation	Meaning
μ_1	Actual value of the mean for population or treatment 1
μ_2	Actual value of the mean for population or treatment 2
$\bar{x}_1$	Sample mean for sample from population or treatment 1
$\bar{x}_2$	Sample mean for sample from population or treatment 2
n_1	Sample size for sample from population or treatment 1
n_2	Sample size for sample from population or treatment 2
s_1	Sample standard deviation for population or treatment 1
s_2	Sample standard deviation for population or treatment 2

The difference in the actual values of the population or treatment means is written as $\mu_1 - \mu_2$. To test hypotheses about the value of $\mu_1 - \mu_2$, you need to know the values of the two sample means, $\bar{x}_1$ and $\bar{x}_2$, the two sample standard deviations, s_1 and s_2, and the sample sizes, n_1 and n_2.

Before using the given expression to calculate the value of the two-sample *t* test statistic, you need to verify that the sample sizes are large enough for the two-sample *t* test to be appropriate. The sample sizes are large enough to use this test if n_1 and n_2 are each greater than or equal to 30. If either sample size is less than 30, this test is only appropriate if it is reasonable to assume that the population distributions are approximately normal. A plot of the data for each sample (dotplot, boxplot, or normal probability plot) can be used to help you decide if this assumption is reasonable.

If you have data from two populations, you also need to confirm that the samples are random samples or that they were selected in a way that makes it reasonable to think that the samples are representative of the populations. If you have data from an experiment, you need to confirm that the experiment used random assignment to the treatment groups.

You will also need to calculate the number of degrees of freedom (*df*) that will be used to determine the appropriate *P*-value, as described earlier in this section.

To calculate the value of the two-sample *t* test statistic, you can substitute the values of the two sample means, $\bar{x}_1$ and $\bar{x}_2$, the two sample standard deviations, s_1 and s_2, and the sample sizes n_1 and n_2 into the expression for the test statistic. The value of $\mu_1 - \mu_2$ in the calculation of the test statistic will be the hypothesized value of this difference that appears in the null hypothesis.

For example, suppose you want to calculate the value of the two-sample *t* test statistic for testing hypotheses about a difference in population means using independent samples. Also suppose that the null hypothesis is $H_0: \mu_1 - \mu_2 = 0$. You have two independent random samples, one from population 1 and one from population 2. The sample from population 1 is size $n_1 = 36$, the sample mean is $\bar{x}_1 = 140$, and the sample standard deviation for this sample is $s_1 = 15$. The sample from population 2 is size $n_2 = 42$, the sample mean is $\bar{x}_2 = 156$, and the sample standard deviation for this sample is $s_2 = 12$.

Substituting these values into the expression for the two-sample *t* test statistic results in the following:

$$t = \frac{(\bar{x}_1 - \bar{x}_2) - (\mu_1 - \mu_2)}{\sqrt{\frac{s_1^2}{n_1} + \frac{s_2^2}{n_2}}} = \frac{(140 - 156) - 0}{\sqrt{\frac{(15)^2}{36} + \frac{(12)^2}{42}}}$$

Begin by evaluating the parts of the expression that appear under the square root. Calculate

$$\frac{s_1^2}{n_1} = \frac{(15)^2}{36} = \frac{225}{36} = 6.250 \quad \text{and} \quad \frac{s_2^2}{n_2} = \frac{(12)^2}{42} = \frac{144}{42} = 3.429$$

Next, add these two values together, and find the square root:

$$\sqrt{6.250 + 3.429} = \sqrt{9.679} = 3.111$$

Substituting this value into the expression for the t test statistics results in

$$t = \frac{(\bar{x}_1 - \bar{x}_2) - (\mu_1 - \mu_2)}{\sqrt{\frac{s_1^2}{n_1} + \frac{s_2^2}{n_2}}} = \frac{(140 - 156) - 0}{\sqrt{\frac{15^2}{36} + \frac{12^2}{42}}} = \frac{(140 - 156)}{3.111} = \frac{-16}{3.111} = -5.14$$

(rounded to two decimal places)

The value of the test statistic for this example is $t = -5.14$.

Check It Out!

a. Suppose that you want to calculate the value of the two-sample t test statistic for testing hypotheses about a difference in population means using independent samples, and that the null hypothesis is $H_0 : \mu_1 - \mu_2 = 0$. You have two independent random samples, one from population Wilma and one from population Betty. The sample from population Wilma is size $n_1 = 63$, the sample mean is $\bar{x}_1 = 33$, and the sample standard deviation for this sample is $s_1 = 14$. The sample from population Betty is size $n_2 = 61$, the sample mean is $\bar{x}_2 = 58$, and the sample standard deviation for this sample is $s_2 = 29$. Calculate the value of the two-sample t test statistic.

b. You are asked to calculate the value of the two-sample t test statistic for testing hypotheses about a difference in population means using independent samples. The null hypothesis is $H_0 : \mu_1 - \mu_2 = 0$. You have two independent random samples, one from population 1 and one from population 2. The sample from population 1 is size $n_1 = 191$, the sample mean is $\bar{x}_1 = 19.6$, and the sample standard deviation for this sample is $s_1 = 6.5$. The sample from population 2 is size $n_2 = 64$, the sample mean is $\bar{x}_2 = 29.3$, and the sample standard deviation for this sample is $s_2 = 4.2$. Calculate the value of the two-sample t test statistic.

Answers

a. Substituting the given sample means and standard deviations and the hypothesized value into the expression for the two-sample t test statistic results in the following:

$$t = \frac{(\bar{x}_1 - \bar{x}_2) - (\mu_1 - \mu_2)}{\sqrt{\frac{s_1^2}{n_1} + \frac{s_2^2}{n_2}}} = \frac{(33 - 58) - 0}{\sqrt{\frac{(14)^2}{63} + \frac{(29)^2}{61}}}$$

Evaluate this expression as shown here:

$$t = \frac{(\bar{x}_1 - \bar{x}_2) - (\mu_1 - \mu_2)}{\sqrt{\frac{s_1^2}{n_1} + \frac{s_2^2}{n_2}}} = \frac{(33 - 58) - 0}{\sqrt{\frac{(14)^2}{63} + \frac{(29)^2}{61}}} = \frac{(33 - 58)}{4.111} = \frac{-25}{4.111} = -6.08$$

(rounded to two decimal places)

The value of the test statistic for this example is $t = -6.08$.

b. Substituting the given sample means and standard deviations and the hypothesized value into the expression for the two-sample t test statistic results in the following:

$$t = \frac{(\bar{x}_1 - \bar{x}_2) - (\mu_1 - \mu_2)}{\sqrt{\frac{s_1^2}{n_1} + \frac{s_2^2}{n_2}}} = \frac{(19.6 - 29.3) - 0}{\sqrt{\frac{(6.5)^2}{191} + \frac{(4.2)^2}{64}}}$$

Evaluate this expression as shown here:

$$t = \frac{(\bar{x}_1 - \bar{x}_2) - (\mu_1 - \mu_2)}{\sqrt{\frac{s_1^2}{n_1} + \frac{s_2^2}{n_2}}} = \frac{(19.6 - 29.3) - 0}{\sqrt{\frac{(6.5)^2}{191} + \frac{(4.2)^2}{64}}} = \frac{(19.6 - 29.3)}{0.705} = \frac{-9.7}{0.705} = -13.76$$

(rounded to two decimal places)

The value of the test statistic for this example is $t = -13.76$.

SECTION 17.1 EXERCISES

17.1 Measurements taken from a random sample of $n_1 = 87$ individuals from population A had a sample mean of $\bar{x}_1 = 41$ and a sample standard deviation of $s_1 = 9.5$. An independent random sample of $n_2 = 32$ individuals from population B produced measurements with sample mean of $\bar{x}_2 = 26$ and sample standard deviation of $s_2 = 4.6$. First calculate the value of the degrees of freedom (*df*), and then calculate the value of the two-sample *t* test statistic for testing the null hypothesis $H_0: \mu_1 - \mu_2 = 0$ using data from these independent samples.

17.2 Individuals participating in an experiment were randomly assigned to one of two treatment groups. Measurements taken from the $n_1 = 90$ individuals assigned to treatment group 1 had a sample mean of $\bar{x}_1 = 9.8$, and a sample standard deviation of $s_1 = 4.1$. Measurements taken from the $n_2 = 60$ individuals assigned to treatment group 2 had a sample mean of $\bar{x}_2 = 9.3$ and a sample standard deviation of $s_2 = 3.8$. Calculate the value of the degrees of freedom (*df*), and then calculate the value of the two-sample *t* test statistic for testing the null hypothesis $H_0: \mu_1 - \mu_2 = 0$.

SECTION 17.2 Guided Practice—Testing Hypotheses About a Difference in Means Using Independent Samples

In this section, you will consider a typical situation that involves using independent samples to carry out a two-sample *t* test for a difference in means. Consider this problem from a statistics textbook:

> The paper "Facebook Use and Academic Performance Among College Students: A Mixed-Methods Study with a Multi-Ethnic Sample" (*Computers in Human Behavior* [2015]: 265–272) describes a survey of a sample of 66 male students and a sample of 195 female students at a large university in Southern California. The authors of the paper believed that these samples were representative of male and female college students in Southern California. For the sample of males, the mean time spent per day on Facebook was 102.31 minutes. For the sample of females, the mean time was 159.61 minutes. The sample standard deviations were not given in the paper, but for purposes of this exercise, suppose that the sample standard deviations were both 100 minutes. Do the data provide convincing evidence that the mean time spent on Facebook is not the same for males and for females? Test the relevant hypotheses using $\alpha = 0.05$.

If you were to encounter a problem like this, you should start by reading the problem carefully. You can use the three-read strategy introduced in Chapter 7.

First Read: Context

1. **What is the general context?**
Researchers wondered if the mean time spent on Facebook is different for male college students and female college students. They collected data by surveying 66 male college students and 195 female college students at a large university in Southern California.

2. **Are there any terms that you don't understand?**
This question is about testing a hypothesis, so if you don't recall the steps required to carry out a hypothesis test, you would want to go to the appropriate section in your textbook to review these steps.

Second Read: Tasks

1. **What question is being asked? What am I being asked to do?**
   This problem asks you carry out a hypothesis test to determine if there is convincing evidence that the mean time spent on Facebook is not the same for males and for females. You are also told you should use a significance level of $\alpha = 0.05$.

Third Read: Details

1. **What data were collected? Were the data collected in a reasonable way?**
   Data were collected from a sample of 66 male college students and 195 female college students. The researchers believed that these samples were representative of male and female college students in Southern California.
2. **Are the actual data provided? What was the sample size (or the sample sizes if there are more than one sample)?**
   With population 1 denoting the population of male college students in Southern California and population 2 denoting the population of female college students in Southern California, $n_1 = 66$ and $n_2 = 195$. The actual data (the actual survey Facebook times for the 66 male students and for the 195 female students) are not provided.
3. **Are any summary statistics or additional information given? If so, which statistics and what are the values of these statistics?**
   The sample means and the sample standard deviations are given. For the sample from population 1, $n_1 = 66$, $\bar{x}_1 = 102.31$, and $s_1 = 100$. For the sample from population 2, $n_2 = 195$, $\bar{x}_2 = 159.61$, and $s_2 = 100$.

Now you are ready to tackle the tasks required to complete this exercise. This problem is asking you to test hypotheses about a difference in population means using data from independent samples. You are asked to determine if there is a difference in the mean time spent on Facebook.

Many textbooks recommend a sequence of steps for carrying out a hypothesis test. When you carry out a hypothesis test, you should follow the recommended sequence of steps. Although the order of the steps may vary somewhat from one text to another, they generally include the following:

Hypotheses: You will need to specify the null and the alternative hypothesis for the hypothesis test and define any symbols that are used in the context of the problem.

Method: You will need to specify what method you plan to use to carry out the test. For example, you might say that you plan to use the two-sample *t* test for a difference in population means if that is a test that would allow you to answer the question posed. You will also need to specify the significance level you will use to reach a conclusion.

Check Conditions: Most hypothesis test procedures are only appropriate when certain conditions are met. Once you have determined the method you plan to use, you will need to verify that any required conditions are met.

Calculate: To reach a conclusion in a hypothesis test, you will need to calculate the value of a test statistic and the associated *P*-value.

Communicate Results: The final step is to use the calculated values of the test statistic and the *P*-value to reach a conclusion and to provide an answer to the question posed. You should always provide a conclusion in words.

These steps are illustrated here for the given example.

> **Hypotheses**: You want to determine if there is convincing evidence that the mean time spent on Facebook is not the same for males and females. You can use μ_1 to represent the mean time spent on Facebook for male college students in Southern California and μ_2 to represent the mean time spent on Facebook for female college students in Southern California. The question being asked usually translates into the alternative hypothesis. Here the question asks if the mean time spent on Facebook is not the same for males and females. This would be the case when $\mu_1 - \mu_2 \neq 0$ (because $\mu_1 - \mu_2 = 0$

says the means are the equal). The alternative hypothesis would be $H_a: \mu_1 - \mu_2 \neq 0$. The null hypothesis would then be obtained by replacing the $\neq$ in the alternative hypothesis with an $=$. This results in hypotheses of

$$H_0: \mu_1 - \mu_2 = 0$$
$$H_a: \mu_1 - \mu_2 \neq 0$$

Method: Because the hypotheses are about a difference in population means and the samples are independent, you would consider the two-sample t test for a difference in population means. The test statistic is

$$t = \frac{(\bar{x}_1 - \bar{x}_2) - (\mu_1 - \mu_2)}{\sqrt{\frac{s_1^2}{n_1} + \frac{s_2^2}{n_2}}}$$

A significance level of $\alpha = 0.05$ was specified.

Check: There are two conditions necessary for the two-sample t test for a difference in population means to be appropriate. You are told that the samples are representative of male college students in Southern California and female college students in Southern California. There are 66 observations in the sample of males and 195 observations in the sample of females. Because both sample sizes are greater than or equal to 30, the sample size condition is met.

Calculations: To complete the calculations for a two-sample t test, you will need to calculate the value for degrees of freedom (df) and the value of the test statistic. The calculation of df is shown below. (For more detail on the calculation of df for the two-sample t test, see Section 17.1.)

$$df = \frac{(V_1 + V_2)^2}{\frac{V_1^2}{n_1 - 1} + \frac{V_2^2}{n_2 - 1}} \text{ where } V_1 = \frac{s_1^2}{n_1} \text{ and } V_2 = \frac{s_2^2}{n_2}$$

$$V_1 = \frac{s_1^2}{n_1} = \frac{(100)^2}{66} = \frac{10{,}000}{66} = 151.515$$

$$V_2 = \frac{s_2^2}{n_2} = \frac{(100)^2}{195} = \frac{10{,}000}{195} = 51.282$$

$$df = \frac{(V_1 + V_2)^2}{\frac{V_1^2}{n_1 - 1} + \frac{V_2^2}{n_2 - 1}} = \frac{(151.515 + 51.282)^2}{\frac{(151.515)^2}{66 - 1} + \frac{(51.282)^2}{195 - 1}} = \frac{(202.797)^2}{\frac{22{,}956.795}{65} + \frac{2629.844}{194}}$$

$$= \frac{41{,}126.623}{353.181 + 13.556} = \frac{41{,}126.623}{366.737} = 112.142$$

Rounding this value down, you would use $df = 112$.

Substituting the values of the two sample means, $\bar{x}_1$ and $\bar{x}_2$, the two sample standard deviations, s_1 and s_2, and the sample sizes n_1 and n_2 into the expression for the two-sample t test statistic results in the following:

$$t = \frac{(\bar{x}_1 - \bar{x}_2) - (\mu_1 - \mu_2)}{\sqrt{\frac{s_1^2}{n_1} + \frac{s_2^2}{n_2}}} = \frac{(102.31 - 159.61) - 0}{\sqrt{\frac{(100)^2}{66} + \frac{(100)^2}{195}}}$$

Begin by evaluating the parts of the expression that appear under the square root. Calculate

$$\frac{s_1^2}{n_1} = \frac{(100)^2}{66} = \frac{10{,}000}{66} = 151.515 \text{ and } \frac{s_2^2}{n_2} = \frac{(100)^2}{195} = \frac{10{,}000}{195} = 51.282$$

Next, add these two values together, and find the square root:

$$\sqrt{151.515 + 51.282} = \sqrt{202.797} = 14.241$$

Substituting this value into the expression for the two-sample t test statistic results in

$$t = \frac{(\bar{x}_1 - \bar{x}_2) - (\mu_1 - \mu_2)}{\sqrt{\frac{s_1^2}{n_1} + \frac{s_2^2}{n_2}}} = \frac{(102.31 - 159.61) - 0}{\sqrt{\frac{(100)^2}{66} + \frac{(100)^2}{195}}}$$

$$= \frac{(102.31 - 159.61)}{14.241} = \frac{-57.30}{14.241} = -4.02$$

(rounded to two decimal places)

The value of the test statistic for this example is $t = -4.02$.

Once you know the value of the test statistic, you need to find the associated P-value. Your statistics text will have an explanation of how the P-value is determined.

This is a two-tailed test (the inequality in H_a is $\neq$), so the P-value is two times an area under a t curve. The degrees of freedom (df) associated with this two-sample t test is $df = 112$. The P-value is the two times area under the t curve with $df = 112$ and to the left of -4.02. Therefore, the P-value is $2(P(t \leq -4.02)) \approx 0$.

Communicate Results: Because the P-value of 0 is less than the selected significance level of 0.05, you reject the null hypothesis. There is convincing evidence that the means for time spent on Facebook are not the same for male college students in Southern California and female college students in Southern California.

SECTION 17.2 EXERCISES

17.3 Each person in a random sample of 228 male teenagers and a random sample of 306 female teenagers was asked how many hours they spent online in a typical week (*Ipsos*, January 25, 2006). For the sample of males, the sample mean was 15.1 hours and the sample standard deviation was 11.4 hours. For females, the sample mean was 14.1 hours and the sample standard deviation was 11.8 hours.

Do the data provided convincing evidence that the mean number of hours spent online in a typical week for the population of male teenagers is not the same as the mean number for the population of female teenagers? Use the three-read strategy to understand the problem, and then test the relevant hypotheses using $\alpha = 0.01$.

17.4 Do faculty and students have similar perceptions of what types of behavior are appropriate in the classroom? This question was examined in the article "Faculty and Student Perceptions of Classroom Etiquette" (*Journal of College Student Development* [1998]: 515–516). Each person in a random sample of 173 students enrolled in general education classes at a large public university was asked to judge various behaviors on a scale from 1 (totally inappropriate) to 5 (totally appropriate). People in a random sample of 98 faculty members also rated the same behaviors. Sample means for three behaviors are included in the following table. Assume that all of the sample standard deviations are equal to 1.0.

Student Behavior	Student Mean Rating	Faculty Mean Rating
Wearing hats in class	2.80	3.63
Addressing instructor by first name	2.90	2.11
Talking on a cell phone	1.11	1.10

Is there sufficient evidence to conclude that the mean "appropriateness" score assigned to wearing a hat in class for the population of students at this university is not the same as the mean "appropriateness" score for the population of faculty at this university? Use the three-read strategy to understand this problem, and then test the relevant hypotheses using $\alpha = 0.05$.

18 Chi-Square Tests—The Math You Need to Know

SECTION 18.1 Evaluating Expressions

In previous chapters, you have considered testing hypotheses about a population proportion using data on a categorical variable with two possible values (like sex, or whether or not a person buys textbooks online). Sometimes you might want to test hypotheses using data on a categorical variable that has more than two categories, or using bivariate categorical data (data consisting of values of two different categorical variables, such as sex and favorite sport). In these situations, a chi-square test might be an appropriate method. To carry out a chi-square test, it might be necessary to calculate the value of a chi-square test statistic as part of the process for carrying out the test.

Note: Depending on your access to technology, such as a graphing calculator or a statistics software package, you may not need to evaluate the expression for the chi-square test statistic by hand. But just in case your statistics class is not relying on technology to do the calculation, this section illustrates how to evaluate this expression.

Test Statistic for Chi-Square Tests

$$X^2 = \sum \frac{(\text{observed count} - \text{expected count})^2}{\text{expected count}}$$

There are three different chi-square tests that are usually covered in an introductory statistics course—the chi-square goodness-of-fit test, the chi-square test of homogeneity, and the chi-square test for independence. All three of these tests involve comparing a set of observed counts to a set of expected counts. The observed counts are the numbers of observations in the data set that take on a particular value of a categorical variable or that take on a particular combination of values of the two categorical variables in a bivariate categorical data set. The expected counts represent what would have been expected if the null hypothesis for the test were true. To calculate the value of the test statistic, you need to know the observed counts and the expected counts, but once these are known, the calculations are not difficult.

Categorical data are often summarized in a one-way table or in a two-way table. For example, suppose that each person in a sample of 100 adults was asked which of four brands of toothpaste they preferred. The resulting data could be summarized in the following one-way table:

Brand	A	B	C	D
Count				

If 32 people said they preferred brand A, 27 said they preferred brand B, 17 said they preferred brand C, and 24 said they preferred brand D, the following table would summarize the data:

Brand	**A**	**B**	**C**	**D**
Count	32	27	17	24

The counts in this table are the observed counts. To calculate the value of the chi-square test statistic for a goodness-of-fit test where the null hypothesis specified that the four brands were preferred equally in the population from which the sample was selected, you would also need to know the expected counts for each of the four brands. For this null hypothesis, the expected counts would be 25, 25, 25, and 25 (if all brands were preferred equally, you would expect ¼ of the people in the sample of 100 to have selected each brand).

There are four different "cells" in the table, corresponding to the four possible values for the categorical variable (Brand A, Brand B, Brand C, and Brand D). For each cell, there is an observed count and an expected count. For example, for brand A, the observed count is 32 and the expected count is 25. To calculate the value of the test statistic, you would calculate the value of

$$\frac{(\text{observed count} - \text{expected count})^2}{\text{expected count}}$$

for each cell and then add those values to obtain the value of the chi-square statistic. For the brand A cell, this would be

$$\frac{(32-25)^2}{25} = \frac{7^2}{25} = \frac{49}{25} = 1.96$$

The values of $\frac{(\text{observed count} - \text{expected count})^2}{\text{expected count}}$ for the other three cells are

$$\frac{(27-25)^2}{25} = \frac{2^2}{25} = \frac{4}{25} = 0.16$$

$$\frac{(17-25)^2}{25} = \frac{(-8)^2}{25} = \frac{64}{25} = 2.56$$

$$\frac{(24-25)^2}{25} = \frac{(-1)^2}{25} = \frac{1}{25} = 0.04$$

The summation sign in the formula for the chi-square test statistic indicates that you should add all the values of $\frac{(\text{observed count} - \text{expected count})^2}{\text{expected count}}$ to obtain the value of the chi-square test statistic, as shown below.

$$X^2 = \sum \frac{(\text{observed count} - \text{expected count})^2}{\text{expected count}} = 1.96 + 0.16 + 2.56 + 0.04 = 4.72$$

The value of the test statistic for this example is $X^2 = 4.72$.

If two or more groups are being compared based on a categorical variable, the data are usually summarized in a two-way table. For example, suppose you are interested in learning about how students, faculty, and staff at a university commute to campus. You might select random samples from each of these three groups—students, faculty, and staff. This would result in three independently selected random samples. If each person in each of the samples was asked how he or she usually travels to campus (car, bus, walk, bicycle, or other), the data could be summarized in a two-way table with rows corresponding to the sample (students, faculty, or staff) and columns corresponding to the different possible values of the categorical variable *method of travel* (car, bus, walk, bicycle, or other). The structure of the table is shown here:

	Car	**Bus**	**Walk**	**Bicycle**	**Other**
Students					
Faculty					
Staff					

In this setting (more than one sample, one categorical variable), the chi-square test of homogeneity might be an appropriate way to analyze the data. This table has 15 cells, with each cell corresponding to a particular sample-method of travel combination. To calculate the value of the test statistic for a chi-square test of homogeneity, you need to know the observed counts and the expected counts for each of the 15 cells. Then you would follow the same process described above to calculate the value of $\frac{(\text{observed count} - \text{expected count})^2}{\text{expected count}}$ for each of the 15 cells and add these values to obtain the value of the chi-square test statistic.

Two-way tables are also used to summarize bivariate categorical data. For example, suppose that each person in a sample of adult California residents was asked where they live (with possible responses of Northern California, Central California, or Southern California). These same people were also asked if they are registered to vote in California (with possible responses registered and not registered). This would result in bivariate categorical data, with a (location of residence, registration status) pair for each person in the sample. The resulting data could be summarized in a two-way table with rows corresponding to the possible responses for the registration status variable and columns corresponding to the location of residence variable. Notice that each of the six cells in this table (see below) corresponds to a particular combination of values for the two variables, such as (Southern California/registered or Northern California/not registered).

	Southern California	Central California	Northern California
Registered			
Not Registered			

In this setting (bivariate categorical data), the chi-square test of independence might be an appropriate way to analyze the data. This table has six cells, with each cell corresponding to a particular location of residence-registration status combination. To calculate the value of the test statistic for a chi-square test of independence, you need to know the observed counts and the expected counts for each of these six cells. Then you would follow the same process described above to calculate the value of $\frac{(\text{observed count} - \text{expected count})^2}{\text{expected count}}$ for each of the six cells and add them to obtain the value of the chi-square test statistic.

Check It Out!

a. An article about the California lottery appeared in the *San Luis Obispo Tribune* (December 15, 1999), and gave the following population age distribution for adults in California: 35% of all California adults at that time were between 18 and 34 years old, 51% were between 35 and 64 years old, and 14% were 65 years old or older.

The article also gave the age distribution for a sample of California adults who purchased lottery tickets. Suppose that the counts (frequencies) for the age groups in the sample were as given in the following table. Assume that the adults in the sample are representative of the population of all California adults.

Age of Purchaser	Frequency
18–34	36
35–64	130
65 and over	34

Calculate the value of the X^2 test statistic that could be used to compare the population age distribution and the observed age distribution for California adults who purchase lottery tickets. The expected cells counts are found using the population age distribution for all California adults, as follows: $0.35 \times 200 = 70$, $0.51 \times 200 = 102$, and $0.14 \times 200 = 28$.

b. The article "Regional Differences in Attitudes Toward Corporal Punishment" (*Journal of Marriage and Family* [1994]: 314–324) included data resulting from a random sample of 978 adults. Each individual in the sample was asked whether they agreed with the statement: "Sometimes it is necessary to discipline a child with a good, hard spanking." Respondents were also classified according to the region of the U.S. in which they lived. The resulting data are summarized in the accompanying table.

Table of Observed Counts

	Response	
Region	**Agree**	**Disagree**
Northeast	130	59
West	146	42
Midwest	211	52
South	291	47

Calculate the value of the χ^2 test statistic that could be used to test for an association between Response (Agree or Disagree) and Region of residence. The expected cell counts are given in the following table:

Table of Expected Counts

	Response	
Region	**Agree**	**Disagree**
Northeast	150.3	38.7
West	149.6	38.4
Midwest	209.2	53.8
South	268.9	69.1

Answers

a. Begin by calculating the following for each cell:

$$\frac{(\text{observed count} - \text{expected count})^2}{\text{expected count}}$$

For the age group 18–34 cell, this would be

$$\frac{(36-70)^2}{70} = \frac{(-34)^2}{70} = \frac{1156}{70} = 16.51$$

The values of $\frac{(\text{observed count} - \text{expected count})^2}{\text{expected count}}$ for the other two cells are

$$\frac{(130-102)^2}{102} = \frac{28^2}{102} = \frac{784}{102} = 7.69$$

$$\frac{(34-28)^2}{28} = \frac{6^2}{28} = \frac{36}{28} = 1.29$$

Add these values to obtain the value of the chi-square test statistic, as shown below.

$$X^2 = \sum \frac{(\text{observed count} - \text{expected count})^2}{\text{expected count}} = 16.51 + 7.69 + 1.29 = 25.5$$

The value of the test statistic for this example is $X^2 = 25.5$.

b. Begin by calculating the following for each cell:

$$\frac{(\text{observed count} - \text{expected count})^2}{\text{expected count}}$$

For the first cell (Northeast and Agree), this is

$$\frac{(130 - 150.3)^2}{150.3} = \frac{(-20.3)^2}{150.3} = \frac{412.09}{150.3} = 2.74$$

The values of $\frac{(\text{observed count} - \text{expected count})^2}{\text{expected count}}$ for the other cells in the table are

$$\frac{(59 - 38.7)^2}{38.7} = \frac{(20.3)^2}{38.7} = \frac{412.09}{38.7} = 10.65$$

$$\frac{(146 - 149.6)^2}{149.6} = \frac{(-3.6)^2}{149.6} = \frac{12.96}{149.6} = 0.09$$

$$\frac{(42 - 38.4)^2}{38.4} = \frac{(3.6)^2}{38.4} = \frac{12.96}{38.4} = 0.34$$

$$\frac{(211 - 209.2)^2}{209.2} = \frac{(1.8)^2}{209.2} = \frac{3.24}{209.2} = 0.02$$

$$\frac{(52 - 53.8)^2}{53.8} = \frac{(-1.8)^2}{53.8} = \frac{3.24}{53.8} = 0.06$$

$$\frac{(291 - 268.9)^2}{268.9} = \frac{(22.1)^2}{268.9} = \frac{488.41}{268.9} = 1.82$$

$$\frac{(47 - 69.1)^2}{69.1} = \frac{(-22.1)^2}{69.1} = \frac{488.41}{69.1} = 7.07$$

Add these values to obtain the value of the chi-square test statistic, as shown below.

$$X^2 = \sum \frac{(\text{observed count} - \text{expected count})^2}{\text{expected count}}$$

$$= 2.74 + 10.65 + 0.09 + 0.34 + 0.02 + 0.06 + 1.82 + 7.07 = 22.8$$

The value of the test statistic for this example is $X^2 = 22.8$.

SECTION 18.1 EXERCISES

18.1 The paper "Cigarette Tar Yields in Relation to Mortality from Lung Cancer in the Cancer Prevention Study II Prospective Cohort" (*British Medical Journal* [2004]: 72–79) included the accompanying data on the tar level of cigarettes smoked for a sample of male smokers who died of lung cancer.

Tar Level	Frequency
Low	103
Medium	378
High	563
Very high	150

Assume that the sample is representative of male smokers who die of lung cancer. Calculate the value of the X^2 test statistic that could be used to test whether the proportions of male smoker lung cancer deaths are not equal for all of the four tar level categories. The expected values are 298.5 for each of the four cells.

18.2 Does including a gift with a request for a donation influence the proportion who will make a donation? This question was investigated in a study described in the report "Gift-Exchange in the Field" (Institute for the Study of Labor, 2007). Letters were sent to a large number of potential donors in Germany. The letter recipients were assigned at random to one of three groups. Those in the first group received a letter with no gift. Those in the second group received a letter that included a small gift (a postcard), and those in the third group received a letter with a larger gift (four postcards). The response of interest was whether or not the letter resulted in a donation.

	Donation	No Donation
No Gift	397	2865
Small Gift	465	2772
Large Gift	691	2656

Calculate the value of the chi-square test statistic that can be used to decide if there is convincing evidence that the proportions for the two donation categories are not the same for all three types of requests. The expected values are given in the table below.

	Donation	No Donation
No Gift	514.5	2747.5
Small Gift	510.6	2726.4
Large Gift	527.9	2819.1

SECTION 18.2 Guided Practice—Chi-Square Tests

In this section, you will consider situations that involve carrying out the three types of chi-square tests usually covered in an introductory statistics course—the chi-square goodness-of-fit test, the chi-square test of homogeneity, and the chi-square test of independence.

Chi-Square Goodness-of-Fit Test

Consider this problem from a statistics textbook:

> The "Global Automotive 2016 Color Popularity Report" (Axalta Coating Systems, axaltacs.com) included data on the colors for a sample of new cars sold in North America. They reported that 25% of the cars in the sample were white, 21% were black, 16% were grey, 11% were silver, 10% were red, and 17% were some other color. Suppose that these percentages were based on a random sample of 1200 new cars sold in North America. Is there convincing evidence that the proportions of new cars sold are not the same for all the six color categories? Carry out a hypothesis test with a significance level of 0.05.

If you were to encounter a problem like this, you should start by reading the problem carefully. You can use the three-read strategy introduced in Chapter 7.

First Read: Context

1. **What is the general context?**
   Data were collected to see if some car colors are more popular than others. The color of each car in a sample of new cars sold in 2016 was recorded. The problem states that the sample was a random sample of 1200 cars.
2. **Are there any terms that you don't understand?**
   This question is about testing a hypothesis, so if you don't recall the steps required to carry out a hypothesis test, you would want to go to the appropriate section in your statistics textbook to review these steps.

Second Read: Tasks

1. **What question is being asked? What am I being asked to do?**
   This problem asks you to decide if the sample data provide convincing evidence that the proportions of new cars sold are not the same for all the six color categories. You are to do this by carrying out a hypothesis test with a significance level of 0.05.

Third Read: Details

1. **What data were collected? Were the data collected in a reasonable way?**
   Data were collected from a random sample of 1200 new cars purchased in 2016. The problem states that the sample is a random sample.

2. **Are the actual data provided? What was the sample size (or the sample sizes if there are more than one sample)?**
The sample size was $n = 1200$. The actual data (the 1200 individual color observations) are not given, but the number of cars that were of each color can be determined from the given percentages and the sample size of 1200.

3. **Are any summary statistics or additional information given? If so, which statistics and what are the values of these statistics?**
The problem states that 25% of the cars in the sample were white, 21% were black, 16% were grey, 11% were silver, 10% were red, and 17% were some other color. You also know that the sample size was $n = 1200$. This allows you to calculate the number of cars in the sample for each color category (the observed counts) as follows:

White: 25% of 1200, so there were 0.25(1200) = 300 white cars in the sample.

Black: 21% of 1200, so there were 0.21(1200) = 252 black cars in the sample.

Grey: 16% of 1200, so there were 0.16(1200) = 192 grey cars in the sample.

Silver: 11% of 1200, so there were 0.11(1200) = 132 silver cars in the sample.

Red: 10% of 1200, so there were 0.10(1200) = 120 red cars in the sample.

Other colors: 17% of 1200, so there were 0.17(1200) = 204 other color cars in the sample.

Now you are ready to tackle the tasks required to complete this exercise. The problem states, "Is there convincing evidence that the proportions of new cars sold are not the same for all the six color categories? Carry out a hypothesis test with a significance level of 0.05." This is asking you to test hypotheses about population proportions for a categorical variable with more than two possible values. The population is all 2016 new cars purchased in North America. You are asked to decide if there is convincing evidence that the color proportions are not all the same for the six color categories. You are also told to use a significance level of 0.05 when carrying out the test.

Many textbooks recommend a sequence of steps for carrying out a hypothesis test. When you carry out a hypothesis test, you should follow the recommended sequence of steps. Although the order of the steps may vary somewhat from one text to another, they generally include the following:

Hypotheses: You will need to specify the null and the alternative hypothesis for the hypothesis test and define any symbols that are used in the context of the problem.

Method: You will need to specify what method you plan to use to carry out the test. For example, you might say that you plan to use the chi-square goodness-of-fit test if that is a test that would allow you to answer the question posed. You will also need to specify the significance level you will use to reach a conclusion.

Check Conditions: Many hypothesis test procedures are only appropriate when certain conditions are met. Once you have determined the method you plan to use, you will need to verify that any required conditions are met.

Calculate: To reach a conclusion in a hypothesis test, you will need to calculate the value of a test statistic and the associated *P*-value.

Communicate Results: The final step is to use the computed values of the test statistic and the *P*-value to reach a conclusion and to provide an answer to the question posed. You should always provide a conclusion in words.

These steps are illustrated here for the given example.

Hypotheses: You want to decide if there is convincing evidence that the color proportions are not all the same for the six color categories. The question being asked usually translates into the alternative hypothesis, so the alternative hypothesis would be

H_a : the six color category proportions are not all equal. The null hypothesis would then specify that the color category proportions are all equal. Since there are six color categories, if the category proportions are all equal, they would each have a value of 1/6. This results in hypotheses of

H_0 : Each of the six color category proportions is equal to 1/6
H_a : The six color category proportions are not all equal

Method: Because the hypotheses are about the distribution of a single categorical variable, you would consider the chi-square goodness-of-fit test. The test statistic is

$$X^2 = \sum \frac{(\text{observed count} - \text{expected count})^2}{\text{expected count}}$$

A significance level of $\alpha = 0.05$ was specified.

Check: There are two conditions necessary for the chi-square goodness-of-fit test to be appropriate. The observed counts must be based on a random sample or a sample that is representative of the population. You also need a large sample, and the sample size is large enough if all expected counts are greater than or equal to five.

The problem states that you can assume that the sample was a random sample. To check the sample size condition, you need to know the expected counts. The expected counts are the counts that would have been expected if the null hypothesis is true. Because the null hypothesis specifies that each color category proportion is 1/6, if the null hypothesis is true, you would expect 1/6 of the 1200 observations to fall in each of the six color categories. This means that the expected count would be $\frac{1}{6}(1200) = 200$ for each color category. Because all the expected counts are greater than or equal to five, the sample size condition is met.

Calculations: The table below gives the observed and expected counts for the six color categories.

Color	Observed Count	Expected Count
White	300	200
Black	252	200
Grey	192	200
Silver	132	200
Red	120	200
Other Colors	204	200

The value of the test statistic is

$$\begin{aligned} X^2 &= \sum \frac{(\text{observed count} - \text{expected count})^2}{\text{expected count}} \\ &= \frac{(300-200)^2}{200} + \frac{(252-200)^2}{200} + \frac{(192-200)^2}{200} \\ &\quad + \frac{(132-200)^2}{200} + \frac{(120-200)^2}{200} + \frac{(204-200)^2}{200} \\ &= 50.00 + 13.52 + 0.32 + 23.12 + 32.00 + 0.08 \\ &= 119.04 \text{ (rounded to two decimal places)} \end{aligned}$$

Once you know the value of the test statistic, you need to find the associated *P*-value. Your statistics textbook will have an explanation of how the *P*-value is determined.

The chi-square goodness-of-fit test is an upper-tailed test, and the P-value is an area under a chi-square curve. The degrees of freedom (df) associated with a chi-square goodness-of-fit test is $k - 1$, where k is the number of different possible values for the categorical variable. Since there are six color categories in this example, $df = 6 - 1 = 5$. The P-value is the area under the chi-square curve with $df = 5$ and to the right of 119.04. For this example, the P-value is $P(X^2 \geq 119.04) \approx 0$.

Communicate Results: Because the P-value is less than the selected significance level of 0.05, you reject the null hypothesis. There is convincing evidence that the color category proportions for the six color categories are not all equal. Looking at the differences between the observed counts and the expected counts, it appears that there is a preference for white and black cars and that silver and red cars are chosen less often.

Chi-Square Test for Homogeneity

Consider this problem from a statistics textbook:

> Some colleges now allow students to pay their tuition using a credit card. The report "Credit Card Tuition Payment Survey 2014" (creditcards.com/credit-card-news/tuition-charge-fee-survey.php) includes data from a survey of 100 public four-year colleges, 100 private four-year colleges, and 100 community colleges. The accompanying table gives information on credit card acceptance for each of these samples of colleges. For this exercise, suppose that these three samples are representative of the populations of public four-year colleges, private four-year colleges, and community colleges in the U.S. Is there convincing evidence that the proportions in each of the two credit card categories is not the same for all three types of colleges? Test the relevant hypotheses using a 0.05 significance level.

	Accepts Credit Cards for Tuition Payment	Does Not Accept Credit Cards for Tuition Payment
Public Four-Year Colleges	92	8
Private Four-Year Colleges	68	32
Community Colleges	100	0

If you were to encounter a problem like this, you should start by reading the problem carefully. You can use the three-read strategy introduced in Chapter 7.

First Read: Context

1. **What is the general context?**
   Data were collected to see if the proportions that allow students to pay tuition using a credit card are not all the same for three types of colleges. The types of colleges compared were public four-year colleges, private four-year colleges, and community colleges. A sample of 100 colleges was selected from each of these populations (public four-year colleges, private four-year colleges, and community colleges). Each college was asked if they accept credit cards for tuition payment. The problem states that the samples are representative of the three populations.
2. **Are there any terms that you don't understand?**
   This question is about testing a hypothesis, so if you don't recall the steps required to carry out a hypothesis test, you would want to go to the appropriate section in your statistics textbook to review these steps.

Second Read: Tasks

1. **What question is being asked? What am I being asked to do?**
   This problem asks you to decide if the sample data provide convincing evidence that the proportions in each of the two credit card categories is not the same for all three types of colleges. You are to do this by carrying out a hypothesis test with a significance level of 0.05.

Third Read: Details

1. **What data were collected? Were the data collected in a reasonable way?**
   The data are from three samples of colleges. One sample consisted of 100 public four-year colleges, one sample consisted of 100 private four-year colleges, and one sample consisted of 100 community colleges. The problem states that the samples are representative of the three types of colleges (the three populations).
2. **Are the actual data provided? What was the sample size (or the sample sizes if there are more than one sample)?**
   Each sample consisted of 100 colleges. The actual data are summarized in the two-way table given in the problem, which provides the observed counts for the "accepts credit cards" and "does not accept credit cards" categories for each of the three samples.
3. **Are any summary statistics or additional information given? If so, which statistics and what are the values of these statistics?**
   The observed counts are provided.

Now you are ready to tackle the tasks required to complete this exercise. This problem is asking you to test hypotheses about distribution of values of a categorical variable and involves comparing two or more populations. There are three populations being compared: public four-year colleges, private four-year colleges, and community colleges. You are asked to decide if there is convincing evidence that the proportions that do and do not accept credit cards are not all the same for the three populations. You are also told to use a significance level of 0.05 when carrying out the test.

Many textbooks recommend a sequence of steps for carrying out a hypothesis test. When you carry out a hypothesis test, you should follow the recommended sequence of steps. Although the order of the steps may vary somewhat from one textbook to another, they generally include the following:

Hypotheses: You will need to specify the null and the alternative hypothesis for the hypothesis test and define any symbols that are used in the context of the problem.

Method: You will need to specify what method you plan to use to carry out the test. For example, you might say that you plan to use the chi-square test for homogeneity if that is a test that would allow you to answer the question posed. You will also need to specify the significance level you will use to reach a conclusion.

Check Conditions: Many hypothesis test procedures are only appropriate when certain conditions are met. Once you have determined the method you plan to use, you will need to verify that any required conditions are met.

Calculate: To reach a conclusion in a hypothesis test, you will need to calculate the value of a test statistic and the associated P-value.

Communicate Results: The final step is to use the calculated values of the test statistic and the P-value to reach a conclusion and to provide an answer to the question posed. You should always provide a conclusion in words.

These steps are illustrated here for the given example.

Hypotheses: You want to decide if there is convincing evidence that the proportions that do and do not accept credit cards are not all the same for the three types of colleges. The question being asked usually translates into the alternative hypothesis, so the alternative hypothesis would be

H_a: The proportions falling into the two credit card response categories are not all the same for the three types of colleges

The null hypothesis would then specify that the credit card response proportions are the same for all three types of colleges. This results in hypotheses of

H_0: The proportions falling into the two credit card response categories are the same for all three types of colleges

H_a: The proportions falling into the two credit card response categories are not all the same for the three types of colleges

Method: Because the hypotheses are about the distribution of a single categorical variable for three populations, you would consider the chi-square test of homogeneity. The test statistic is

$$X^2 = \sum \frac{(\text{observed count} - \text{expected count})^2}{\text{expected count}}$$

A significance level of $\alpha = 0.05$ was specified.

Check: There are two conditions necessary for the chi-square test of homogeneity to be appropriate. The observed counts must be based on independent random samples or samples that are representative of the populations. You also need large samples, and the sample sizes are large enough if all expected counts are greater than or equal to 5.

The problem states that you can assume that the samples are representative of the three types of colleges (the three populations of interest). To check the sample size condition, you need to know the expected counts. The expected counts are the counts that would have been expected if the null hypothesis is true.

For a chi-square test of homogeneity, the expected counts for the cells in the two-way table are found by calculating

$$\frac{(\text{row total})(\text{column total})}{\text{grand total}}$$

for each cell in the table. The grand total is the total number of observations in all the samples combined. The row totals, column totals and the grand total have been added to the two-way table, as shown below.

	Accepts Credit Cards for Tuition Payment	Does Not Accept Credit Cards for Tuition Payment	Total
Public Four-Year Colleges	92	8	**100**
Private Four-Year Colleges	68	32	**100**
Community Colleges	100	0	**100**
Total	**260**	**40**	**300**

For the cell that corresponds to public four-year colleges and accepts credit cards, the row total is 100, the column total is 260, and the grand total is 300. The expected cell count for this cell can then be calculated as

$$\frac{(\text{row total})(\text{column total})}{\text{grand total}} = \frac{(100)(260)}{300} = \frac{26{,}000}{300} = 86.67$$

The other expected counts are calculated in a similar way, and they have been entered in parentheses just below the observed counts in the following table.

	Accepts Credit Cards for Tuition Payment	Does Not Accept Credit Cards for Tuition Payment	Total
Public Four-Year Colleges	92 (86.67)	8 (13.33)	**100**
Private Four-Year Colleges	68 (86.67)	32 (13.33)	**100**
Community Colleges	100 (86.67)	0 (13.33)	**100**
Total	**260**	**40**	**300**

Because all the expected counts are greater than or equal to 5, the sample size condition is met.

Calculations: The value of the test statistic is

$$X^2 = \sum \frac{(\text{observed count} - \text{expected count})^2}{\text{expected count}}$$

$$= \frac{(92 - 86.67)^2}{86.67} + \frac{(68 - 86.67)^2}{86.67} + \frac{(100 - 86.67)^2}{86.67}$$

$$+ \frac{(8 - 13.33)^2}{13.33} + \frac{(32 - 13.33)^2}{13.33} + \frac{(0 - 13.33)^2}{13.33}$$

$$= 0.39 + 4.02 + 2.05 + 2.13 + 26.15 + 13.33$$

$$= 48.07 \text{ (rounded to two decimal places)}$$

Once you know the value of the test statistic, you need to find the associated *P*-value. Your statistics textbook will have an explanation of how the *P*-value is determined.

The chi-square test of homogeneity is an upper-tailed test, and the *P*-value is an area under a chi-square curve. The degrees of freedom (df) associated with a chi-square test of homogeneity is $(r - 1)(c - 1)$, where r is the number of rows in the two-way table and c is the number of columns in the two-way table (not counting the total row and column). Since there are 3 rows and 2 columns in the table for this problem, $df = (3 - 1)(2 - 1) = (2)(1) = 2$. The *P*-value is the area under the chi-square curve with $df = 2$ and to the right of 48.07. Therefore, the *P*-value is $P(X^2 \geq 48.07) \approx 0$.

Communicate Results: Because the *P*-value is less than the selected significance level of 0.05, you reject the null hypothesis. There is convincing evidence that the proportions that do and do not accept credit cards are not all the same for the three types of colleges. Looking at the differences between the observed counts and the expected counts, it appears that private four-year colleges are less likely to accept credit cards for payment than the other two types of colleges.

Chi-Square Test for Independence

Consider this problem from a statistics textbook:

Each person in a representative sample of 445 college students age 18 to 24 was classified according to age and to the response to the following question: "How often have you used a credit card to buy items knowing you wouldn't have money to pay the bill when it arrived?" Possible responses were never, rarely, sometimes, or frequently ("Majoring in Money: How American College Students Manage Their Finances," June 28, 2016, salliemae.newshq.businesswire.com). The responses are summarized in the accompanying table. Do these data provide evidence that there is an association between age group and the response to the question? Test the relevant hypotheses using $\alpha = 0.01$.

	Age 18 to 20	Age 21 to 22	Age 23 to 24
Never	72	62	29
Rarely	36	34	32
Sometimes	30	42	40
Frequently	12	24	32

If you were to encounter a problem like this, you should start by reading the problem carefully. You can use the three-read strategy introduced in Chapter 7.

First Read: Context

1. **What is the general context?**
   Data were collected to learn about the way in which college students use credit cards. College students were asked about how often they use a credit card to buy items knowing that they would not have money to pay the bill when it arrived. Responses were obtained from a representative sample of 445 college students age 18 to 24.
2. **Are there any terms that you don't understand?**
   This question is about testing for an association, so if you don't recall what it means for there to be an association between two categorical variables or if you don't recall the steps required to carry out a hypothesis test, you would want to go to the appropriate sections in your statistics textbook to review the definition of association and the steps in a hypothesis test.

Second Read: Tasks

1. **What question is being asked? What am I being asked to do?**
   This problem asks you to decide if the data provide evidence that there is an association between age group and the response to the question. You are to do this by carrying out a hypothesis test with a significance level of 0.01.

Third Read: Details

1. **What data were collected? Were the data collected in a reasonable way?**
   Data were collected from a representative sample of 445 college students age 18 to 24. The resulting data set is a bivariate categorical data set because two variables were recorded for each student—age and response to the question about credit card use. The problem states that the sample is representative of college students age 18 to 24.
2. **Are the actual data provided? What was the sample size (or the sample sizes if there are more than one sample)?**
   The sample size was $n = 445$. The data set is a bivariate categorical data set because two variables—age group and credit card use response—were recorded for each student in the sample. The actual data are summarized in the two-way table given in the problem, which provides the observed counts for the 12 different age group/response category combinations (such as "age 18 to 20"/"never").
3. **Are any summary statistics or additional information given? If so, which statistics and what are the values of these statistics?**
   The observed counts are provided.

Now you are ready to tackle the tasks required to complete this exercise. This problem is asking you to test hypotheses about an association between two categorical variables. The two variables are age group (with categories 18 to 20, 21 to 22, and 23 to 24) and response to the question about how often credit cards are used to make purchases knowing that there would not be money to pay the bill when it arrived (with categories never, rarely, sometimes, and frequently). You are asked to decide if there is evidence that there is an association between age group and response. You are also told to use a significance level of 0.01 when carrying out the test.

Many textbooks recommend a sequence of steps for carrying out a hypothesis test. When you carry out a hypothesis test, you should follow the recommended sequence of steps. Although the order of the steps may vary somewhat from one textbook to another, they generally include the following:

Hypotheses: You will need to specify the null and the alternative hypothesis for the hypothesis test and define any symbols that are used in the context of the problem.

Method: You will need to specify what method you plan to use to carry out the test. For example, you might say that you plan to use the chi-square test for independence if that is a test that would allow you to answer the question posed. You will also need to specify the significance level you will use to reach a conclusion.

Check Conditions: Many hypothesis test procedures are only appropriate when certain conditions are met. Once you have determined the method you plan to use, you will need to verify that any required conditions are met.

Calculate: To reach a conclusion in a hypothesis test, you will need to calculate the value of a test statistic and the associated *P*-value.

Communicate Results: The final step is to use the computed values of the test statistic and the *P*-value to reach a conclusion and to provide an answer to the question posed. You should always provide a conclusion in words.

These steps are illustrated here for the given example.

Hypotheses: You want to decide if there is evidence of an association between age group and response to the credit card use question. The question being asked usually translates into the alternative hypothesis, so the alternative hypothesis would be

H_a: There is an association between age group and response to the credit card use question

The null hypothesis would then specify that there is no association between age group and credit card use response (which is equivalent to saying that age group and credit card use response are independent). This results in hypotheses of

H_0: There is no association between age group and response to the credit card use question

H_a: There is an association between age group and response to the credit card use question

Method: Because the hypotheses are about association between two categorical variables, you would consider the chi-square test of independence. The test statistic is

$$X^2 = \sum \frac{(\text{observed count} - \text{expected count})^2}{\text{expected count}}$$

A significance level of $\alpha = 0.01$ was specified.

Check: There are two conditions necessary for the chi-square test of independence to be appropriate. The observed counts must be based on a random sample or a sample that is representative of the population. You also need large sample, and the sample size is large enough if all expected counts are greater than or equal to 5.

The problem states that you can assume that the sample is representative of the population of college students age 18 to 24. To check the sample size condition, you need to know the expected counts. The expected counts are the counts that would have been expected if the null hypothesis is true.

For a chi-square test of independence, the expected counts for the cells in the two-way table are found by calculating

$$\frac{(\text{row total})(\text{column total})}{\text{grand total}}$$

for each cell in the table. The grand total is the total number of observations in the sample. The row totals, column totals, and the grand total have been added to the two-way table, as shown below.

	Age 18 to 20	Age 21 to 22	Age 23 to 24	Total
Never	72	62	29	**163**
Rarely	36	34	32	**102**
Sometimes	30	42	40	**112**
Frequently	12	24	32	**68**
Total	**150**	**162**	**133**	**445**

For the cell that corresponds to age 18 to 20 and never, the row total is 163, the column total is 150, and the grand total is 445. The expected cell count for this cell can then be calculated as

$$\frac{\text{(row total)(column total)}}{\text{grand total}} = \frac{(163)(150)}{445} = \frac{24{,}445}{445} = 54.94$$

The other expected counts are calculated in a similar way, and they have been entered in parentheses just below the observed counts in the following table.

	Age 18 to 20	Age 21 to 22	Age 23 to 24	Total
Never	72 (54.94)	62 (59.34)	29 (48.72)	**163**
Rarely	36 (34.38)	34 (37.13)	32 (30.49)	**102**
Sometimes	30 (37.75)	42 (40.77)	40 (33.47)	**112**
Frequently	12 (22.92)	24 (24.76)	32 (20.32)	**68**
Total	**150**	**162**	**133**	**445**

Because all the expected counts are greater than or equal to five, the sample size condition is met.

Calculations: The value of the test statistic is

$$\begin{aligned} X^2 &= \sum \frac{(\text{observed count} - \text{expected count})^2}{\text{expected count}} \\ &= \frac{(72-54.94)^2}{54.94} + \frac{(62-59.34)^2}{59.34} + \frac{(29-48.72)^2}{48.72} \\ &\quad + \frac{(36-34.38)^2}{34.38} + \frac{(34-37.13)^2}{37.13} + \frac{(32-30.49)^2}{30.49} \\ &\quad + \frac{(30-37.75)^2}{37.75} + \frac{(42-40.77)^2}{40.77} + \frac{(40-33.47)^2}{33.47} \\ &\quad + \frac{(12-22.92)^2}{22.92} + \frac{(24-24.76)^2}{24.76} + \frac{(32-20.32)^2}{20.32} \\ &= 5.30 + 0.12 + 7.98 + 0.08 + 0.26 + 0.07 + 1.59 + 0.04 \\ &\quad + 1.27 + 5.20 + 0.02 + 6.71 \\ &= 28.64 \text{ (rounded to two decimal places)} \end{aligned}$$

Once you know the value of the test statistic, you need to find the associated P-value. Your statistics textbook will have an explanation of how the P-value is determined.

The chi-square test of independence is an upper-tailed test, and the P-value is an area under a chi-square curve. The degrees of freedom (df) associated with a chi-square test of independence is $(r - 1)(c - 1)$, where r is the number of rows in the two-way table and c is the number of columns in the two-way table (not counting the Total row and Total column). Since there are 4 rows and 3 columns in the table for this problem, $df = (4 - 1)(3 - 1) = (3)(2) = 6$. The P-value is the area under the chi-square curve with $df = 6$ and to the right of 28.64. The P-value is $P(X^2 \geq 28.64) \approx 0$.

Communicate Results: Because the P-value is less than the selected significance level of 0.01, you reject the null hypothesis. There is convincing evidence that there is an association between age group and the response to the credit card use question.

SECTION 18.2 EXERCISES

18.3 Use the three-read strategy to understand the following exercise from a statistics textbook, and then proceed to complete the exercise.

A popular urban legend is that more babies than usual are born during certain phases of the lunar cycle, especially near the full moon. The paper "The Effect of the Lunar Cycle on Frequency of Births and Birth Complications" (*American Journal of Obstetrics and Gynecology* [2005]: 1462–1464) classified births according to the lunar cycle. Data for a sample of randomly selected births, consistent with summary quantities in the paper, are given in the accompanying table.

Lunar Phase	Number of Days	Number of Births
New Moon	24	7,680
Waxing Crescent	152	48,442
First Quarter	24	7,579
Waxing Gibbous	149	47,814
Full Moon	24	7,711
Waning Gibbous	150	47,595
Last Quarter	24	7,733
Waning Crescent	152	48,230

Calculate the value of the X^2 goodness-of-fit test statistic that could be used to determine if the distribution of Number of Births in each lunar phase is what would be expected based on the number of days in each phase. Use the value of the test statistic to test the relevant hypotheses using a 0.10 significance level.

18.4 Use the three-read strategy to understand the following exercise from a statistics textbook, and then proceed to complete the exercise.

In a study to determine if hormone therapy increases risk of venous thrombosis in menopausal women, each person in a sample of 579 women who had been diagnosed with venous thrombosis was classified according to hormone use. Each woman in a sample of 2243 women who had not been diagnosed with venous thrombosis was also classified according to hormone use. Data from the study are given in the accompanying table (*Journal of the American Medical Association* [2004]: 1581–1587). The women in each of the two samples were selected at random from patients at a large HMO in the state of Washington.

	Current Hormone Use		
	None	Esterified Estrogen	Conjugated Equine Estrogen
Venous Thrombosis	372	86	121
No Venous Thrombosis	1,439	515	289

Is there convincing evidence that the proportions of those who would fall into each of the hormone use categories are not the same for women who have been diagnosed with venous thrombosis and those who have not? Carry out a hypothesis test with a significance level of 0.05.

18.5 Use the three-read strategy to understand the following exercise from a statistics textbook, and then proceed to complete the exercise.

The paper "Overweight Among Low-Income Preschool Children Associated with the Consumption of Sweet Drinks" (*Pediatrics* [2005]: 223–229) described a study of children who were underweight or normal weight at age 2. Children in the sample were classified according to the number of sweet drinks consumed per day and whether or not the child was overweight one year after the study began. Is there evidence of an association between whether or not children are overweight after 1 year and the number of sweet drinks consumed? Assume that the sample of children in this study is representative of 2- to 3-year-old children. Test the appropriate hypotheses using a 0.01 significance level.

		Overweight?	
		Yes	No
Number of Sweet Drinks Consumed per Day	0	22	930
	1	73	2,074
	2	56	1,681
	3 or more	102	3,390

19 Estimating and Testing Hypotheses About the Slope of a Population Regression Line—The Math You Need to Know

SECTION 19.1 Evaluating Expressions

There are several expressions that you will encounter as you learn about for estimating the slope of a population regression line and testing hypotheses about the slope of a population regression line.

Note: Depending on your access to technology, such as a graphing calculator or a statistics software package, you may not need to evaluate these expressions by hand. But just in case your statistics class is not relying on technology to do these calculations, this section looks at how to evaluate the expressions.

Chapter 4 of this book includes related material on how to calculate deviations from a line, the slope and intercept of the least-squares line, the sum of squared residuals (SSR), the coefficient of determination (r^2), and the standard deviation about the least-squares line (s_e). It is a good idea to review the material in Sections 4.4 and 4.6 before continuing with this chapter.

Standard Deviation of the Slope of the Least-Squares Line

$$s_b = \frac{s_e}{\sqrt{\sum(x - \bar{x})^2}}$$

The value of the slope of the least-squares regression line, b, depends on the sample data used to calculate the equation of the least-squares line. Different samples from the same population will result in different values of the slope simply because of sampling variability. The standard deviation of the slope of the least-squares regression line, s_b, tells you how much sample to sample variability you can expect in the value of the slope. The standard deviation of the slope of the least-squares regression line is used in the calculation of confidence intervals for the slope of a population regression line and also in calculating the value of the test statistic used to test hypotheses about the slope of a population regression line.

To calculate the value of s_b, you need to know the value of s_e, which is the standard deviation about the least-squares line. To review how s_e is calculated, see "Standard Deviation About the Least-Squares Line, s_e" in Section 4.6. Once you know or have calculated the value of s_e, you can proceed to evaluating the expression for the standard deviation of the slope of the least-squares regression line, $s_b = \frac{s_e}{\sqrt{\Sigma(x - \bar{x})^2}}$. Begin by evaluating expression that appears in the denominator under the square root symbol. Notice that this is the sum of the squared deviations from the sample mean for the x values in the data set. This sum also appears in the expressions used to calculate the variance and standard deviation of a data set (See Section 3.4 for a review of calculating the sum of squared deviations from the sample mean). This sum is calculated by:

1. Calculating the mean of the x values, $\bar{x}$
2. Subtracting $\bar{x}$ from each x value in the sample to get the deviations from the mean
3. Squaring each deviation to get the squared deviations from the mean
4. Adding up all the squared deviations from the mean

For example, suppose that the standard deviation about the least-squares line for a bivariate data set is $s_e = 21.45$ and that the x values in the data set are

10 26 8 14 17

The mean of these values is $\bar{x} = \frac{\Sigma x}{n} = \frac{10 + 26 + 8 + 14 + 17}{5} = \frac{75}{5} = 15.$

Subtracting 15 from each of the x values results in the following deviations from the mean:

$$10 - 15 = -5$$
$$26 - 15 = 11$$
$$8 - 15 = -7$$
$$14 - 15 = -1$$
$$17 - 15 = 2$$

Next, square each deviation and add them to get the sum of squared deviations:

$$\sum(x - \bar{x})^2 = (-5)^2 + (11)^2 + (-7)^2 + (-1)^2 + (2)^2$$
$$= 25 + 121 + 49 + 1 + 4 = 200$$

Then

$$\sqrt{\sum(x - \bar{x})^2} = \sqrt{200} = 14.142$$

Now it is easy to complete the calculation of s_b.

$$s_b = \frac{s_e}{\sqrt{\sum(x - \bar{x})^2}} = \frac{21.45}{14.142} = 1.517$$

Check It Out!

a. Suppose that you know that the standard deviation about the least-squares line for a bivariate data set is $s_e = 7.4$, and that the six x values in the data set are

5.5 1.2 6.4 0.8 9.9 8.3

Calculate the value of s_b, the standard deviation of the slope of the least-squares regression line.

b. Suppose that you know that the standard deviation about the least-squares line for a bivariate data set is $s_e = 33$, and that the four x values in the data set are

47 36 14 21

Calculate the value of s_b, the standard deviation of the slope of the least-squares regression line.

Answers

a. The mean of these values is $\bar{x} = \frac{\Sigma x}{n} = \frac{5.5 + 1.2 + 6.4 + 0.8 + 9.9 + 8.3}{6} = \frac{32.1}{6} = 5.35$. Subtracting 5.35 from each of the x values results in the following deviations from the mean:

$$5.5 - 5.35 = 0.15$$
$$1.2 - 5.35 = -4.15$$
$$6.4 - 5.35 = 1.05$$
$$0.8 - 5.35 = -4.55$$
$$9.9 - 5.35 = 4.55$$
$$8.3 - 5.35 = 2.95$$

Square each deviation and add them to get the sum of squared deviations:

$$\begin{aligned}\sum(x - \bar{x})^2 &= (0.15)^2 + (-4.15)^2 + (1.05)^2 + (-4.55)^2 + (4.55)^2 + (2.95)^2\\ &= 0.022 + 17.223 + 1.102 + 20.703 + 20.703 + 8.702\\ &= 68.46\end{aligned}$$

Then

$$\sqrt{\sum(x - \bar{x})^2} = \sqrt{68.46} = 8.27$$

Complete the calculation of s_b as shown here:

$$s_b = \frac{s_e}{\sqrt{\sum(x - \bar{x})^2}} = \frac{7.4}{8.27} = 0.89$$

b. The mean of these values is $\bar{x} = \frac{\Sigma x}{n} = \frac{47 + 36 + 14 + 21}{4} = \frac{118}{4} = 29.5$. Subtracting 29.5 from each of the x values results in the following deviations from the mean:

$$47 - 29.5 = 17.5$$
$$36 - 29.5 = 6.5$$
$$14 - 29.5 = -15.5$$
$$21 - 29.5 = -8.5$$

Square each deviation and add them to get the sum of the squared deviations:

$$\begin{aligned}\sum(x - \bar{x})^2 &= (17.5)^2 + (6.5)^2 + (-15.5)^2 + (-8.5)^2\\ &= 306.25 + 42.25 + 240.25 + 72.25\\ &= 661\end{aligned}$$

Then

$$\sqrt{\sum(x - \bar{x})^2} = \sqrt{661} = 25.710$$

Complete the calculation of s_b:

$$s_b = \frac{s_e}{\sqrt{\sum (x - \bar{x})^2}} = \frac{33}{25.710} = 1.28$$

Confidence Interval for the Slope of a Population Regression Line

$$b \pm (t \text{ critical value})s_b$$

To calculate a confidence interval for the slope of a population regression line, you need to know the value of the slope of the least-squares regression line, b, the standard deviation of the slope of the least-squares line, s_b, and the sample size n. You also need to know the confidence level, which determine the t critical value that is used to calculate the confidence interval.

Recall that an interval is defined by its endpoints—it has a lower endpoint and an upper endpoint. Notice that the expression for the confidence interval includes the symbol $\pm$, which is read as "plus and minus." This means that you will do two calculations, once using $+$ and once using $-$. Each one of these calculations results in one of the two interval endpoints.

To calculate the endpoints of the confidence interval, you can start by determining the appropriate t critical value for the specified confidence level. Your statistics textbook should have instructions for how this is done. Once you have determined the appropriate t critical value, you can substitute the values of b and s_b into the expression for the confidence interval.

For example, suppose you want to calculate a 95% confidence interval for the slope of a population regression line, and that you have already used sample data from $n = 20$ observations to calculate the value of the slope of the least-squares regression line $b = 6.85$ (see Section 4.6 if you want to review how to do this) and the standard deviation of the slope of the least-squares line, $s_b = 0.521$. To determine the t critical value, you first must calculate the degrees of freedom (df) associated with this interval. For a confidence interval for the slope of a population regression line, $df = n - 2$, which for this example would be $df = 20 - 2 = 18$. A table of t critical values or technology could then be used to find the t critical value. For a 95% confidence level and $df = 18$, the t critical value is 2.10.

Substituting values into the expression for the confidence interval results in the following:

$$b \pm (t \text{ critical value})s_b$$
$$6.85 \pm (2.10)(0.521)$$
$$6.85 \pm 2.621$$

To calculate the endpoints of the interval, you would evaluate

$$6.85 \pm 2.621$$

The lower endpoint of the interval is $6.85 - 2.621 = 4.229$. The upper endpoint of the interval is $6.85 + 2.621 = 9.471$.

The 95% confidence interval estimate of the slope of the population regression line is (4.229, 9.471). You would interpret this interval by saying that based on the data from the sample, you can be 95% confident that the actual value of the slope of the population regression line is somewhere between 4.229 and 9.471.

Check It Out!

a. Suppose that you are asked to calculate a 90% confidence interval for the slope of a population regression line, and that you have already used sample data from $n = 28$ observations to calculate the value of the slope of the least-squares regression line $b = 62.0$ and the standard deviation of the slope of the least-squares line, $s_b = 9.7$. Calculate the confidence interval and interpret the confidence interval in context.

b. Suppose that you wish to calculate a 95% confidence interval for the slope of a population regression line, and that you have already used sample data from $n = 41$ observations to calculate the value of the slope of the least-squares regression line $b = 7.32$ and the standard deviation of the slope of the least-squares line, $s_b = 1.58$. Calculate the confidence interval and interpret the confidence interval in context.

Answers

a. For a confidence interval for the slope of a population regression line, $df = n - 2$, which for this example would be $df = 28 - 2 = 26$. A table of t critical values or technology could then be used to find the t critical value. For a 90% confidence level and $df = 26$, the t critical value is 1.71.

Substituting values into the expression for the confidence interval results in the following:

$$b \pm (t \text{ critical value})s_b$$
$$62.0 \pm (1.71)(9.7)$$
$$62.0 \pm 16.587$$

The lower endpoint of the interval is $62.0 - 16.587 = 45.41$ (rounded to two decimal places). The upper endpoint of the interval is $62.0 + 16.587 = 78.59$.

The 90% confidence interval estimate of the slope of the population regression line is (45.41, 78.59). You would interpret this interval by saying that based on the data from the sample, you can be 90% confident that the actual value of the slope of the population regression line is somewhere between 45.41 and 78.59.

b. For a confidence interval for the slope of a population regression line, $df = n - 2$, which for this example would be $df = 41 - 2 = 39$. A table of t critical values or technology could then be used to find the t critical value. For a 95% confidence level and $df = 39$, the t critical value is 2.02.

Substituting values into the expression for the confidence interval results in the following:

$$b \pm (t \text{ critical value})s_b$$
$$7.32 \pm (2.02)(1.58)$$
$$7.32 \pm 3.192$$

The lower endpoint of the interval is $7.32 - 3.192 = 4.13$ (rounded to two decimal places). The upper endpoint of the interval is $7.32 + 3.192 = 10.51$.

The 95% confidence interval estimate of the slope of the population regression line is (4.13, 10.51). You would interpret this interval by saying that based on the data from the sample, you can be 95% confident that the actual value of the slope of the population regression line is somewhere between 4.13 and 10.51.

Test Statistic for Testing Hypotheses About the Slope of a Population Regression Line

$$t = \frac{b - \beta_0}{s_b}$$

A hypothesis test about the slope of a population regression line uses data from a sample to decide between two competing hypotheses—a null hypothesis (denoted by H_0) and an alternative hypothesis (denoted by H_a). The null hypothesis specifies a particular hypothesized value of the population slope, denoted by β_0. The null hypothesis will be of the form $H_0 : \beta = \beta_0$. For example, you might want to carry out a hypothesis test where the null hypothesis states that the slope of the population regression line is 0. The null hypothesis would then be $H_0 : \beta = 0$. For purposes of calculating the value of the test statistic, the value of β_0 for this null hypothesis is $\beta_0 = 0$.

To calculate the value of the test statistic, you need to know the values of the slope of the least-squares regression line, b, and the standard deviation of the slope of the least-squares regression line, s_b. Sometimes these values will be given, but other times you might need to calculate them from given data. For a review of how to calculate the value of b, see Section 4.6. The calculation of the standard deviation of the slope of the least-squares regression line appears earlier in this section.

Suppose that you want to test the null hypothesis $H_0 : \beta = 0$ and that a sample of size $n = 25$ resulted in a slope of $b = 3.4$ and a standard deviation for the slope of the least-squares line of $s_b = 1.69$. To calculate the value of the test statistic, substitute the values of b (the value of the slope of the least-squares line), s_b (the value of the standard deviation of the slope of the least-squares line), and β_0 (the hypothesized value from the null hypothesis) into the formula for the test statistic to obtain:

$$t = \frac{b - \beta_0}{s_b} = \frac{3.4 - 0}{1.69} = \frac{3.4}{1.69} = 2.01 \text{ (rounded to two decimal places)}$$

The value of the test statistic for this example is $t = 2.01$.

Check It Out!

a. Suppose that you want to test the null hypothesis $H_0 : \beta = 0$ and that a sample of size $n = 65$ resulted in a slope of $b = 56.9$ and a standard deviation for the slope of the least-squares line of $s_b = 11.8$. Calculate the value of the test statistic for testing this null hypothesis.

b. Suppose that you want to test the null hypothesis $H_0 : \beta = 0$ and that a sample of size $n = 37$ resulted in a slope of $b = -0.68$ and a standard deviation for the slope of the least-squares line of $s_b = 0.41$. Calculate the value of the test statistic for testing this null hypothesis.

Answers

a. To calculate the value of the test statistic, substitute the values of b (the value of the slope of the least-squares line), s_b (the value of the standard deviation of the slope of the least-squares line), and β_0 (the hypothesized value from the null hypothesis) into the formula for the test statistic to obtain

$$t = \frac{b - \beta_0}{s_b} = \frac{56.9 - 0}{11.8} = \frac{56.9}{11.8} = 4.82 \text{ (rounded to two decimal places)}$$

The value of the test statistic for this example is $t = 4.82$.

b. To calculate the value of the test statistic, substitute the values of b (the value of the slope of the least-squares line), s_b (the value of the standard deviation of the slope of the least-squares line), and β_0 (the hypothesized value from the null hypothesis) into the formula for the test statistic to obtain

$$t = \frac{b - \beta_0}{s_b} = \frac{-0.68 - 0}{0.41} = \frac{-0.68}{0.41} = -1.66 \text{ (rounded to two decimal places)}$$

The value of the test statistic for this example is $t = -1.66$.

SECTION 19.1 EXERCISES

19.1 The effects of grazing animals on grasslands have been the focus of numerous investigations by ecologists. One such study, reported in "The Ecology of Plants, Large Mammalian Herbivores, and Drought in Yellowstone National Park" (*Ecology* [1992]: 2043–2058), proposed using a simple linear regression model to relate y = green biomass concentration (g/cm^3) to x = elapsed time since snowmelt (in days). Suppose that the sum of the squared deviations is $\sum(x - \bar{x})^2 = 3.86$. The sample size is $n = 58$.

Also, suppose that the slope of the least-squares regression line is 0.640 g/cm^3 per day, and that the standard deviation about the least-squares line for this bivariate data set is $s_e = 0.41$.

a. Calculate the value of s_b, the standard deviation of the slope of the least-squares regression line.
b. Calculate a 95% confidence interval for the slope of a population regression line, and give an interpretation of the confidence interval, in context.

19.2 The article "Snow Cover and Temperature Relationships in North America and Eurasia" (*Journal of Climate and Applied Meteorology* [1983]: 460–469) explored the relationship between October–November continental snow cover (x, in millions of square km) and December–February temperature (y, in °C). There are $n = 13$ time periods represented by the Eurasia data (1969–1970, 1970–1971, … , 1981–1982).

Suppose that the sum of the squared deviations is $\sum(x - \bar{x})^2 = 0.172$.

Also, suppose that the slope of the least-squares regression line is -0.228 °C per million square km, and that the standard deviation about the least-squares line for this bivariate data set is $s_e = 0.047$.

a. Calculate the value of s_b, the standard deviation of the slope of the least-squares regression line.
b. Calculate the value of the test statistic that could be used to test the null hypothesis: $H_0 : \beta = 0$.

SECTION 19.2 Guided Practice—Confidence Interval for the Slope of a Population Regression Line

Consider this problem from a statistics textbook:

Acrylamide is a chemical that is sometimes found in cooked starchy foods and which is thought to increase the risk of certain kinds of cancer. The paper "A Statistical Regression Model for the Estimation of Acrylamide Concentrations in French Fries for Excess Lifetime Cancer Risk Assessment" (*Food and Chemical Toxicology* [2012]: 3867–3876) describes a study to investigate the effect of frying time (in seconds) and acrylamide concentration (in micrograms per kilogram) in batches of French fries. The data in the accompanying table are approximate values read from a graph that appeared in the paper for six batches of French fries.

Frying Time	Acrylamide Concentration
150	155
240	120
240	190
270	185
300	140
300	270

For these data, the estimated regression line for predicting y = acrylamide concentration based on x = frying time is $\hat{y} = 87 + 0.359x$, and the standard deviation of the slope of the least-squares line is $s_b = 0.438$. Calculate and interpret a 95% confidence interval for the slope of the population regression line.

If you were to encounter a problem like this, you should start by reading the problem carefully. You can use the three-read strategy introduced in Chapter 7.

First Read: Context

1. **What is the general context?**
   A study was conducted to see if there is a relationship between the frying time for French fries and the concentration of a chemical that is thought to increase cancer risk. The researchers recorded frying times and then measured the concentration of the chemical in the fries for each of six batches of French fries.
2. **Are there any terms that you don't understand?**
   This question is about constructing a confidence interval, so if you aren't comfortable with the meaning of this term you would want to go to the appropriate section in your textbook to review the definition. The term "acrylamide" is probably not familiar, but this term is defined in the problem.

Second Read: Tasks

1. **What question is being asked? What am I being asked to do?**
   This problem asks you to construct a 95% confidence interval estimate of the slope of the population regression line and to provide an interpretation of the interval.

Third Read: Details

1. **What data were collected? Was the data collected in a reasonable way?**
   Data were collected from six batches of French fries. For each batch, the frying time and the acrylamide concentration were measured. This results in bivariate data. The problem does not say how the batches were selected, so you would need to assume that the batches are representative of French fries that have been fried for various amounts of time.
2. **Are the actual data provided? What was the sample size (or the sample sizes if there are more than one sample)?**
   The sample size was $n = 6$ (corresponding to the six batches). The actual data, in six (frying time, concentration) pairs, are provided.
3. **Are any summary statistics or additional information given? If so, which statistics and what are the values of these statistics?**
   The slope of the equation of the least-squares regression line and the standard deviation of the slope of the least-squares regression line are both given. The equation of the least-squares regression line was $\hat{y} = 87 + 0.359x$, so the slope of the line is $b = 0.359$. The standard deviation of the slope of the least-squares regression line was $s_b = 0.438$.

Now you are ready to tackle the tasks required to complete this exercise. This problem asks you to construct a 95% confidence interval estimate of the slope of the least-squares regression line. You are also asked to interpret the interval.

If the assumptions of the linear regression model are reasonable, it is appropriate to use the given expression to calculate a confidence interval. To determine the t critical value, you first must calculate the degrees of freedom (df) associated with this interval. For a confidence interval for the slope of a population regression line, $df = n - 2$, which for this example would be $df = 6 - 2 = 4$. A table of t critical values or technology could then be used to find the t critical value. For a 95% confidence level and $df = 4$, the t critical value is 2.78. Substituting values into the expression for the confidence interval results in the following:

$$b \pm (t \text{ critical value})s_b$$
$$0.359 \pm (2.78)(0.438)$$
$$0.359 \pm 1.218$$

To calculate the endpoints of the interval, you would evaluate

$$0.359 \pm 1.218$$

The lower endpoint of the interval is $0.359 - 1.218 = -0.859$. The upper endpoint of the interval is $0.359 + 1.218 = 1.577$.

The 95% confidence interval estimate of the slope of the population regression line is $(-0.859, 1.577)$. You would interpret this interval by saying that based on the data from the sample, you can be 95% confident that the actual value of the slope of the population regression line is somewhere between -0.859 and 1.577. Because 0 is included in this interval, 0 is a plausible value for the slope of the population regression line. This is of interest because a slope of 0 indicates no linear relationship between frying time and acrylamide concentration.

SECTION 19.2 EXERCISES

19.3 The accompanying data were read from a plot (and are a subset of the complete data set) given in the article "Cognitive Slowing in Closed-Head Injury" (*Brain and Cognition* [1996]: 429–440). The data represent the mean response times for a group of individuals with closed-head injury (CHI) and a matched control group without head injury on 10 different tasks. Each observation was based on a different study and used different subjects, so it is reasonable to assume that the observations are independent.

Study	Control	CHI
1	250	303
2	360	491
3	475	659
4	525	683
5	610	922
6	740	1,044
7	880	1,421
8	920	1,329
9	1,010	1,481
10	1,200	1,815

For these data, the estimated regression line for predicting y = CHI based on x = Control is $\hat{y} = -96.7 + 1.59x$, and the standard deviation of the slope of the least-squares line is $s_b = 0.059$. Use the three-read strategy to understand the problem, and then calculate and interpret a 99% confidence interval for the slope of the population regression line.

19.4 The article "Photocharge Effects in Dye Sensitized Ag[Br,I] Emulsions at Millisecond Range Exposures" (*Photographic Science and Engineering* [1981]: 138–144) gave the accompanying data on x = % light absorption and y = peak photovoltage.

x	4.0	8.7	12.7	19.1	21.4	24.6	28.9	29.8	30.5
y	0.12	0.28	0.55	0.68	0.85	1.02	1.15	1.34	1.29

For these data, the estimated regression line for predicting y = peak photovoltage based on x = % light absorption is $\hat{y} = -0.08 + 0.045x$, and the standard deviation of the slope of the least-squares line is $s_b = 0.0022$. Use the three-read strategy to understand the problem, and then calculate and interpret a 95% confidence interval for the slope of the population regression line.

SECTION 19.3 Guided Practice—Testing Hypotheses About the Slope of a Population Regression Line

In this section, you will consider a typical situation that involves testing hypotheses about the slope of a population regression line.

Consider this problem from a statistics textbook:

> The accompanying data are a subset of data from the report "Great Jobs, Great Lives" (Gallup-Purdue Index 2015 Report, gallup.com/reports/197144/gallup-purdue-index-report-2015.aspx). The values are approximate values read from a scatterplot. Students at each of a number of universities were asked if they agreed that their education was worth the cost. One variable in the table is the *U.S. News and World Report* ranking of the university in 2015. The other variable in the table is the percentage of students at the university who responded, "strongly agree."

University Ranking	Percentage of Alumni Who Strongly Agree
28	53
29	58
30	62
37	55
45	54
47	62
52	55
54	62
57	70
60	58
65	66
66	55
72	65
75	57
82	67
88	59
98	75

For these data, the equation of the least-squares regression line is $\hat{y} = 51.305 + 0.163x$, where $\hat{y}$ is the predicted percentage of alumni who strongly agree that their education was worth the cost, and x is the 2015 university ranking. The standard deviation of the slope of the least-squares regression line was calculated to be $s_b = 0.064$. You may assume that the assumptions needed for the linear regression model to be appropriate are met.

Do the sample data support the hypothesis that there is a useful linear relationship between the percentage of alumni who strongly agree that their education was worth the cost and 2015 university ranking? Test the appropriate hypotheses using a significance level of $\alpha = 0.05$.

If you were to encounter a problem like this, you should start by reading the problem carefully. You can use the three-read strategy introduced in Chapter 7.

First Read: Context

1. **What is the general context?**
   Data were collected to investigate whether there is a relationship between university ranking and the percentage of students who think that their education was worth the cost. Students at each of 17 universities were asked if they thought their education was worth the cost, and the percentage who said yes was reported. The university ranking was also reported for each of these 17 universities.
2. **Are there any terms that you don't understand?**
   This question is about testing a hypothesis, so if you don't recall the steps required to carry out a hypothesis test, you would want to go to the appropriate section in your statistics textbook to review these steps. If you are interested in how university ranking was determined, you might look at the actual report to find out.

Second Read: Tasks

1. **What question is being asked? What am I being asked to do?**
   This problem asks you to decide if the sample data provide support for the hypothesis that there is a useful linear relationship between the percentage of alumni who strongly agree that their education was worth the cost and 2015 university ranking. A useful linear relationship is characterized by a slope that is not equal to 0, so you are being asked to carry out a test to see if there is convincing evidence that the slope of the population regression line is not equal to 0. You will use a significance level of 0.05.

Third Read: Details

1. **What data were collected. Were the data collected in a reasonable way?**
   The data are from 17 universities. For each university, the percentage in a sample of students who responded that they strongly agree that their education was worth the cost and the university ranking were recorded. From the problem description, it is not clear that it is reasonable to consider this group of universities as representative of universities in general. For this reason, you should be cautious about generalizing conclusions to a larger population.
2. **Are the actual data provided? What was the sample size (or the sample sizes if there are more than one sample)?**
   The sample size was $n = 17$ (the 17 universities for which data are given). The actual data (the 17 (university ranking, percentage who strongly agree) pairs) are given.
3. **Are any summary statistics or additional information given? If so, which statistics and what are the values of these statistics?**
   The slope of the equation of the least-squares regression line and the standard deviation of the slope of the least-squares regression line are both given. The equation of the least-squares regression line was $\hat{y} = 51.305 + 0.163x$, so the slope of the line is $b = 0.163$. The standard deviation of the slope of the least-squares regression line was $s_b = 0.064$.

Now you are ready to tackle the tasks required to complete this exercise. The problem states "Do the sample data support the hypothesis that there is a useful linear relationship between the percentage of alumni who strongly agree that their education was worth the cost and 2015 university ranking? Test the appropriate hypotheses using a significance level of $\alpha = 0.05$." This is asking you to test hypotheses about the slope of a population regression line, β. You are asked to decide if there is convincing evidence that there is a useful linear relationship between the percentage of alumni who strongly agree that their education was worth the cost and university ranking. You are also told to use a significance level of 0.05 when carrying out the test.

Many textbooks recommend a sequence of steps for carrying out a hypothesis test. When you carry out a hypothesis test, you should follow the recommended sequence of steps. Although the order of the steps may vary somewhat from one text to another, they generally include the following:

Hypotheses: You will need to specify the null and the alternative hypothesis for the hypothesis test and define any symbols that are used in the context of the problem.

Method: You will need to specify what method you plan to use to carry out the test. For example, you might say that you plan to use the *t* test for the slope of a population regression line, if that is a test that would allow you to answer the question posed. You will also need to specify the significance level you will use to reach a conclusion.

Check Conditions: Many hypothesis test procedures are only appropriate when certain conditions are met. Once you have determined the method you plan to use, you will need to verify that any required conditions are met.

Calculate: To reach a conclusion in a hypothesis test, you will need to calculate the value of a test statistic and the associated *P*-value.

Communicate Results: The final step is to use the computed values of the test statistic and the *P*-value to reach a conclusion and to provide an answer to the question posed. You should always provide a conclusion in words.

These steps are illustrated here for the given example.

> **Hypotheses**: You want to determine if there is convincing evidence that there is a useful linear relationship between the percentage of alumni who strongly agree that their education was worth the cost and university ranking. You can use β to represent the slope of the population regression line. The question being asked usually translates into the alternative hypothesis, so the alternative hypothesis would be $H_a: \beta \neq 0$ because a

slope of 0 indicates no linear relationship. The null hypothesis would then be obtained by replacing the ≠ in the alternative hypothesis with an =. This results in hypotheses of

$$H_0: \beta = 0$$
$$H_a: \beta \neq 0$$

Method: Because the hypotheses are about the slope of a population regression line, you would consider the t test for the slope of a population regression line. The test statistic is

$$t = \frac{b - 0}{s_b}$$

A significance level of $\alpha = 0.05$ was specified.

Check: The problem specified that you can assume that the assumptions of the linear regression model are met.

Calculations: The value of the slope of the least-squares regression line is $b = 0.163$, and the value of the standard deviation of the slope of the least-squares line is $s_b = 0.064$. You also know that the sample size was $n = 17$. The value of the test statistic is

$$t = \frac{b - 0}{s_b} = \frac{0.163 - 0}{0.064} = \frac{0.163}{0.064} = 2.55 \text{ (rounded to two decimal places)}$$

Once you know the value of the test statistic, you need to find the associated P-value. Your statistics text will have an explanation of how the P-value is determined.

This is a two-tailed test (the inequality in H_a is ≠), so the P-value is two times an area under a t curve. The degrees of freedom (df) associated with a t test for the slope is $n - 2$, so $df = 17 - 2 = 15$. The P-value is 2 times the area under the t curve with $df = 15$ and to the right of 2.55. The P-value is $2 \cdot P(t \geq 2.55) = 0.022$.

Communicate Results: Because the P-value of 0.022 is less than the significance level of 0.05, we reject H_0. There is convincing evidence of a useful linear relationship between the percentage of alumni who strongly agree that their education was worth the cost and the 2015 university ranking.

SECTION 19.3 EXERCISES

19.5 The paper "Physiological Characteristics and Performance of Top U.S. Biathletes" (*Medicine and Science in Sports and Exercise* [195]: 1302–1310) describes a study of the relationship between cardiovascular fitness, as measured by x = time (in minutes) to exhaustion running on a treadmill and y = performance (in minutes) on a 20 km ski race. Data on 11 athletes were collected, and are displayed in the accompanying table:

Treadmill	Ski Time
7.7	71.0
8.4	71.4
8.7	65.0
9.0	68.7
9.6	64.4
9.6	69.4
10.0	63.0
10.2	64.6
10.4	66.9
11.0	62.6
11.7	61.7

For these data, the equation of the least-squares regression line is $\hat{y} = 88.80 - 2.334x$, where $\hat{y}$ is the predicted finishing time in a 20 km ski race, and x is the time to exhaustion running on a treadmill. The standard deviation of the slope of the least-squares regression line was calculated to be $s_b = 0.591$. You may assume that the assumptions needed for the linear regression model to be appropriate are met.

Do the sample data support the hypothesis that there is a useful linear relationship between time to exhaustion running on a treadmill and finishing time in a 20 km ski race? Use the three-read strategy to understand the problem, and then test the appropriate hypotheses using a significance level of $\alpha = 0.05$.

19.6 The authors of the paper "Weight-Bearing Activity During Youth Is a More Important Factor for Peak Bone Mass than Calcium Intake" (*Journal of Bone and Mineral Research* [1994], 1089–1096) studied a number of variables they thought might be related to bone mineral density (BMD). The accompanying data on x = weight at age 13 and y = bone mineral density at age 27 are consistent with summary quantities for women given in the paper.

Weight (kg)	BMD (g/cm³)
54.4	1.15
59.3	1.26
74.6	1.42
62.0	1.06
73.7	1.44
70.8	1.02
66.8	1.26
66.7	1.35
64.7	1.02
71.8	0.91
69.7	1.28
64.7	1.17
62.1	1.12
68.5	1.24
58.3	1.00

For these data, the equation of the least-squares regression line is $\hat{y} = 0.558 + 0.0094x$, where $\hat{y}$ is the predicted bone mineral density at age 27, and x is the weight at age 13. The standard deviation of the slope of the least-squares regression line was calculated to be $s_b = 0.0071$. You may assume that the assumptions needed for the linear regression model to be appropriate are met.

Do the sample data support the hypothesis that there is a useful linear relationship between weight at age 13 and bone mineral density at age 27? Use the three-read strategy to understand the problem, and then test the appropriate hypotheses using a significance level of $\alpha = 0.10$.

20 Testing Hypotheses About More than Two Means—The Math You Need to Know

SECTION 20.1 Evaluating Expressions

There are many different expressions that you will encounter as you learn about methods for testing hypotheses about more than two population or treatment means.

Note: Depending on your access to technology, such as a graphing calculator or a statistics software package, you may not need to evaluate most of these expressions by hand. But just in case your statistics class is not relying on technology to do these calculations, this section looks at how to evaluate the expressions.

The following data set will be used to illustrate how the expressions considered in this section are evaluated. These data are measured responses from an experiment in which 27 people were assigned at random to one of three experimental treatment groups. The overall mean listed in the table is the mean of all 27 responses, and is denoted by $\bar{\bar{x}}$.

Treatment Group	Response									Sample Mean	Sample Standard Deviation
1	4	7	7	4	5	3	4	7	6	$\bar{x}_1 = 5.2$	$s_1 = 1.56$
2	4	1	3	2	6	2	5	3	4	$\bar{x}_2 = 3.3$	$s_2 = 1.58$
3	3	4	5	6	5	4	2	4	4	$\bar{x}_3 = 4.1$	$s_3 = 1.17$
										Overall mean $= \bar{\bar{x}} = 4.2$	

Sums of Squares

There are three different sums of squares that you might need to calculate when carrying out a hypothesis test about more than two means—Sum of Squares for Treatments (*SSTr*), Sum of Squared Error (*SSE*), and Total Sum of Squares (*SSTo*), which is just the sum of $SSTr$ and SSE. The expressions for these sums of squares use the following notation:

Notation	Meaning	Example
k	Number of samples or treatments	For the given example, there are three treatment groups, so $k = 3$.
$\bar{x}_1, \bar{x}_2, \cdots, \bar{x}_k$	The k sample or treatment means. $\bar{x}_1$ is the mean for sample or treatment 1, $\bar{x}_2$ is the mean for sample or treatment 2, and so on.	For the given example, there are three treatment groups, and the three treatment means were given in the table: $\bar{x}_1 = 5.2$, $\bar{x}_2 = 3.3$, $\bar{x}_3 = 4.1$
N	The total number of observations in all the samples or treatment groups combined.	For the given example, there were three treatment groups and each group has 9 observations. The total number of observations in all groups combined is $N = 27$.
$\bar{\bar{x}}$	The mean of all the observations in all the samples or treatments combined. $\bar{\bar{x}}$ is calculated by adding all the data values and dividing by the total number of observations.	For the given example, there are a total of 27 observations. The sum of all 27 observations is $\sum x = 114$, so $\bar{\bar{x}} = \frac{114}{27} = 4.2$ (rounded to one decimal place).
$s_1, s_2, \cdots, s_k$	The k sample or treatment standard deviations. s_1 is the standard deviation for sample or treatment 1, s_2 is the standard deviation for sample or treatment 2, and so on.	For the given example, there are three treatment groups, and the three treatment standard deviations were given in the table: $s_1 = 1.56$, $s_2 = 1.58$, $s_3 = 1.17$

Sum of Squares for Treatments (*SSTr*)

$$SSTr = n_1(\bar{x}_1 - \bar{\bar{x}})^2 + n_2(\bar{x}_2 - \bar{\bar{x}})^2 + \cdots + n_k(\bar{x}_k - \bar{\bar{x}})^2$$

To calculate the value of $SSTr$, you need to know the values of the sample or treatment means, $\bar{x}_1, \bar{x}_2, \cdots, \bar{x}_k$, and the value of the overall mean, $\bar{\bar{x}}$. It is helpful to use a table to organize the calculation of $SSTr$, so begin by setting up a table like the one below that includes the sample or treatment means and the sample sizes.

Sample or Treatment	$\bar{x}$	$(\bar{x} - \bar{\bar{x}})$	$(\bar{x} - \bar{\bar{x}})^2$	n	$n(\bar{x} - \bar{\bar{x}})^2$
1	5.2			9	
2	3.3			9	
3	4.1			9	

You can then begin to fill in the other columns in the table. The overall mean for the given data set was $\bar{\bar{x}} = 4.2$, so you can fill in the $(\bar{x} - \bar{\bar{x}})$ column by subtracting $\bar{\bar{x}} = 4.2$ from each of the three treatment means.

Sample or Treatment	$\bar{x}$	$(\bar{x} - \bar{\bar{x}})$	$(\bar{x} - \bar{\bar{x}})^2$	n	$n(\bar{x} - \bar{\bar{x}})^2$
1	5.2	$5.2 - 4.2 = 1.0$		9	
2	3.3	$3.3 - 4.2 = -0.9$		9	
3	4.1	$4.1 - 4.2 = -0.1$		9	

Next, calculate the values for the $(\bar{x} - \bar{\bar{x}})^2$ column by squaring each of the $(\bar{x} - \bar{\bar{x}})$ values.

Sample or Treatment	$\bar{x}$	$(\bar{x} - \bar{\bar{x}})$	$(\bar{x} - \bar{\bar{x}})^2$	n	$n(\bar{x} - \bar{\bar{x}})^2$
1	5.2	$5.2 - 4.2 = 1.0$	$(1.0)^2 = 1.00$	9	
2	3.3	$3.3 - 4.2 = -0.9$	$(-0.9)^2 = 0.81$	9	
3	4.1	$4.1 - 4.2 = -0.1$	$(-0.1)^2 = 0.01$	9	

The values for the last column in the table can now be calculated by multiplying the sample sizes by the corresponding $(\bar{x} - \bar{\bar{x}})^2$ values in the table, as shown here:

Sample or Treatment	$\bar{x}$	$(\bar{x} - \bar{\bar{x}})$	$(\bar{x} - \bar{\bar{x}})^2$	n	$n(\bar{x} - \bar{\bar{x}})^2$
1	5.2	$5.2 - 4.2 = 1.0$	$(1.0)^2 = 1.00$	9	$9(1.00) = 9.00$
2	3.3	$3.3 - 4.2 = -0.9$	$(-0.9)^2 = 0.81$	9	$9(0.81) = 7.29$
3	4.1	$4.1 - 4.2 = -0.1$	$(-0.1)^2 = 0.01$	9	$9(0.01) = 0.09$

The expression for *SSTr* is

$$SSTr = n_1(\bar{x}_1 - \bar{\bar{x}})^2 + n_2(\bar{x}_2 - \bar{\bar{x}})^2 + \cdots + n_k(\bar{x}_k - \bar{\bar{x}})^2$$

This is the sum of the values in the last column of the table you used to organize the calculations, so

$$\begin{aligned} SSTr &= n_1(\bar{x}_1 - \bar{\bar{x}})^2 + n_2(\bar{x}_2 - \bar{\bar{x}})^2 + \cdots + n_k(\bar{x}_k - \bar{\bar{x}})^2 \\ &= 9.00 + 7.29 + 0.09 \\ &= 16.38 \end{aligned}$$

Check It Out!

A study was conducted by randomly assigning 24 human subjects to four treatment groups of six subjects each. A score was measured for the subjects at the conclusion of the study.

Treatment Group	Scores						Sample Mean	Sample Standard Deviation
1	34.0	43.1	33.0	34.5	36.7	37.1	$\bar{x}_1 = 36.42$	$s_1 = 3.65$
2	37.9	28.6	27.8	25.2	28.7	29.0	$\bar{x}_2 = 29.53$	$s_2 = 4.31$
3	24.4	22.4	28.2	23.4	15.9	31.7	$\bar{x}_3 = 24.34$	$s_3 = 5.37$
4	35.9	31.7	28.2	29.9	27.1	30.4	$\bar{x}_4 = 30.52$	$s_4 = 3.11$
							Overall mean $= \bar{\bar{x}} = 30.20$	

Calculate the value of the sum of squares for treatments (SSTr).

Answer

Begin by filling in the table below. The overall mean for the given data set was $\bar{\bar{x}} = 30.20$, so you can fill in the $(\bar{x} - \bar{\bar{x}})$ column by subtracting $\bar{\bar{x}} = 30.20$ from each of the treatment means.

Group	$\bar{x}$	$(\bar{x} - \bar{\bar{x}})$	$(\bar{x} - \bar{\bar{x}})^2$	n	$n(\bar{x} - \bar{\bar{x}})^2$
1	36.42			6	
2	29.53			6	
3	24.34			6	
4	30.52			6	

The completed table is shown below.

Group	$\bar{x}$	$(\bar{x} - \bar{\bar{x}})$	$(\bar{x} - \bar{\bar{x}})^2$	n	$n(\bar{x} - \bar{\bar{x}})^2$
1	36.42	$36.42 - 30.20 = 6.22$	$6.22^2 = 38.688$	6	$6(38.688) = 232.128$
2	29.53	$29.53 - 30.20 = -0.67$	$(-0.67)^2 = 0.449$	6	$6(0.449) = 2.694$
3	24.34	$24.34 - 30.20 = -5.86$	$(-5.86)^2 = 34.340$	6	$6(34.340) = 206.040$
4	30.52	$30.52 - 30.20 = 0.32$	$0.32^2 = 0.102$	6	$6(0.102) = 0.612$

The expression for *SSTr* is

$$SSTr = n_1(\bar{x}_1 - \bar{\bar{x}})^2 + n_2(\bar{x}_2 - \bar{\bar{x}})^2 + \cdots + n_k(\bar{x}_k - \bar{\bar{x}})^2$$

This is the sum of the values in the last column of the table you used to organize the calculations, so

$$\begin{aligned} SSTr &= n_1(\bar{x}_1 - \bar{\bar{x}})^2 + n_2(\bar{x}_2 - \bar{\bar{x}})^2 + \cdots + n_k(\bar{x}_k - \bar{\bar{x}})^2 \\ &= 232.128 + 2.694 + 206.040 + 0.612 \\ &= 441.47 \end{aligned}$$

Sum of Squared Error (*SSE*)

$$SSE = (n_1 - 1)s_1^2 + (n_2 - 1)s_2^2 + \cdots + (n_k - 1)s_k^2$$

To calculate the value of *SSE*, you need to know the values of the sample or treatment standard deviations, $s_1, s_2, \ldots, s_k$, and the sample sizes. It may be helpful to use a table to organize the calculation of *SSE*, so begin by setting up a table like the one below that includes the sample or treatment standard deviations and the sample sizes.

Sample or Treatment	s	s^2	n	$n-1$	$(n-1)s^2$
1	1.56		9		
2	1.58		9		
3	1.17		9		

You can then begin to fill in the other columns in the table. Squaring each of the standard deviations will produce the entries for the s^2 column. The values for the $n - 1$ column are obtained by subtracting 1 from each of the sample sizes. The values for the last column in the table can now be calculated by multiplying the value in the s^2 column by the corresponding value in the $n - 1$ column, as shown here:

Sample or Treatment	s	s^2	n	$n-1$	$(n-1)s^2$
1	1.56	2.434	9	8	8(2.434) = 19.472
2	1.58	2.496	9	8	8(2.496) = 19.968
3	1.17	1.369	9	8	8(1.369) = 10.952

The expression for *SSE* is

$$SSE = (n_1 - 1)s_1^2 + (n_2 - 1)s_2^2 + \cdots + (n_k - 1)s_k^2$$

This is the sum of the values in the last column of the table you used to organize the calculations, so

$$\begin{aligned} SSE &= (n_1 - 1)s_1^2 + (n_2 - 1)s_2^2 + \cdots + (n_k - 1)s_k^2 \\ &= 19.472 + 19.968 + 10.952 \\ &= 50.392 \end{aligned}$$

Check It Out!

Use the following table to calculate the value of the Sum of Squared Error (*SSE*), and check your work.

Group	s	s^2	n	$n-1$	$(n-1)s^2$
1	3.65		6		
2	4.31		6		
3	5.37		6		
4	3.11		6		

Answer

The completed table is shown below.

Group	s	s^2	n	$n - 1$	$(n - 1)s^2$
1	3.65	$3.65^2 = 13.322$	6	5	$5(13.322) = 66.610$
2	4.31	$4.31^2 = 18.576$	6	5	$5(18.576) = 92.880$
3	5.37	$5.37^2 = 28.837$	6	5	$5(28.837) = 144.185$
4	3.11	$3.11^2 = 9.672$	6	5	$5(9.672) = 48.360$

The expression for SSE is

$$SSE = (n_1 - 1)s_1^2 + (n_2 - 1)s_2^2 + \cdots + (n_k - 1)s_k^2$$

This is the sum of the values in the last column of the table you used to organize the calculations, so

$$\begin{aligned} SSE &= (n_1 - 1)s_1^2 + (n_2 - 1)s_2^2 + \cdots + (n_k - 1)s_k^2 \\ &= 66.610 + 92.880 + 144.185 + 48.360 \\ &= 352.04 \end{aligned}$$

Analysis of Variance *F* Test Statistic

$$F = \frac{MSTr}{MSE} = \frac{SSTr/(k-1)}{SSE/(N-k)}$$

An easy way to calculate the value of the analysis of variance F statistic is to first calculate the mean square for treatments ($MSTr$) and the mean square error (MSE). These values are the numerator and the denominator of the expression for the F statistic. Each mean square is a sum of squares divided by an associated degrees of freedom. For the example being used to illustrate the calculations in this section, you now already have the following values (N is the overall sample size when all samples are combined):

$$\begin{aligned} SSTr &= 16.38 \\ SSE &= 50.392 \\ N &= 27 \end{aligned}$$

You can use these values to calculate the two mean squares needed for the F statistic:

$$MSTr = \frac{SSTr}{k-1} = \frac{16.38}{3-1} = \frac{16.38}{2} = 8.19$$

$$MSE = \frac{SSE}{N-k} = \frac{50.392}{27-3} = \frac{50.392}{24} = 2.10$$

The value of the F statistic is then

$$F = \frac{MSTr}{MSE} = \frac{8.19}{2.10} = 3.90 \text{ (rounded to two decimal places)}$$

Check It Out!

Calculate the value of the F statistic, using the values of the $SSTr$ and SSE that you found in the previous Check It Out! examples. The overall sample size when all of the groups are combined is $N = 24$.

Answer

From the previous Check it Out! examples, $SSTr = 441.47$ and $SSE = 352.04$. Begin by calculating the mean square for treatments ($MSTr$) and the mean square error (MSE):

$$MSTr = \frac{SSTr}{k-1} = \frac{441.47}{4-1} = \frac{441.47}{3} = 147.157$$

$$MSE = \frac{SSE}{N-k} = \frac{352.04}{24-4} = \frac{352.04}{20} = 17.60$$

The value of the F statistic is then

$$F = \frac{MSTr}{MSE} = \frac{147.157}{17.60} = 8.36 \text{ (rounded to two decimal places)}$$

SECTION 20.1 EXERCISES

20.1 Suppose that a random sample of size $n = 5$ was selected from the vineyard properties for sale in Sonoma County, California, in each of 3 years. The following data are consistent with summary information on price per acre (in dollars, rounded to the nearest thousand) for disease-resistant grape vineyards in Sonoma County (*Wines and Vines*, November 1999):

Year	Observations				
1996	30,000	34,000	36,000	38,000	40,000
1997	30,000	35,000	37,000	38,000	40,000
1998	40,000	41,000	43,000	44,000	50,000

Calculate the mean square for treatments ($MSTr$), the mean square error (MSE), and the value for the F statistic.

20.2 Consider the accompanying data on plant growth after the application of five different types of growth hormone.

Hormone	Observations			
1	13	17	7	14
2	21	13	20	17
3	18	14	17	21
4	7	11	18	10
5	6	11	15	8

Calculate the mean square for treatments ($MSTr$), the mean square error (MSE), and the value for the F statistic.

SECTION 20.2 Guided Practice—One-Way Analysis of Variance

In this section, you will consider a typical situation that involves using independent samples to test hypotheses about more than two means.

Consider this problem from a statistics textbook:

Do people feel hungrier after sampling a healthy food? The authors of the paper "When Healthy Food Makes You Hungry" (*Journal of Consumer Research* [2010]: S34–S44) carried out a study to answer this question. They randomly assigned volunteers into one of three groups. The people in the first group were asked to taste a snack that was billed as a new health bar containing high levels of protein, vitamins, and fiber. The people in the second group were asked to taste the same snack, but were told it was a tasty chocolate bar with a raspberry center. After tasting the snack, participants were asked to rate their hunger level on a scale from 1 (not at all hungry) to 7 (very hungry). The people in the third group were asked to rate their hunger but were not given a snack.

The data in the accompanying table are consistent with summary quantities given in the paper (although the sample sizes in the actual study were larger). Do these data provide evidence that the mean hunger rating is not the same for all three treatments ("healthy" snack, "tasty" snack, no snack)? Test the relevant hypotheses using a significance level of 0.05.

Treatment Group	Hunger Rating									Sample Mean	Sample Standard Deviation
Healthy	4	7	7	4	5	3	4	7	6	$\bar{x}_1 = 5.2$	$s_1 = 1.56$
Tasty	4	1	3	2	6	2	5	3	4	$\bar{x}_2 = 3.3$	$s_2 = 1.58$
No Snack	3	4	5	6	5	4	2	4	4	$\bar{x}_3 = 4.1$	$s_3 = 1.17$
										Overall mean $= \bar{\bar{x}} = 4.2$	

If you were to encounter a problem like this, you should start by reading the problem carefully. You can use the three-read strategy introduced in Chapter 7.

First Read: Context

1. **What is the general context?**
   Researchers wondered if people feel hungrier after eating a food that they think is healthy. They collected data by carrying out an experiment with three experimental groups (treatments). One group ate a snack that was described as healthy, one group ate the same snack, but it was described as a tasty chocolate bar, and the third group did not eat a snack. Participants rated their hunger on a scale from 1 to 7.
2. **Are there any terms that you don't understand?**
   This question is about testing a hypothesis, so if you don't recall the steps required to carry out a hypothesis test, you would want to go to the appropriate section in your statistics textbook to review these steps.

Second Read: Tasks

1. **What question is being asked? What am I being asked to do?**
   This problem asks you carry out a hypothesis test to determine if the data provide evidence that the means for hunger rating are not all the same for the three treatments ("healthy" snack, "tasty" snack, no snack). You are also told you should use a significance level of $\alpha = 0.05$.

Third Read: Details

1. **What data were collected? Were the data collected in a reasonable way?**
   There were 27 volunteers who participated in the experiment. These people were randomly assigned to one of the three treatment groups.
2. **Are the actual data provided? What was the sample size (or the sample sizes if there are more than one sample)?**
   The sample size for treatment group 1 (healthy snack) was $n_1 = 9$. The sample size for treatment group 2 (tasty snack) was $n_2 = 9$. The sample size for treatment group 3 (no snack) was $n_3 = 9$. The actual data (the nine hunger ratings for each treatment group) are provided.
3. **Are any summary statistics or additional information given? If so, which statistics and what are the values of these statistics?**
   The treatment means and the treatment sample standard deviations are given. For these samples, $\bar{x}_1 = 5.2$, $\bar{x}_2 = 3.3$, and $\bar{x}_3 = 4.1$, and $s_1 = 1.56$, $s_2 = 1.58$, and $s_3 = 1.17$. The overall mean, $\bar{\bar{x}} = 4.2$, is also given.

Now you are ready to tackle the tasks required to complete this exercise. This problem is asking you to test hypotheses about a difference in population means using data from independent samples. You are asked to determine if there are differences among the means for hunger rating for the three snack treatment groups.

Many statistics textbooks recommend a sequence of steps for carrying out a hypothesis test. When you carry out a hypothesis test, you should follow the recommended sequence of steps. Although the order of the steps may vary somewhat from one text to another, they generally include the following:

Hypotheses: You will need to specify the null and the alternative hypothesis for the hypothesis test and define any symbols that are used in the context of the problem.

Method: You will need to specify what method you plan to use to carry out the test. For example, you might say that you plan to use the analysis of variance F test if that is a test that would allow you to answer the question posed. You will also need to specify the significance level you will use to reach a conclusion.

Check Conditions: Many hypothesis test procedures are only appropriate when certain conditions are met. Once you have determined the method you plan to use, you will need to verify that any required conditions are met.

Calculate: To reach a conclusion in a hypothesis test, you will need to calculate the value of a test statistic and the associated P-value.

Communicate Results: The final step is to use the computed values of the test statistic and the P-value to reach a conclusion and to provide an answer to the question posed. You should always provide a conclusion in words.

These steps are illustrated here for the given example.

Hypotheses: You want to determine if there is convincing evidence that the means for hunger rating are not all the same for the three treatments ("healthy" snack, "tasty" snack, no snack). You can use μ_1 to represent the mean hunger rating for treatment 1 (healthy snack), μ_2 to represent the mean hunger rating for treatment 2 (tasty snack), and μ_3 to represent the mean hunger rating for treatment 3 (no snack). The question being asked usually translates into the alternative hypothesis. Here the question asks if the means for hunger rating are not all equal, so the alternative hypothesis would be H_a : not all three treatment means are equal. The null hypothesis would specify that all three treatment means are equal, so $H_0 : \mu_1 = \mu_2 = \mu_3$. This results in hypotheses of

$H_0 : \mu_1 = \mu_2 = \mu_3$
H_a : at least two of the treatment means are different

Method: Because the hypotheses are about more than two treatment means, you would consider the analysis of variance F test. The test statistic is

$$F = \frac{MSTr}{MSE} = \frac{SSTr/(k-1)}{SSE/(N-k)}$$

A significance level of $\alpha = 0.05$ was specified.

Check: There are three conditions necessary for the analysis of variance F test. The volunteers in the experiment were randomly assigned to treatment groups, so the random sample or random assignment condition is met. The treatment sample standard deviations are not very different and the largest standard deviation (1.58) is not more than twice as large as the smallest (1.17), so the assumption of equal standard deviations seems reasonable. The treatment sample sizes are not greater than or equal to 30, so you must be willing to assume that the population response distributions are approximately normal for each of the three treatments. Boxplots (shown below) of the data for each group are roughly symmetric and there are no outliers, so it is reasonable to think that the normality condition is also plausible.

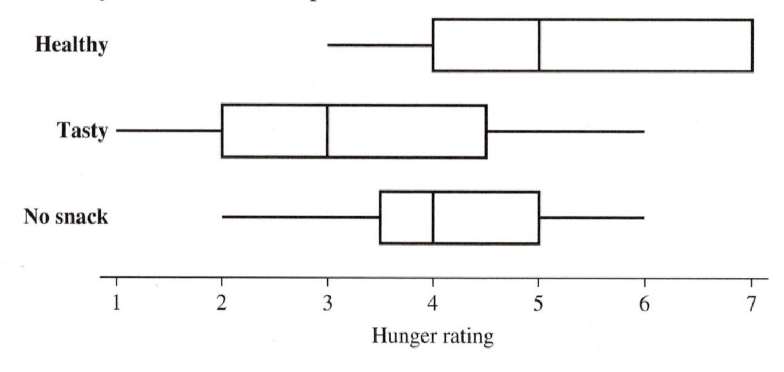

Calculations: To complete the calculations for an analysis of variance F test, you will need to calculate the values of *SSTr*, *SSE*, *MSTr*, and *MSE*. The data for this example were used in Section 20.1 to illustrate these calculations. This is the place in the hypothesis testing process where you would need to carry out those calculations. From Section 20.1, you know the following:

$$SSTr = 16.38$$
$$SSE = 50.392$$
$$k = 3$$

$N = 27$ (N is the overall sample size when all samples are combined)

$$MSTr = \frac{SSTr}{k-1} = \frac{16.38}{3-1} = \frac{16.38}{2} = 8.19$$

$$MSE = \frac{SSE}{N-k} = \frac{50.392}{27-3} = \frac{50.392}{24} = 2.10$$

$$F = \frac{MSTr}{MSE} = \frac{8.19}{2.10} = 3.90 \text{ (rounded to two decimal places)}$$

Once you know the value of the test statistic, you need to find the associated P-value. Your statistics textbook will have an explanation of how the P-value is determined. The analysis of variance F test is an upper-tailed test, so the P-value is an area under a F curve. The numerator degrees of freedom associated with this test is numerator $df = k - 1 = 2$. The denominator degrees of freedom associated with this test is denominator $df = N - k = 27 - 2 = 24$. The P-value is the area under the F curve with numerator $df = 2$, denominator $df = 24$, and to the right of 3.90. For this test, the P-value is $P(F \geq 3.90) = 0.034$.

Communicate Results: Because the P-value of 0.034 is less than the selected significance level of 0.05, you reject the null hypothesis. There is convincing evidence that the means for hunger rating are not all the same for the three treatments groups—"healthy" snack, "tasty" snack, and no snack. What you can't tell yet is which treatments were different—you just know they are not all the same. The next thing that the researchers would want to do is to investigate where the meaningful differences were. It is interesting to note that the greatest mean hunger rating was not for those who did not eat a snack, but rather for those who ate what they thought was a healthy snack.

SECTION 20.2 EXERCISES

20.3 The article "Compression of Single-Wall Corrugated Shipping Containers Using Fixed and Floating Text Platens" (*Journal of Testing and Evaluation* [1992]: 318–320) described an experiment in which several different types of boxes were compared with respect to compression strength (in pounds). The accompanying data resulted from an experiment involving $k = 4$ types of boxes. Do these data provide evidence to support the claim that the means for compression strength are not all the same for the four box types? Use the three-read strategy to understand the problem, and then test the relevant hypothesis using a significance level of 0.05.

Type of Box	Compression Strength (lb)
1	655.5 788.3 734.3 721.4 679.1 699.4
2	789.2 772.5 786.9 686.1 732.1 774.8
3	737.1 639.0 696.3 671.7 717.2 727.1
4	535.1 628.7 542.4 559.0 586.9 520.0

20.4 The accompanying data on calcium content of wheat are consistent with summary quantities that appeared in the article "Mineral Contents of Cereal Grains as Affected by Storage and Insect Infestation" (*Journal of Stored Products Research* [1992]: 147–151). Four different storage times were considered.

Storage Time	Observations
0 months	58.75 57.94 58.91 56.85 55.21 57.30
1 month	58.87 56.43 56.51 57.67 59.75 58.48
2 months	59.13 60.38 58.01 59.95 59.51 60.34
4 months	62.32 58.76 60.03 59.36 59.61 61.95

Do these data provide evidence that the means for calcium content are not all the same for the four storage periods? Use the three-read strategy to understand the problem, and then test the relevant hypotheses using a significance level of 0.01.

Answers to the Odd Exercises

CHAPTER 1

Getting Ready for Statistics

SECTION 1.1

1.1 2

1.3 4

1.5 −2

1.7 3

1.9 −17

1.11 6

1.13 47

1.15 −64

1.17 1

1.19 36

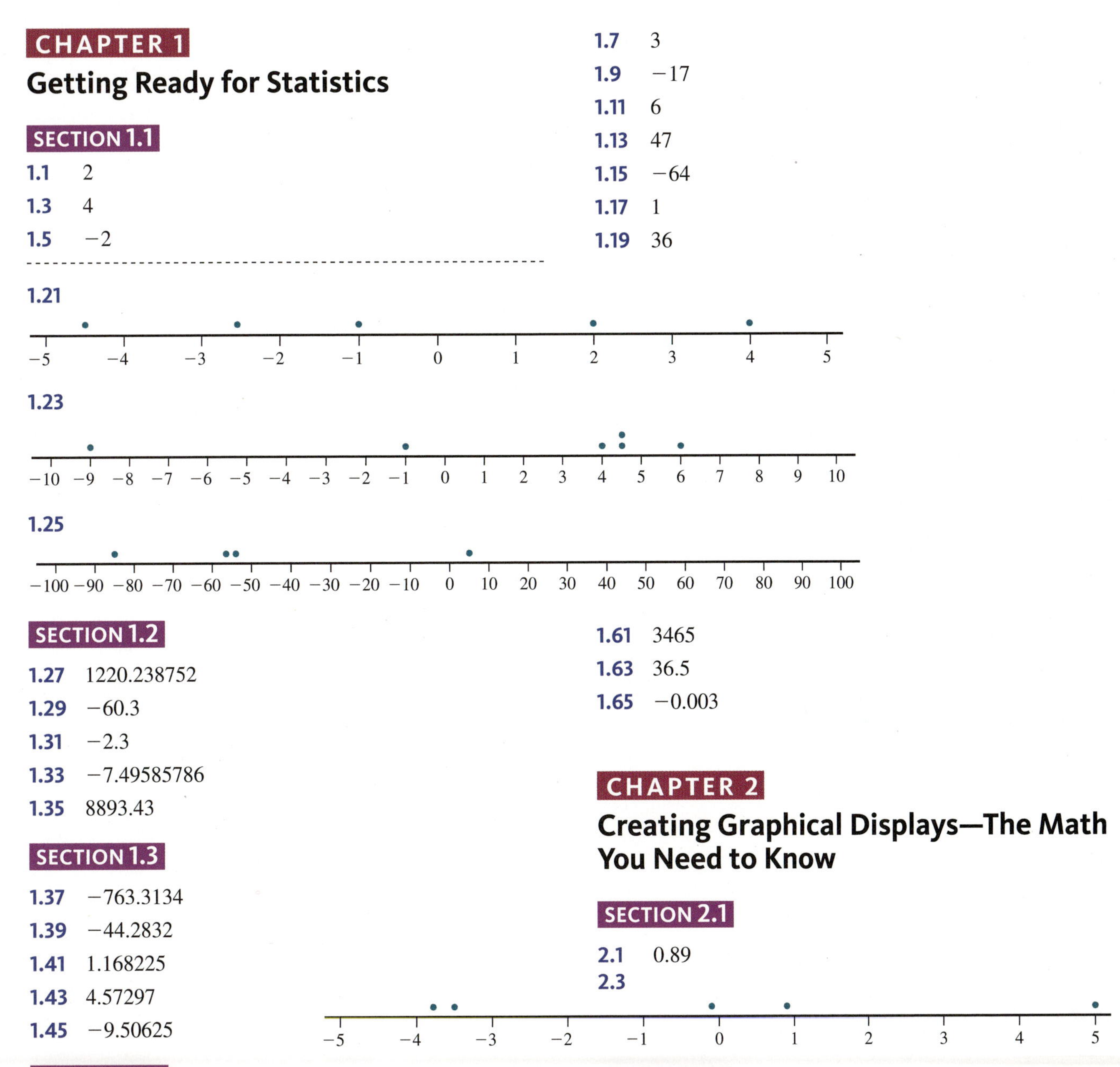

1.21

1.23

1.25

SECTION 1.2

1.27 1220.238752

1.29 −60.3

1.31 −2.3

1.33 −7.49585786

1.35 8893.43

SECTION 1.3

1.37 −763.3134

1.39 −44.2832

1.41 1.168225

1.43 4.57297

1.45 −9.50625

SECTION 1.4

1.47 186,624

1.49 1.91

1.51 −292.47

1.53 4

1.55 −71.91

1.57 −966.24

1.59 −0.34

1.61 3465

1.63 36.5

1.65 −0.003

CHAPTER 2

Creating Graphical Displays—The Math You Need to Know

SECTION 2.1

2.1 0.89

2.3

SECTION 2.2

2.5 Answers may vary, but one problem is that all of the dots seem to be rounded to the nearest scale value ending in 0 or 5.

2.7 Answers may vary, but one problem is that the dots are equally spaced with frequency one from 6 through 14, but the data values do not correspond to this pattern.

2.9 Answers may vary, but one problem is that the negative data values are not represented on the dotplot.

2.11

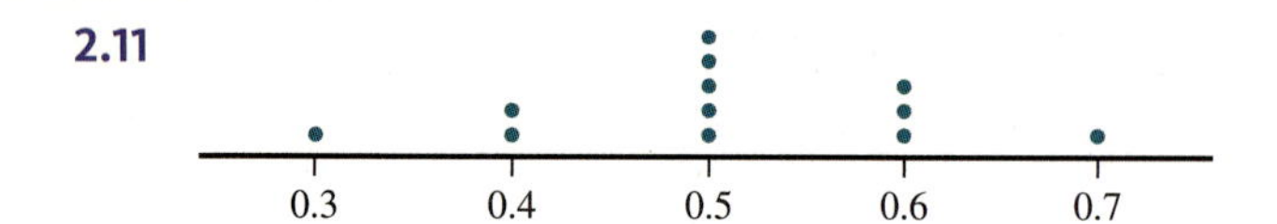

2.13

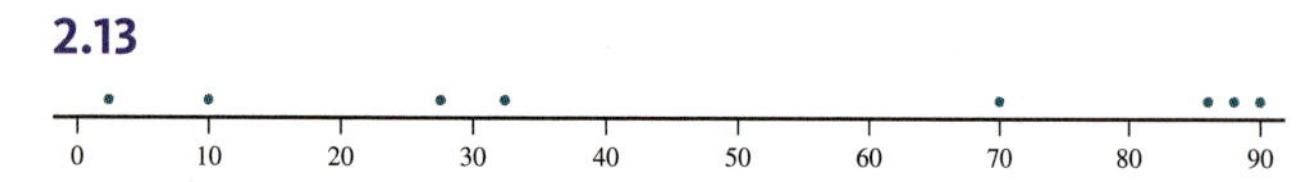

2.15 Answers may vary. Here is one choice:

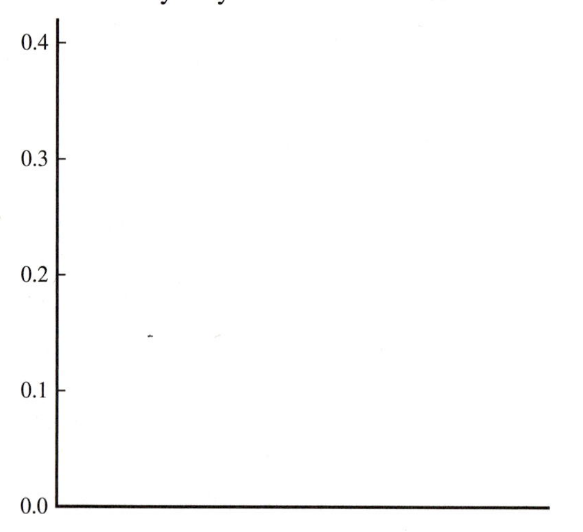

SECTION 2.3

2.17 9.0, and all of the numbers between 7.5 and 9.0

2.19 (9, 82]

2.21 15.8

2.23 16

SECTION 2.4

2.25 0.333

2.27 0.333

2.29 47.4%

2.31 91%

2.33 0.001

2.35 0.043

SECTION 2.5

2.37

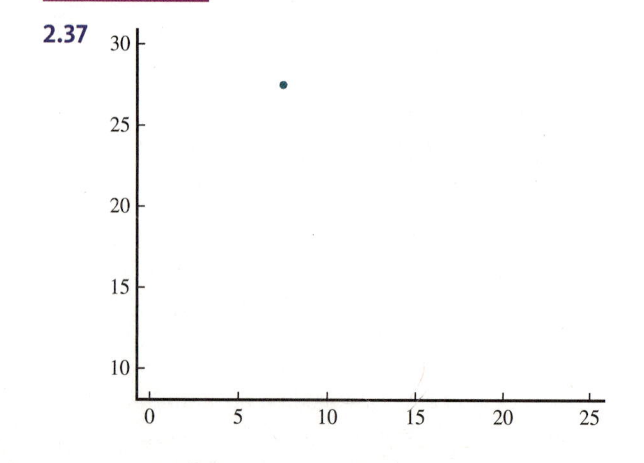

2.39

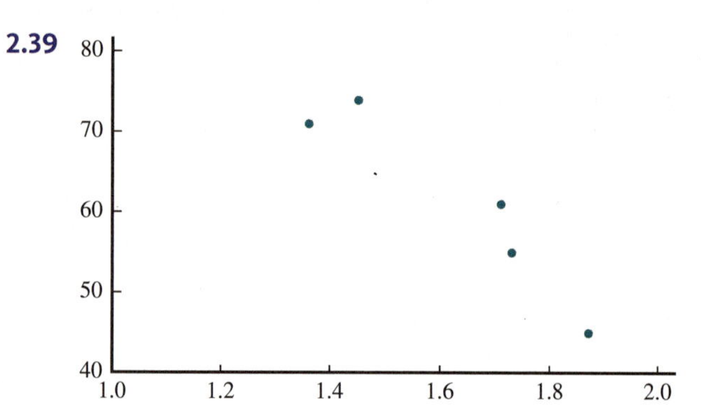

2.41 As x increases, the values of y tend to decrease. That is, there is a negative linear pattern.

2.43 There is no linear pattern in the plot. However, there is less spread in the y direction for lower values of x than there is for larger values of x.

2.45 Positive linear pattern

2.47 No linear pattern

SECTION 2.6

2.49 0.613

2.51 0.130

2.53 0.4

2.55 24

2.57 0.103; 0.037; 0.002; 0.010

CHAPTER 3

Measures of Center and Variability—The Math You Need to Know

SECTION 3.1

3.1 −69, −62, −56, −28, −24, 35, 40, 49, 49

3.3 −51.4, −3.1, 3.91, 57.9, 66.8

3.5 22.69

3.7 1.67

3.9 56

3.11 49

SECTION 3.2

3.13 3.5

3.15 −156

3.17 Score on the first test, and Score on the final exam

3.19 Medication taken, and Whether hives developed

3.21 −1, 75, 24, 18, and 49

3.23 −9.6, and −6.3

SECTION 3.3

3.25 419; 24,695; 175,561

3.27 −10.0, 580.480, 100.0

SECTION 3.4

3.29 41.9, 7138.89

3.31 −1.11, 569.36

SECTION 3.5

3.33 3.99, 10.481, 3.237

3.35 1.2, 33.70, 5.81

3.37 −0.60

3.39 0.13

CHAPTER 4

Describing Bivariate Numerical Data—The Math You Need to Know

SECTION 4.1

4.1 Independent is Variable 2, Dependent is Variable 1

4.3 Independent is Variable 2, Dependent is Variable 1

4.5 Scatterplot A, because there is no visible pattern

4.7 Scatterplot A

4.9 Nonlinear

4.11 Positive, because more sunlight should promote more growth

4.13 No association, because Grade point average should not be associated with Height

4.15 −7.45, and −1.83

SECTION 4.2

4.17 −10, and 98.9

4.19 38, and 71

4.21 $y = 3 + 4.1x$

4.23 14.7

4.25 7.82

4.27 Answers may vary. For $x = 1$, $y = 8$.

4.29 Positive

4.31 Negative

4.33 Line B has a slope of 90, and Line A has a slope of 45.

4.35 1.13

4.37 31, (9, 31)

4.39 120.06, (4.1, 120.06)

4.41 (2, 37.6), (4, 56.2)

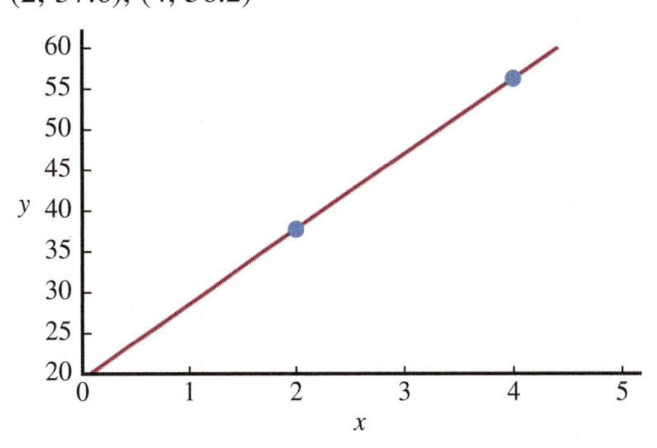

SECTION 4.3

4.43 No, because the pattern in the scatterplot is not a line.

4.45 Represents a linear model, because it contains only a y-intercept and a slope.

4.47 −27.31

4.49 Not appropriate for Point 1, due to extrapolation. Appropriate for Point 2.

4.51 For each additional day since snow melt, we predict a decrease of 0.64 units of Green biomass concentration, in g/cm^3.

4.53 Answers may vary. The value of the y-intercept is 90 minutes, which is a potentially possible 20 km ski time. It would be for a person who is immediately exhausted on a treadmill, which is not realistic for competitive skiers.

SECTION 4.4

4.55 0.406

SECTION 4.5

4.57 Answers may vary. Line B, because it better captures the steepness of the positive linear association.

SECTION 4.6

4.59 0.313, and 0.356

4.61 0.155

CHAPTER 5

Probability—The Math You Need to Know

SECTION 5.1

5.1 42%

5.3 0.889

5.5 0.09

5.7 0.60

SECTION 5.2

5.9 {2, 3, 4, 5, 7, 8, 9, 10, 11, 14}, {2}

5.11 Answers may vary. $A = \{1, 2, 3, 4, 5, 6, 7, 8, 9, 10\}$, $B = \{3, 5, 7, 8, 10, 15, 19\}$

5.13 {0, 2, 6, 7, 9}

5.15 $\{E, H, K, T\}$

SECTION 5.3

5.17 0.740

5.19 0.883

CHAPTER 6

Random Variables and Probability Distributions—The Math You Need to Know

SECTION 6.1

6.1 6561

6.3 7.14

6.5 28.6%

6.7 0.612

6.9 10.4%

6.11 −0.94

6.13 2.44

SECTION 6.2

6.15 Answers may vary. One random variable is Whether they exercised the previous week, and another is The number of minutes spent exercising.

6.17 Select a person at random from those working at the car dealership, and a random variable is their Job satisfaction rating, from 1 to 10.

SECTION 6.3

6.19

0 10 20 30 40 50

6.21 (0.741, 0.871)

6.23 (82, 89)

80 82 84 86 88 90

SECTION 6.4

6.25 =, <

6.27 <

6.29 $x < 6.0$

6.31 The value of x is between 5.6 and 7.6.

6.33 The value of m is no larger than 165/19 = 8.68.

6.35 Answers may vary. The value of c is greater than 2.86, and the number 2.86 is less than the value of c.

6.37 The value of w is about 3.16.

SECTION 6.5

6.39 $P(8.2 < x < 8.4)$

6.41 $P(x \geq 5.0)$

6.43

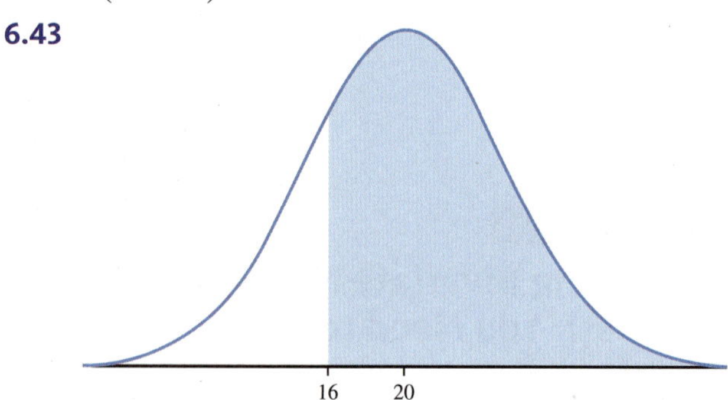

6.45 The probability that a normal random variable x takes a value between 15 and 30, and $P(15 < x < 30)$

6.47 The probability that a normal random variable y takes a value greater than 77, and $P(y > 77)$

SECTION 6.6

6.49 0.3

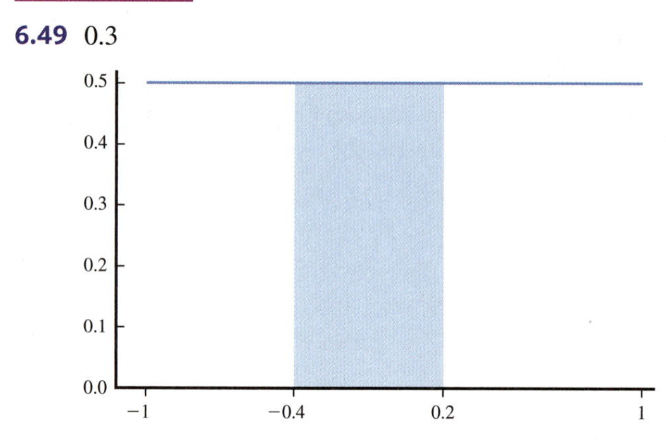

6.51 0.438

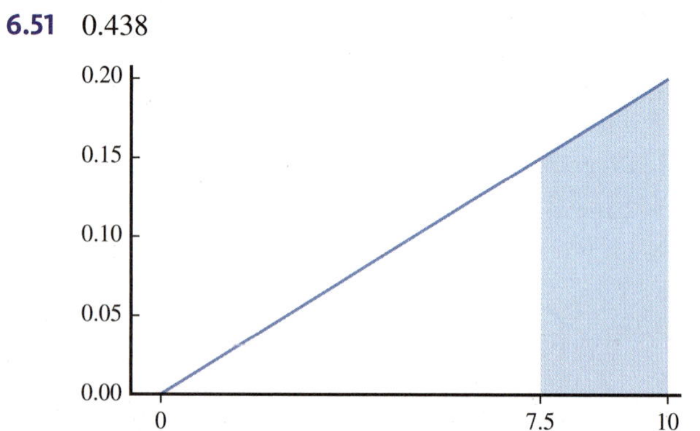

SECTION 6.7

6.53 98.6

SECTION 6.8

6.55 120

6.57 3,628,800

6.59 42

6.61 1

SECTION 6.9

6.63 1.53, 1.24

6.65 0.273

CHAPTER 7

How to Read a Statistics Problem

SECTION 7.4

7.1

Context:	Estimate the population mean following distance for taxi drivers while they are using a phone.
Terms:	"Following distance" is how far behind the lead car the taxi was. A "driving simulator" is an environment representing a taxi, for drivers to practice and be tested.
Questions:	Construct and interpret a confidence interval, and state assumptions.
Data collection:	Following distance; Collected in a reasonable way; Limited to taxi drivers.
Sample size:	Actual data not provided; Sample size is 50 taxi drivers.
Statistics:	Sample mean is 3.20 m; Sample standard deviation is 1.11 m.

7.3

Context:	Compare the proportions of adults and teens who believe in reincarnation.
Terms:	"Reincarnation" means after death, being reborn as another living thing. A "proportion" is a number between 0 and 1, the decimal version of a percentage.
Questions:	Calculate and interpret a confidence interval.
Data collection:	A poll was conducted; Conducted reasonably.
Sample size:	We don't know individual responses; Samples consisted of 2463 adults and 510 teens.
Statistics:	19% of teens and 26% of adults indicated that they believe in reincarnation.

CHAPTER 8

Estimating a Population Proportion—The Math You Need to Know

SECTION 8.1

8.1 Conditions satisfied; 0.115; If 0.48 is the estimate of the population proportion based on this sample of size 73, it is unlikely that this estimate differs from the actual value of the population proportion by more than 0.115.

8.3 Conditions satisfied; (0.257, 0.503); Based on the data from the sample, you can be 90% confident that the actual value of the population proportion is somewhere between 0.258 and 0.504.

8.5 601

SECTION 8.2

8.7

Context:	Researchers studied Internet use by college students.
Terms:	"Computer-mediated" means communicating with another person but through a device.
Questions:	Estimate and interpret a margin of error.
Data collection:	Students indicated if they use the Internet more than 3 hours per day.
Sample size:	Individual responses are not provided; 7421 students were surveyed.
Statistics:	The value of the estimated proportion is 0.404.

(a) 0.404; (b) Conditions satisfied; (c) 0.009; (d) You would expect the estimated proportion to fall within 0.009 of the actual value of the population proportion.

SECTION 8.3

8.9

Context:	Researchers considered potential jurors' exposure to crime-scene television series.
Terms:	"CSI" means crime-scene investigation.
Questions:	Construct and interpret a 90% confidence interval.
Data collection:	Potential jurors were asked if they are regular watchers of at least one crime series.
Sample size:	Individual responses are not provided; 500 potential jurors were surveyed.
Statistics:	The value of the sample proportion is 0.7.

(0.666, 0.734); You can be 90% confident that the actual value of the population proportion falls between 0.666 and 0.734.

SECTION 8.4

8.11

Context:	A survey explored the percentage of U.S. consumers who would choose a diesel car.
Terms:	"Diesel" is an alternative automobile fuel type.
Questions:	Determine the sample size required to estimate a population proportion to a specified margin of error.
Data collection:	U.S. consumers were asked if they would choose a diesel car.
Sample size:	No data are provided, because the question is about planning a sample size.
Statistics:	No statistics are provided, because the question is about planning a sample size.

Required sample size = 7572

CHAPTER 9

Testing Hypotheses About a Population Proportion—The Math You Need to Know

SECTION 9.1

9.1 0.374; −1.46; Appropriate

SECTION 9.2

9.3

Context:	The Associated Press surveyed adult Americans to determine their preference between watching movies at home or in a theater.
Terms:	A "majority" is more than half of a group.
Questions:	Test for convincing evidence that a majority prefer to watch movies at home.
Data collection:	A survey was conducted of randomly selected adult Americans, which is reasonable.
Sample size:	Individual responses are not available; 1000 adult Americans were surveyed.
Statistics:	The sample proportion of those who prefer to watch movies at home is 0.730.

$z = 14.55$; P-value < 0.001; There is convincing evidence that a majority of adult Americans prefer to watch movies at home.

CHAPTER 10

Estimating a Difference in Two Proportions—The Math You Need to Know

SECTION 10.1

10.1 (−0.629, −0.130); Based on the data from the samples, you can be 90% confident that the actual value of the difference in population proportions (Group 1 − Group 2) is somewhere between −0.629 and −0.130. The sample size condition is not satisfied.

SECTION 10.2

10.3

Context:	Research was conducted to compare sunburn incidence between college graduates and people without a high school diploma.
Terms:	The "incidence" is the proportion of people who have a condition, such as sunburn.
Questions:	Estimate the difference between two population proportions of those getting sunburnt, for college graduates and those without a high school diploma.
Data collection:	Independent random samples were collected from the two populations; Appropriate.
Sample size:	Individual responses are not provided; 200 were sampled from each population.
Statistics:	Sample percentages sunburnt were 43% for the college graduate group and 25% for those without a high school diploma.

Conditions satisfied; (0.104, 0.256); You can be 90% confident that the difference in proportions falls between 0.104 and 0.256.

CHAPTER 11

Testing Hypotheses About a Difference in Two Proportions—The Math You Need to Know

SECTION 11.1

11.1 −7.16; Conditions satisfied

SECTION 11.2

11.3

Context:	Research was conducted to compare the rates of listening to music at high volume for boys and girls.

Terms:	"Precautions" are steps taken to prevent undesirable outcomes.
Questions:	Test the appropriate hypotheses to compare two population proportions, and check conditions.
Data collection:	Dutch boys and girls were surveyed using random samples; Reasonable.
Sample size:	Individual responses are not provided; 764 boys, and 748 girls.
Statistics:	Counts are provided, to be used in calculating the two sample proportions.

$z = 3.00$; P-value $= 0.001$; Reject H_0; There is convincing evidence that the proportion of boys who listen to music at high volumes is greater than the proportion for girls; Conditions satisfied.

CHAPTER 12

Estimating a Population Mean—The Math You Need to Know

SECTION 12.1

12.1 (a) 0.330, It is unlikely that the sample mean time spent studying differs from the actual value of the population mean by more than 0.330 hours. (b) (7.410, 8.070), You can be 95% confident that the actual value of the population mean time spent studying is somewhere between 7.410 and 8.070 hours; 711.

SECTION 12.2

12.3

Context:	Researchers estimated the population mean time spent studying for university business students.
Terms:	No unusual terms.
Questions:	Calculate and interpret the margin of error for an estimate of a population mean.
Data collection:	Students logged time spent studying; Supposedly a random sample.
Sample size:	The individual data values are not provided; 212 business students.
Statistics:	Sample mean = 9.66 hours; Sample standard deviation = 6.62 hours.

0.896; $df = 211$; It is unlikely that the estimate differs from the actual value of the population mean by more than 0.896 hours.

SECTION 12.3

12.5

Context:	The population mean reaction time while driving and using a cell phone was estimated.
Terms:	"Reaction time" is the time it takes to react to a stimulus; "Simulated driving" means that the people studied were assessed in a driving simulator.
Questions:	Calculate and interpret a 95% confidence interval for a population mean; State an assumption.
Data collection:	Little information is provided about the sample or the individuals.
Sample size:	No individual data values are provided; 548 individuals participated.
Statistics:	Sample mean reaction time = 530 msec; Sample standard deviation = 70 msec.

(524.1, 535.9); $df = 547$; You can be 95% confident that the actual value of the population mean falls between 524.1 and 535.9 msec; Representative sample.

SECTION 12.4

12.7

Context:	Lead levels in California wines will be tested for consumer safety.
Terms:	"BATF" is a government office that enforces laws about alcohol, tobacco, and guns. Lead is a chemical element that is dangerous if consumed by humans.
Questions:	Provide a recommended sample size for wine specimens, given the value of a margin of error and a confidence level.
Data collection:	The data are not yet collected.
Sample size:	You will recommend a sample size.
Statistics:	You will recommend a sample size, and the data will be used to estimate the population mean lead level.

Required sample size = 1015

CHAPTER 13

Testing Hypotheses About a Population Mean—The Math You Need to Know

SECTION 13.1

13.1 −9.32

SECTION 13.2

13.3

Context: Researchers compared elapsed time to perception for abstaining smokers.

Terms: "Abstain" means to not complete a task, such as smoking cigarettes, for some period of time. "Elapsed time" is how much measured time has passed until a task is completed.

Questions: Carry out an appropriate test of hypothesis to determine if the population mean time perceived by abstaining smokers differs from the actual elapsed time of 45 seconds.

Data collection: Perceived elapsed times were assessed for abstaining smokers; The sample is not very large, and it is not clearly random.

Sample size: Individual data values are not provided; 20 abstaining smokers were studied.

Statistics: Sample mean = 59.3 seconds; Sample standard deviation = 9.84 seconds.

$t = 6.50$; $df = 19$; P-value < 0.001; Reject H_0; There is convincing evidence that the mean perceived elapsed time is different from 45 seconds.

CHAPTER 14

Estimating a Difference in Means Using Paired Samples—The Math You Need to Know

SECTION 14.1

14.1 (6.47, 6.93); You can be 90% confident that the population mean difference in reaction time (Before – After) falls between 6.47 milliseconds and 6.93 milliseconds.

SECTION 14.2

14.3

Context: A study compared the differences in reported height and actual height for men with online dating profiles.

Terms: "Online dating sites" report user-provided information, which may not always be accurate.

Questions: Use a confidence interval to estimate the population mean difference in height (Profile – Actual), and interpret the interval in context.

Data collection: Reported and actual heights were collected for men with online dating profiles; Not clearly a random sample.

Sample size: Individual data values are not provided; 40 men with online dating profiles.

Statistics: The sample mean difference is 0.57 inches, and the sample standard deviation is 0.81 inches.

(0.223, 0.917); $df = 39$; You can be 99% confident that the mean difference in heights falls between 0.223 and 0.917 inches.

CHAPTER 15

Testing Hypotheses About a Difference in Means Using Paired Samples—The Math You Need to Know

SECTION 15.1

15.1 44.77

SECTION 15.2

15.3

Context: Researchers tested for an effect of diesel exhaust exposure on brain activity.

Terms: "Diesel" is an alternative type of petroleum-based fuel. "Median power frequency" (MPF) is a measure of brain activity.

Questions: Test whether the mean difference in MPF before and after diesel exposure is different from 0.

Data collection: MPF before and after diesel exhaust exposure; Volunteer sample.

Sample size: The data values are provided; 10 men.

Statistics: Statistics must be calculated from the data.

$t = -3.11$; $df = 9$; P-value $= 0.013$; Reject H_0; There is convincing evidence that the mean difference is different from zero.

CHAPTER 16

Estimating a Difference in Two Means Using Independent Samples—The Math You Need to Know

SECTION 16.1

16.1 61.04; (−5.63, −0.37)

SECTION 16.2

16.3

Context: A study estimated the difference in mean wait times in coffee shops for males and females.

Terms: No unusual terms.

Questions: Construct and interpret a 99% confidence interval for the difference in mean wait times in coffee shops for males and females.

Data collection: Representative samples of waiting times were collected; Reasonable.

Sample size: Individual data values are not provided; 145 males and 141 females.

Statistics: Sample mean for males is 85.2 seconds, with sample standard deviation 50 seconds. Sample mean for females is 113.7 seconds, with sample standard deviation 75 seconds.

$(-48.1, -8.9)$; $df = 243.0$; You can be 99% confident that the difference in mean wait times in coffee shops for males and females falls between -48.1 and -8.9 seconds.

CHAPTER 17

Testing Hypotheses About a Difference in Means Using Independent Samples—The Math You Need to Know

SECTION 17.1

17.1 108.4; 11.51

SECTION 17.2

17.3

Context: A report on a survey compared hours spent online per week for male and female teenagers.

Terms: "Typical week" means a week that is not substantially unusual compared to other weeks.

Questions: Conduct a hypothesis test to determine if there is convincing evidence that the means for hours spent online per week are not the same for males and for females.

Data collection: Teenagers were surveyed, which may lead to response bias.

Sample size: The data values are not provided; 228 males and 306 females were surveyed.

Statistics: The sample means and the sample standard deviations are provided for both male and female teenagers.

$t = 0.99$; $df = 497.9$; P-value $= 0.324$; Fail to reject H_0; There is insufficient evidence that the means for males and females are different.

CHAPTER 18

Chi-Square Tests—The Math You Need to Know

SECTION 18.1

18.1 457.46

SECTION 18.2

18.3

Context: Researchers examined the numbers of births occurring during different phases of the moon.

Terms: The "lunar cycle" is the pattern of the phases of the moon, from full moon to new moon, and back again.

Questions: Test whether the proportions of births during each phase of the moon are different from the expected rates based on the lengths of the lunar phases.

Data collection: A random sample of births was selected; Reasonable, although perhaps not over more than one month.

Sample size: Individual data values are not provided; 222,784 births are summarized.

Statistics: The necessary proportions and expected counts are not provided, but can be calculated.

6.31; $df = 7$; P-value $= 0.504$; Fail to reject H_0; There is insufficient evidence to conclude that the proportions of births are different from expected rates based on lengths of phases.

18.5

Context: Researchers explored the association between children being overweight and daily sweet drinks consumption.

Terms: "Sweet drinks" are beverages with high sugar content.

Questions: Test for evidence of an association in children between drinking various numbers of sweet drinks per day and being overweight.

Data collection: It is not stated whether the sample was random; The results may not generalize.

Sample size: Individual data values are not provided; 8328 children were included.

Statistics: The necessary proportions and expected counts are not provided, but can be calculated.

3.03; $df = 3$; P-value $= 0.387$; Fail to reject H_0; There is insufficient evidence to conclude that there is an association between children being overweight and daily sweet drinks consumption.

CHAPTER 19

Estimating and Testing Hypotheses About the Slope of a Population Regression Line—The Math You Need to Know

SECTION 19.1

19.1 (a) 0.209; (b) (0.221, 1.059; You can be 95% confident that the actual value for the slope of the least squares regression line falls between 0.221 and 1.059 g/cm³ per day.

SECTION 19.2

19.3

Context: A least squares regression line was used to relate mean response time for individuals with closed-head injury (CHI) and matched controls.

Terms: A "closed-head injury" is an injury inside the skull. One example is a concussion. "Mean response time" is how long it takes an individual to complete a task, on average.

Questions: Calculate and interpret a 99% confidence interval for the slope of the least squares regression line.

Data collection: Individuals without CHI were matched to those with CHI; The samples were not random.

Sample size: There are 10 observations in the study, and the data values are provided.

Statistics: The least squares regression line and the standard deviation of the slope are provided.

$df = 8$; (1.39, 1.79); You can be 99% confident that the slope of the population regression line falls between and 1.39 and 1.79.

SECTION 19.3

19.5

Context: A least squares regression line was used to predict finishing time in a 20 km ski race to time to exhaustion running on a treadmill.

Terms: "Time to exhaustion" means that the athlete ran on the treadmill so long that they were not capable of running any longer.

Questions: Test where there is a useful linear relationship between time to exhaustion running on a treadmill and finishing time in a 20 km ski race.

Data collection: Data for time to exhaustion and a 20 km ski race were collected for 11 athletes; The sample was not random, but it may be representative.

Sample size: The actual data are provided for 11 athletes.

Statistics: The equation of the least squares regression line and the standard deviation of the slope are provided.

$df = 9$; $t = 3.95$; P-value $= 0.003$; Reject H_0; There is evidence of a statistically significant useful linear relationship.

CHAPTER 20

Testing Hypotheses About More than Two Means—The Math You Need to Know

SECTION 20.1

20.1 101,600,000; 14,866,667; 6.83

SECTION 20.2

20.3

Context: Compare the mean compression strength for four different types of boxes.

Terms: "Compression strength" (in pounds) measures how much weight a box can bear before it collapses.

Questions: Conduct a one-way analysis of variance, to test whether the mean compression strengths of at least two of the box types are different.

Data collection: Compression strengths were measured for six boxes of each type; Not a random sample, but perhaps representative.

Sample size: Actual data are provided for the compression strengths of six boxes of each type.

Statistics: The data can be used to calculate an ANOVA table and carry out the test.

H_0: All four means are equal; $F = 25.09$; P-value < 0.001; Reject H_0; Convincing evidence that not all four Type means are the same for mean Strength.

Index

T

U